INFECTIOUS LIBERTY

LIT Z

Sara Guyer and Brian McGrath, series editors

Lit Z embraces models of criticism uncontained by conventional notions of history, periodicity, and culture, and committed to the work of reading. Books in the series may seem untimely, anachronistic, or out of touch with contemporary trends because they have arrived too early or too late. Lit Z creates a space for books that exceed and challenge the tendencies of our field and in doing so reflect on the concerns of literary studies here and abroad.

At least since Friedrich Schlegel, thinking that affirms literature's own untimeliness has been named romanticism. Recalling this history, Lit Z exemplifies the survival of romanticism as a mode of contemporary criticism, as well as forms of contemporary criticism that demonstrate the unfulfilled possibilities of romanticism. Whether or not they focus on the romantic period, books in this series epitomize romanticism as a way of thinking that compels another relation to the present. Lit Z is the first book series to take seriously this capacious sense of romanticism.

In 1977, Paul de Man and Geoffrey Hartman, two scholars of romanticism, team-taught a course called Literature Z that aimed to make an intervention into the fundamentals of literary study. Hartman and de Man invited students to read a series of increasingly difficult texts and through attention to language and rhetoric compelled them to encounter "the bewildering variety of ways such texts could be read." The series' conceptual resonances with that class register the importance of recollection, reinvention, and reading to contemporary criticism. Its books explore the creative potential of reading's untimeliness and history's enigmatic force.

INFECTIOUS LIBERTY

Biopolitics between Romanticism and Liberalism

Robert Mitchell

Fordham University Press

New York 2021

Publication of this open monograph was the result of Duke University's participation in TOME (Toward an Open Monograph Ecosystem), a collaboration of the Association of American Universities, the Association of University Presses, and the Association of Research Libraries. TOME aims to expand the reach of long-form humanities and social science scholarship including digital scholarship. Additionally, the program looks to ensure the sustainability of university press monograph publishing by supporting the highest quality scholarship and promoting a new ecology of scholarly publishing in which authors' institutions bear the publication costs. Funding from Duke University Libraries made it possible to open this publication to the world.

Fordham University Press has no responsibility for the persistence or accuracy of URLs for external or third-party Internet websites referred to in this publication and does not guarantee that any content on such websites is, or will remain, accurate or appropriate.

Fordham University Press also publishes its books in a variety of electronic formats. Some content that appears in print may not be available in electronic books.

Visit us online at www.fordhampress.com.

Library of Congress Cataloging-in-Publication Data available online at https://catalog.loc.gov.

Printed in the United States of America

23 22 21   5 4 3 2 1

First edition

Contents

Preface

When I first began working on the chapters of the project that eventually became this book, I did not intend to write a book about liberalism. Liberalism was not an intellectual movement that had ever particularly interested me, and though as an undergraduate and graduate student I had dutifully read classics of liberal thought by authors such as John Locke, David Hume, Adam Smith, John Stuart Mill, Isaiah Berlin, John Rawls, and Richard Rorty, I often battled against a sense of greyness and boredom as I worked through their texts. In comparison with the tradition of continental philosophy that excited me—authors such as Martin Heidegger, Hannah Arendt, Michel Foucault, Jacques Derrida, and Gilles Deleuze—the "theorists" of liberalism seemed cramped, provincial, and reactionary. Moreover, the aspects of liberal theory that interested me—for example, Hume's and Smith's reflections on sympathy—seemed like points at which their liberalisms revealed theoretical contradictions. And yet I have ended up writing a book that is, at least from my perspective, quite fundamentally about liberalism. It thus may be useful to explain how my interest in liberalism developed, as this narrative clarifies the stakes of my project.

Michel Foucault's recently published lectures from the 1970s on both liberalism and biopolitics were, without a doubt, the key vector that linked my interest in continental philosophy with liberalism. Roberto Esposito's rereading of Foucault in terms of what Esposito calls "the immunitary paradigm" further cemented that link. As a consequence, the understanding of liberalism articulated in this book is rather different from what I encountered as a student, and I am not sure that the classical authors of liberalism noted above would fully agree with my reading, heavily influenced by Foucault and Esposito, of liberalism as a mode of biopolitics. Or, to put this another way, I see the traditional understanding of liberalism that I encountered as a student as failing to grasp fully what was at stake in the work of authors such as Locke, Smith, Mill, and Hayek.

Yet even as Foucault provided an important theoretical lens with which to reenvision liberalism, his work was not the origin of this project but rather a means for me to think further the relationship of my research interests to a series of existential commitments and hopes. To begin with my research interests, this book represents the intersection and convergence of two quite different research projects. On the one hand, it extends parts of my earlier monograph *Experimental Life: Vitalism in Romantic Science and Literature.* I had explored there ways that Romantic literary authors appropriated the concept of "experiment" from the sciences, often transforming the very meaning of the concept of experimentation by means of their projects. Edmund Burke, for example, worried that the French Revolution represented an unprecedented social experiment, while William Wordsworth and Samuel Taylor Coleridge's collection *Lyrical Ballads* (1798) initiated a new genre of "experimental" poetry. I stressed in *Experimental Life* that Romantic-era experimentation with the concept of experiment was also the attempt to create new relationships among experiments, the sciences, the arts, and collective social life. Because this connection between experimentation and ways of living was at the core of *Experimental Life*, I had originally hoped to discuss there John Stuart Mill's post-Romantic defense of liberalism as committed to what he called "experiments in living." However, time and space constraints prevented me from considering either Mill specifically or the relationships between experimentation and liberalism more generally. The present book thus began with the goal of considering how experimentation and Romantic-era liberalism were bound up with each other in the eighteenth and nineteenth centuries.

On the other hand, I had developed a quite separate research trajectory focused on late-twentieth- and early-twenty-first-century relationships between the health sciences and economics. This research project, which was cast more in the mode of critical sociology than literary criticism, drew me to the topic of biobanks—that is, collections of biological material and related information drawn from many members of a population—and other forms of population-level data collection. These contemporary biomedical projects tend to alter the relationship of the medical sciences to the general public via, for example, the aspiration of running experiments on entire national populations, the use of technical data procedures that eliminate informed consent, and hopes that algorithmic data-mining methods will enable purely empirical, "theory-free" science. This research focus led me in turn (in part via the inimitable work of the historian of economics Philip Mirowski) to focus on relationships between the sciences and *neo-liberalism*: that is, a form of liberalism first formulated in the 1940s and 1950s by authors such as Friedrich Hayek and Milton Friedman and that became dominant in policy circles in the 1970s (and continues to gain strength in the present). My work on biobanks thus led me to consider relationships between contemporary neoliberalism and earlier forms of liberalism, and the roles of experimentation and the sciences in each of these liberalisms. And so even as my interest in Romantic-era experimentation led me forward to Mill's postulated linkage between liberalism and experimentation, my interest in contemporary neoliberal transformations of the health sciences led me back to Mill and then further back into the eighteenth century. These links were in turn reinforced by the connections Foucault drew between eighteenth-century liberalism and mid- to late-twentieth-century neoliberalism.

The result is a book that, like *Experimental Life*, focuses primarily on the eighteenth and nineteenth centuries but that also emphasizes the relationship of the Romantic past to our present. Though these reflections on our present are limited, my hope is nevertheless to illuminate significant resonances between the eighteenth- and nineteenth-century concepts and topics upon which I have focused here—for example, genius, population, environmental transformation, regulation, and social experimentation—and many of the key categories of contemporary neoliberalism (for example, information, "smartness," spontaneous order, and regulation).

Though I have described my research interests as though they emerged and proceeded of their own accord, such interests are always bound up—albeit in ways often inscrutable to the author—with what I will call existential commitments: that is, hopes and aspirations for the best forms of relationships among humans and their environments (and also nonhuman living beings). My own existential commitments have become increasingly bound up and looped through both "abstract" concerns about global warming and consequent environmental disaster—abstract in the sense that these processes can only be grasped through scientific lenses—and the more concrete and often quite messy concerns that have emerged as my wife and I have sought to raise two children in this world.

No doubt because my first daughter was born about the time I began working on this book, it was hard to miss the many ways that parenting is engaged with biopolitics. For me, the birth of my daughter brought, in addition to new forms of love and joy, a terrifying feeling of absolute responsibility for another human being. This responsibility felt absolute in the sense that I feel responsible for *everything* that happens to this human being, even in circumstances in which I have little or no influence over what in fact happens to her. (The weight and terror of this feeling of absolute responsibility has certainly not been lessened by my sense that I lack anything close to adequate knowledge or ability to discharge even those tasks that in principle fall within my sphere of action.) The biopolitical decisions one is asked to make as a parent in the United States in the second decade of the twenty-first century often direct this feeling of responsibility toward corporeal concerns. The promise and peril of population-level disease inoculation, for example, shifted from a theoretical to an embodied register every time I was asked by a medical professional to hold my terrified daughter while she was injected with multiple inoculation cultures, some of which seemed to me of questionable utility. (Do children really need to be inoculated against chicken pox? Having suffered through chicken pox as a child, I am happy my daughter will not need to do so. Yet I am unsure how to assess the tradeoffs involved in an inoculation for a condition that is generally not fatal, and unsure if there can be any logical end to the ever-increasing number of inoculations that become mandatory.) And whereas disease inoculation is a form of biopolitics over which parents have some minimal control, children—especially poor children and children who are not white—are part of neoliberal biopolitical programs over which parents have no control at all, as the 2015 revelations about lead poisoning of drinking water in Flint, Michigan, highlighted. Biopolitics is not just a matter of "making live" through technical means such as inoculation but is also a matter of decisions about where *not* to intervene—a pernicious kind of laissez-faire that ensures that some groups will suffer and die in much greater percentages than others. The fact that my final revisions of this book were written during the "social-distancing" protocols of the 2019–2020 COVID-19 pandemic—and the fact that the burdens of social distancing are inequitably borne by the poor—simply underscored for me the ever-increasing relevance of these concerns.[1]

Since the provision of safe drinking water is a biopolitical measure—and since, in my ideal communities of the future, humans will likely still need to be inoculated against known diseases and will have to deal collectively with emerging diseases such as COVID-19—the answer cannot be to "get rid of" biopolitics, whatever that might mean.[2] Rather, the answer—or at least part of the answer—has to be the development of what Roberto Esposito calls an "affirmative" biopolitics. By this he means a form of biopolitics that does not presume that the cost of ensuring the survival of purportedly "worthy" lives is the letting die of those deemed less worthy. The development and coordination of an affirmative biopolitics with just ways of collective living seems to me a vital task for our present and future. Though the Romantics do not have all the answers to the questions raised by such a project, the premise of this book is that they provide us with important resources for raising and answering such questions.

INFECTIOUS LIBERTY

Introduction

Many of the basic elements of recognizably modern European and American literature emerged in the late eighteenth and early nineteenth centuries—that is, the Romantic period—and many of these focused readers on *differences* among individuals. It was during this period, for example, that poets argued against older modes of poetry that involved "personifications of abstract ideas" and in favor of the use of "the real language of men in a state of vivid sensation" to focus on wildly eccentric individuals or on the poet's own idiosyncratic emotional experiences.[1] In keeping with this focus on individual idiosyncrasies, the quests of action-oriented heroes across external geographies that characterized earlier literature were "internalized" into dramas of the individualized psyche.[2] It was also during this period that novelists began to develop what would eventually be called realistic novelistic techniques, such as an emphasis on "round" (versus flat) characters; free indirect discourse, which brought readers into contact with the fugitive thoughts and feelings of those round characters; and the construction of plots that involved increasingly large populations of ordinary, yet also individualized, characters.[3] At this same time, the very term "literature" was narrowed down to include solely imaginative texts, and these kinds of texts began to be described as both "experimental" and potentially world changing.[4] This period also witnessed the consolidation of the Romantic understanding of authorship—the understanding of the literary author as a unique genius who experiences the world more intensely than others but who is able to pass on that experience to readers via literature—and this premise about literary production not only continues to determine our understanding of authorship but structures more generally our current intellectual property regime.[5] The Romantic era saw the development, especially with respect to novels, of a large literary market segmented by different genres, along with the consolidation of a journal reviewing system that allowed a reader to locate those genres and novels that best corresponded to her particular taste.[6] Though each of these developments has been refined, altered, or contested in subsequent literary periods, each nevertheless continues to determine our own sense of what counts as good (and bad) literature.

Earlier literary criticism has tended to approach this explosion of difference—differences among the individuals of increasingly large and diverse populations of characters in novels, differences among literary genres, and differences in literary preferences among an ever-expanding number of readers—as a surface-level illusion that covered up more subterranean processes of homogenization and sameness. For critics working in a Marxist literary tradition, these various literary developments were part of the more general bourgeois construction of the cultural capital necessary to remake social relations to advance their class interests. The forms of difference among characters, texts, and readers are thus understood as ideological illusions of variety that function in the service of enforcing very narrow forms of normative behavior among readers.[7] For literary critics more influenced by Michel Foucault's work on the rise of the disciplinary society, these differences are also treated as illusory and in this case as distractions that obscured the ways that literature contributed to an increasing disciplining and policing of individuals.[8]

Despite significant theoretical differences between these two camps of literary criticism, both have approached literature as, in essence, a technology of normativity. In the case of accounts of the rise and development of the new literary form of the novel, for example, whether it is the "monitory image" that Ian Watt suggested a novel such as *Robinson Crusoe* established for modern society; or the symbolic acts, ideologemes, and assumptions about genre that, according to Fredric Jameson, establish the limits of our utopian "collective thinking and our collective fantasies about history and reality"; or the proleptic semiotic construction of "the ordered space we now recognize as the household," which Nancy Armstrong describes as serving as "the context for representing normal behavior"; or the disciplinary "spiritual exercise[s]" that D. A. Miller suggests were provided by the Victorian novel, all of these critics understand the ultimate effect of eighteenth- and nineteenth-century novels as the inculcation of "liberal" normative beliefs and practices, a drive toward sameness disguised under the cover of the celebration of individual difference.[9] Or, as Miller puts it, the function of nineteenth-century novels and all their attendant forms of apparent difference was to

confirm the novel-reader in his identity as "liberal subject," a term with
which I allude not just to the subject whose private life, mental or domestic,
is felt to provide constant inarguable evidence for his constitutive "freedom,"
but also to . . . the political regime that sets store by this subject. (x)

For critics from both Marxist and Foucauldian literary-critical camps, literature is "liberal" in the ideological sense that it produces sameness out of
difference and does so by naturalizing normative beliefs and behaviors and
encouraging readers to conform to these social norms.

While I draw on much of this earlier literary-critical work in this book,
I nevertheless approach the relationship of modern literature to difference
in a fundamentally different way. I also read the elements and institutions
of modern literature that I have sketched as liberal, that is, as fundamentally
connected to the emergence of a liberal political-economic philosophy that
valorized the unique individual and her freedom but purported that a
capitalist order was the only guarantee of such liberty. However, I read
liberalism as fundamentally a mode of what Foucault called biopolitics.
Foucault used the term "biopolitics" to refer to the development of techniques focused not on training and disciplining individuals but rather
addressed to "a multiplicity of men"—that is, to populations—"to the
extent that they form . . . a global mass that is affected by overall processes
characteristic of birth, death, production, illness and so on."[10] From this
perspective, liberalism was one of several eighteenth-century attempts to
see the world biopolitically, in terms of populations and the regularities
that occur within these collective bodies, and to use such knowledge to
alter those regularities.

Foucault's approach allows us to see eighteenth-century liberalism as
part of a wider biopolitical effort to use the sciences to limit sovereign
political and legal power. Biopolitics limited political and legal power by
"scientizing" a concept such as population and arguing that this collective
body had immanent and natural dynamics that could be illuminated, and
partially regulated, by sciences such as political economy or medicine.
Biopolitical techniques were premised on the position that these immanent
and natural dynamics took place beyond, or below, the direct reach of law
and political power. Legal authorities, for example, could not prevent
smallpox outbreaks by commanding members of a population to avoid
contracting this disease, but outbreaks could be limited by employing new
developments in inoculation (and, later, vaccination). In similar fashion,
liberal theorists such as John Locke, Adam Smith, and Thomas Malthus

argued that the science of political economy revealed population-level economic behavior that could not be politically commanded or forbidden in the traditional sense but that could be partially regulated by adapting law to the findings of political economy. Inoculation campaigns, efforts to regulate population growth, liberal political-economic theory: These were all expressions of a more general biopolitical approach to collective relations.

Foucault's approach allows us to see in a new light the link between liberalism's stress on individuals and liberty, on the one hand, and the collective entity that became known as population, on the other. Commentators have sometimes sought to distinguish between "political" liberalism, which is primarily focused on securing individual liberty, and "economic" liberalism, which "is focused on the general prosperity of the society, not on individual advantage."[11] Yet this distinction between two modes of liberalism obscures the fact that key "political" liberal theorists such as John Locke and John Stuart Mill were also economic theorists (a point to which I return in Chapters 5 and 6) and overlooks the political implications of the epistemological shift stressed by Foucault. Foucault's approach helps us see the liberal defense of the unique individual as a means for securing certain kinds of population-level regularities. That is, liberals wish individuals to be "free to choose" within certain areas, such as religion or market relations, *because* such individual autonomy and difference purportedly produces regulated order at the level of the population in a way that enforcing more rigid norms would not.[12] Adam Smith's claims about the "invisible hand" of the market, which comes into being when each individual is free to focus on his or her own interest, is one well-known example of this link between individual freedom and population-level regularities. In this sense, the individual uniqueness prized by liberal theorists—as well as the legal and ideological means for protecting individual uniqueness, such as individual property rights—was a biopolitical means and not an end in itself.[13]

These points allow us to see relationships among late-eighteenth- and nineteenth-century literature, difference, and norms in a significantly different way. If we think late-eighteenth- and nineteenth-century literature within the milieu of biopolitics and understand liberalism as one particularly important expression of biopolitical logic, then we can understand modern literature less as a normalizing technology that encouraged readers to internalize specific norms and instead as a technology that encouraged readers to see the world biopolitically, especially in terms of populations. Encouraging readers to see the world in terms of populations is itself a certain kind of normalization, since it may discourage readers from understanding the world in terms of other kinds of collectives (for example, classes or "the multitude"). Yet it is a very peculiar mode of normalization, since population concepts aim at *altering* norms. Moreover, as I document in this book, there were multiple eighteenth- and nineteenth-century concepts of population, and so understanding the world in terms of populations meant that readers were always implicitly or explicitly choosing one concept or model of population over others. Some population models, such as Malthus's "principle of population," implied that there were few important differences between members of a population.[14] However, other models of population stressed the importance of individual uniqueness, the multitude of norms that such uniqueness enables, the regularities that could emerge immanently on the basis of those differences, and the capacity to use the knowledge of regularities to change those regularities (that is, to create new norms). This does not mean that the literary critics I have cited here were wrong, for modern literature may indeed have disciplined some readers to conform to a small number of bourgeois or liberal norms. My point, though, is that modern literature could do this only because it was more fundamentally a technology able to establish new norms, and this latter capacity could never be fully contained within the frame of those bourgeois or liberal norms.[15]

Approaching the liberalism of modern literature in this way changes our relationship, as literary critics, to literature. The literary-critical methodologies I have noted approach literature by means of a hermeneutic suspicion that seeks out the ways that literature constrained the imaginative and political potential of readers. Approaching the institutions of modern literature through the biopolitical frame I have sketched out allows us instead—or at least in addition—to understand late-eighteenth- and nineteenth-century literary texts as proposing a multitude of different models and norms for the relationship of readers to their natural, political, and economic milieux. This encourages us to consider both the ground of possibility for such norm creation and to focus on how this ground of possibility enables the creation of new norms, which in turn allows us to see continuities between past literary texts and our own efforts to discover new norms that can help create a better world.[16] The critical task to which I hope this book in part contributes is, in other words, that of pushing our understanding of biopolitics beyond the frame of liberalism and toward a positive, affirmative, and just version of biopolitics.

I exemplify the potential of this approach with six chapters. The first three focus on classic "literary" concepts: genius, character-systems, and free indirect discourse. I refer to these as literary concepts to emphasize that we now tend to associate these terms primarily with the production and reading of literary texts. I suggest, however, that each concept is better understood as a means for aligning literary and biopolitical strategies. The next three chapters take up three more general concepts—global flows, collective experiments, and self-regulation—that were central to the alignment of the sciences, biopolitics, and liberalism in the eighteenth and early nineteenth centuries and developed both in literary texts and nonliterary texts. These chapters also take up literary concepts: The final chapter, for example, reads the contest between different eighteenth-century concepts of regulation as the enabling frame for a key mid-eighteenth-century debate about whether there was a "standard of taste." However, these final chapters focus on concepts that were not specific to literary or artistic endeavors and that linked scientific debates with liberal and biopolitical policy proposals.

In the first chapter, I demonstrate that a biopolitical lens illuminates the extent to which mid- and late-eighteenth-century debates about genius were not primarily about the "nature" of genius, as earlier critics have suggested, but more fundamentally about how to *maximize* the number of geniuses in a population. I stress that this effort depended on the cultivation of two kinds of worry: the worry of losing potential geniuses and the worry of losing past geniuses in the torrent of new kinds of print publications. Significantly, the cultivation of these worries occurred not in treatises focused on genius but in poetry about genius, including that of Thomas Gray and William Wordsworth.

In the second chapter, I focus on the development of a "population imaginary" in both the political philosophies of Thomas Malthus and William Godwin and the fiction of Mary Shelley. Despite Malthus's and Godwin's public antagonism toward each other, both assumed that population-level analyses could disregard individual differences. Hence, the true rivals of Malthus—or rather, of the Malthus-Godwin couple—were authors such as Mary Shelley, who presumed that the individuals who make up a population *differ* from one another in innumerable ways. As I note, some of Shelley's first readers—namely, critics of her novel who wrote for periodicals—demonstrated that her encouragement to view the world in terms of populations could be spread even via criticisms of her novel.

The third chapter focuses on two literary devices central to the development of nineteenth-century novels: the massive expansion of *character-systems* in nineteenth-century novels and the emergence and development of *free indirect discourse*. I present the emergence of free indirect discourse as a tool through which nineteenth-century novelists developed "probes" that allowed them to search, in quasi-scientific ways, for forces that determined the characteristics of populations and that included nondiscursive causes such as habitual comportments and evolutionary endowments. Yet understanding free indirect discourse in this way requires that we keep this device connected to character-systems, since free indirect discourse could only function as a probe when it was tied to the territory of a specific character-system.

The next three chapters focus on more general concepts, ones central to the development of both biopolitics and liberalism in the eighteenth and early nineteenth centuries but that also help us rethink the nature and possible functions of literature. The fourth chapter focuses on the concept of global flow. I use this term to capture a mode of thinking, exemplified in multiple Romantic-era sciences, including physics, meteorology, and political economy, which presumed that, since humans were inextricably situated on a globe, every movement in one direction would eventually return to its point of origin. Seeing the world in terms of flows across the surface of a globe contributed to the tendency of liberal theorists to understand liberalism as a necessarily global phenomenon that would eventually encompass all of humanity. However, as I demonstrate through readings of Erasmus Darwin's and Percy Bysshe Shelley's reflections on possible human transformations of the physical characteristics of the globe, the concept of global flow also gave rise to reflections on globality that exceeded that liberal frame and did so by focusing attention on the extent to which humans can *alter* their global natural environment. Emphasizing the resonance of these Romantic-era reflections with our own concerns about global warming, I suggest that Shelley's approach is especially useful for us in our era of the Anthropocene.

The fifth chapter focuses on the concept of the "collective experiment." My starting point is what the Victorian theorist of liberalism John Stuart Mill described as the need for collective "experiments in living," and I trace the genealogy of Mill's concept back into eighteenth-century liberalism and forward into twentieth-century neoliberalism. As earlier literary criticism has noted, the concept of experiment was central for Romantic literary practice and political theory, the former exemplified by William Wordsworth and Samuel Taylor Coleridge's description of many of the poems in *Lyrical Ballads* as "experiments" and the latter by Edmund Burke's description of French revolutionaries as desiring that "the whole fabric [of the French government] should be at once pulled down, and the area cleared for the erection of a theoretic, experimental edifice in its place."[17] My argument is that liberalism has been, since its eighteenth-century origins, intrinsically bound to the concept of the collective experiment and that the latter concept is a key way that specific liberalisms express their biopolitical visions. Liberalisms from the eighteenth to twentieth centuries employ the concept of collective experimentation as a way of mediating among individual freedom, collective "learning," and institutional stability. To put my point polemically, the concept of the collective experiment seems to me more central to liberalism than the principle of individual liberty, which is generally understood as the core of liberal theory. To put my point less polemically, the specific way that a liberal theorist understands collective experimentation determines how he or she understands the nature and extent of individual liberties. Recognizing this allows us to understand why advocates of liberalism turn consistently not only to categories of individual choice, property, and liberty but also to concepts of population and immunity.

The final chapter takes up one of the central concepts—arguably, *the* central concept—of both biopolitics and liberalism, namely, "self-regulation." Aspirations for self-regulation appear in many late-eighteenth- and early-nineteenth-century literary, political, scientific, and philosophical texts. However, as my focused readings of Thomas Malthus and Immanuel Kant demonstrate, such efforts tended to oscillate back and forth between two different models of regulation. The first model was grounded in a schema of conformity to an invariable, sovereign-issued standard; the other model of regulation focused on a fluctuating, hidden, and variable standard that revealed itself in the interactions of a collective. I contend that these Romantic efforts to puzzle out what self-regulation might mean are important for a variety of eighteenth- and nineteenth-century debates about the ways that "standards" (including the standard of taste) could guide and regulate the collective. Equally significant, though, these debates remain relevant for our own "Anthropocenic" moment, since these same difficulties in understanding the nature of self-regulation reappear in contemporary reflections on how to deal with the impact of a large global population on an increasingly endangered global ecosystem.

In each chapter, I have focused on texts that were important within the Romantic era but that also—and perhaps even more importantly—had long post-Romantic afterlives. I focus on Thomas Gray's *Elegy* in Chapter 1, for example, in large part because it became one of the most heavily anthologized English poems during the nineteenth century and into the mid–twentieth century. Mary Shelley's *Frankenstein*, the main subject of Chapter 2, more or less instantly became a cultural myth and is now likely the most read Romantic-era novel in the world, while Thomas Malthus's *An Essay on the Principle of Population*, another key text in Chapter 2 and also in Chapters 3 and 6, continues to exert a strong influence on our own reflections about global population control. I focus on Percy Bysshe Shelley's *Queen Mab* in Chapters 4 and 6 in part because it had a long influence on the labor movements in multiple European countries throughout the nineteenth and early twentieth centuries. My corpus of eighteenth-century and nineteenth-century texts in this book is thus intended less to be representative of Romanticism as a whole—though these were all important texts in that period—and more to underscore the continuing persistence of Romantic-era approaches, framings, dilemmas, and considerations into the present.

As a consequence, while these chapters are focused primarily on eighteenth- and early-nineteenth-century texts and debates, each chapter also traces the reverberations and resonances of these debates within the twentieth-century transformation of liberalism into neoliberalism, which latter is characterized by the efforts of economists and policy makers to force every aspect of life into economically oriented models of population experimentation. I explore these resonances between Romantic and contemporary texts and debates with the hope that careful exploration of Romantic-era efforts to coordinate liberalism, biopolitics, and literature can provide us with resources for our own struggles in the era of neoliberalism, population-oriented "smart" biopolitical technologies, and ecological disaster. Though my analysis will point to liberal dimensions of Romanticism, my point is not, fundamentally, to critique Romanticism as a mode of liberal ideology, but rather to understand Romanticism as an attempt to steer the biopolitical techniques of liberalism toward more liberatory shores.

Liberalism and Biopolitics: A Few More Words

Though each chapter of this book is largely self-sufficient, and so the chapters could be read in any order, my hope is that readers will read them in the order presented, as the stakes of the last three chapters are more evident when read after the first three. However, since some readers may choose to approach the book differently, I will explain here how I approach the terms liberalism and biopolitics, so that I can avoid repeating this discussion in each chapter.

For scholars of Romanticism, my focus on liberalism may seem peculiar. Liberalism is not often associated with Romanticism, and there is, accordingly, very little work focused specifically on their relationship.[18] This ought to surprise us, since many of the key authors cited in intellectual histories of liberalism—Thomas Hobbes, John Locke, David Hume, Adam Smith, Jean-Jacques Rousseau, and Immanuel Kant—are also widely acknowledged as key reference points for many Romantic authors.[19] Nor is it hard to see parallels between the twin premises of most varieties of liberalism—namely, that if individuals are "freed" to determine and follow their own sense of their best interests, then an immanent form of collective order will emerge—and the stress of Romantic authors on individual uniqueness, which was frequently linked to the hope that this would lead, in some often unspecified way, to new, more equitable forms of social order. Such parallels between liberalism and Romanticism were evident to the Victorian-era theorist of liberalism John Stuart Mill, who explicitly acknowledged the importance of Romantic-era poets such as William Wordsworth for his overall intellectual development and also cited the German Romantic-era author Wilhelm von Humboldt prominently in *Of Liberty*, Mill's classic articulation of liberalism.[20] Yet one of the peculiarities of contemporary literary-critical scholarship is that even as liberalism is a central category for Victorianist literary criticism, it is more or less absent as a term from Romantic literary criticism.[21]

The relative disinterest of scholars of Romanticism in liberalism may be a function of a long-standing tendency within Romantic literary criticism to parse discussions of Romantic-era politics through the binary of "radicalism" versus "conservatism."[22] Liberalism is hard to situate within such a binary, since it is both radical in its aim to eliminate traditional social relationships in favor of new modes of human relations and conservative insofar as these changes tend to benefit an existing bourgeois class. It was for this reason that Isaac Kramnick coined the term "bourgeois radicalism" to describe the liberal political positions of authors such as Joseph Priestley and Tom Paine, and a similar rationale underlies Saree Makdisi's distinction between Blake's (true) radicalism and the (merely bourgeois) radicalism of authors such as Thomas Paine and Mary Wollstonecraft. Yet the risk run by such coined terms and distinctions is that we lose sight of a longer history that connects Romantic-era authors both to earlier building blocks of liberalism, such as seventeenth-century political arithmetic (which was not clearly "bourgeois"), and to post-Romantic developments in liberalism, such as John Stuart Mill's critique of capitalism and, conversely, the tight neoliberal embrace of capitalism represented by twentieth-century authors such as Friedrich von Hayek. In this book, I explicitly employ the term "liberalism" in order to keep this longer history in view.

My use of the term "liberalism" also draws heavily from Foucault's approach, which differs in emphasis from more traditional philosophical and political-theoretical accounts of liberalism. As many historians and theorists of liberalism have noted, it is difficult in practice to establish unambiguously the core tenets of liberalism. Or, as the political theorist Alan Ryan puts it, it is "easy to list famous liberals; it is hard to say what they have in common."[23] However, the brief definition of liberalism that I offered—the premise that if the resources of the state, such as law (especially that which ensures private property and contracts), are oriented toward "freeing" individuals to determine and follow their own sense of their best interests, then an immanent form of collective order will emerge—seems to me to be a fairly uncontentious summary of key aspects of what Ryan calls "classical" liberalism, exemplified by authors such as John Locke and Adam Smith.[24]

Foucault's account of classical liberalism does not so much contest this definition as refine it. This is evident if we consider Ryan's own definition of classical liberalism. Ryan argues that classical liberalism was

> focuse[d] on the idea of limited government, the maintenance of the rule
> of law, the avoidance of arbitrary and discretionary power, the sanctity
> of private property and freely made contracts, and the responsibility of indi-
> viduals for their own fates. (24)

Rather than simply noting that liberal theorists such as Locke or Smith advocated for limited government and against arbitrary and discretionary power, Foucault focused on the transformation of epistemology that enabled the success of this attack on sovereign power. He stressed that this occurred because authors such as Locke, Hume, Smith, and the French physiocrats, among others, convinced their readers that there were "natural" dynamics of economic order that escaped the reach of sovereign power but that nevertheless could be channeled and regulated by new forms of knowledge, such as the new sciences of political economy and physiocracy. Foucault's focus on liberalism as dependent upon an epistemological transformation led him in turn to recognize the importance of eighteenth-century concepts such as "population" and "regulation" for liberalism, for these were among the new concepts that enabled theorists of liberalism to alter the relationship between epistemology and political power.[25]

Foucault's stress on the importance of concepts such as population and regulation for liberalism underscores the extent to which classical liberalism was a form of biopolitics, that is, an approach to power focused not on micromanaging individual behavior but instead interested in locating, and thereby enabling some kind of steering of, regularities within the collective "body" of the population. Romantic literary criticism has been more attentive to the development of biopolitics than to liberalism.[26] However, within most literary-critical accounts of Romantic-era biopolitics, the latter has been assumed to be pernicious, and literary authors are cast either as the heroes who fight against biopolitics or the dupes who furthered its reach.[27] Though this approach is not surprising when it comes to assessing, for example, the rather nasty implications Thomas Malthus drew from his account of the "principle of population," biopolitics also covers developments such as smallpox inoculation and various other forms of disease prevention that developed from that. It is often not clear in Romantic literary criticism whether those latter biopolitical technologies are also understood as inherently oppressive and problematic. It is thus worth considering more closely three rationales that seem to underpin this more general literary-critical tendency to understand biopolitics as something that should be opposed at all costs, even if these rationales have often remained implicit in Romantic literary criticism.

First, insofar as biopolitics requires some form of political coercion over individuals—for example, via strong suggestions or even legal requirements that individuals be inoculated against smallpox—such control can be opposed in the name of individual freedom and autonomy. I am tempted to call this the "liberal" critique of biopolitics, since it opposes biopolitical developments in the name of the individual autonomy that has been so important to classical liberal theorists. Yet if, as I argue here, liberalism is itself a mode of biopolitics, to critique biopolitics in the name of individual autonomy will likely result in a paradoxical and self-defeating position. Part of the intellectual appeal of liberalism is that it does not *oppose* individual autonomy and population-level interests; it instead suggests that individual differences and autonomy *lead to* collective improvement, and it strikes me as unlikely that any position that begins by valuing individual autonomy will not end up finding itself led back to the more fundamental body of the population.

A second argument against biopolitics contests its apparent premise that there are corporeal aspects of collective living—for example, the bare minimum of food an individual requires to survive or an individual's resistance to a disease such as smallpox—that cannot be altered by means of collective rational discussion. It was precisely this critique that structured the Romantic-era debate between William Godwin and Thomas Malthus that I consider in Chapters 2 and 6. Godwin argued that all aspects of human existence, including reproduction, disease, and human lifespan, could be altered through rational decision making; Malthus responded that reproductive drives and food production would always escape such control. Karl Marx embraced Godwin's side of this debate, and a vaguely Marxist vision of humanity's complete control over nature lies, I suspect, behind much of the recent animus toward biopolitics. Yet such an animus raises the question of whether the classless society of the future would also engage in fundamentally biopolitical practices such as disease vaccination and leaves obscure the question of how the vision of human mastery over nature comports with our increasing awareness of our embeddedness within ecological systems that cannot be controlled in this way.

Finally, some theorists have argued that a seemingly life-oriented bio-politics either contingently or necessarily leads to its opposite, what the political theorist Roberto Esposito calls a "thanatopolitics," in which bio-political policies are employed to deliberately kill (or let die) certain groups of people for the sake of saving other, purportedly more valuable lives. This line of argument has been developed most extensively by Giorgio Agamben and Esposito. Agamben argues that biopolitics necessarily employs what he calls the "sovereign exception," by means of which sovereign political power is constituted around an ancient Greek distinction between properly political life (*zoe*) and the "mere" life (*bios*) of those who are excluded from properly political life and who can be killed without legal or moral consequence.[28] Agamben traces the political-legal structure of the sovereign exception back to ancient Roman law and sees modern biopol-itics as simply the continuation of this logic within a modern context. What distinguishes our modern context, though, are technologies that enable those excluded from properly political life to be killed en masse, and Agamben sees National Socialist concentration and extermination camps as the logical terminus of modern biopolitics.

Esposito, by contrast, argues that biopolitics leads to thanatopolitics only when the former is combined with what he calls the "immunitary para-digm."[29] Building on Foucault's work on biopolitics, Esposito argues that modernity—that is, those political concepts, institutions, and practices that first emerged in Europe in the seventeenth century and continue to deter-mine national and international political relations—is best understood as the unfolding and increasing intensification of the paradigm of immuni-zation. This paradigm has three key elements. First, politics is understood as the solution to the problem of human survival (rather than, for example, the collective pursuit of excellence, as politics was understood in the classical Greek tradition). Second, what purportedly threatens collective survival is not an external threat, but rather something immanent to human relations. Third, the latter problem is to be solved by "shelter[ing] life in the same powers that interdict its development"—that is, by "immunizing" social relations against the internal threat that places them in jeopardy, most generally by sacrificing lives understood as less worthy for those considered more worthy.[30] For Esposito, when biopolitical techniques are subordinated to the immunitary paradigm, they indeed lead to the thanatopolitical outcome of Nazi Germany to which Agamben points. However, because biopolitics and the immunitary paradigm are analytically separate for Esposito, he also holds out hope for what he calls an "affirmative" biopolitics, which would not understand politics as oriented primarily toward human survival and hence would not engage in the immunitary solution of sacrificing some lives for the sake of others.

Of the critiques of biopolitics I have outlined here, I find Esposito's the most convincing and also the most useful for understanding the relationship between biopolitics and liberalism.[31] Esposito argues compellingly, for example, that the modern, liberal concept of liberty—that is, the explicit centerpiece of liberalism in all its forms—developed fully within the immunitary paradigm. Echoing a long lineage of liberal theorists, Esposito distinguishes between an "ancient" and a "modern" understanding of liberty.[32] He contends that "the concept of liberty, in its [ancient] germinal nucleus, alludes to a connective power that grows and develops according to its own internal law, and to an expansion or to a deployment that unites its members in a shared dimension."[33] That ancient concept of liberty was connected to the awarding of a "privilege" that was granted to some individuals by a collective body, such as a city-state, so that the individual could more fully pursue his or her (usually his) ends within that collective. However, Esposito contends that, beginning in the seventeenth century, liberty began to be understood in a "negative" fashion, that is, *not* as a mode of becoming, such as the ability to grow and change within a collective framework, but rather as the absence of any obstacle between oneself and one's will (71).[34] Modern liberty is thus fundamentally a concept of security of the individual against outside threats, not a concept of privilege that allows one to do something (72). Esposito contends that, because of this emphasis on security, modern authors never limit themselves "to the simple enunciation of the imperative of liberty" but also necessarily "implicat[e] the organization of conditions that make this effectively possible," namely, strong law and police forces (74). Hence, even though liberty is understood as freedom from external constraints and compulsion, this goal apparently can only be pursued when it is supported by external constraints and compulsion. For Esposito, this emphasizes in dramatic fashion what he calls the self-destructive nature of the immunitary paradigm, which, he argues, always ends up destroying precisely what it had aimed to save. Or, as Esposito writes, "it isn't possible to determine or define liberty except by contradicting it" (75).

Esposito's approach provides a way of understanding liberalism as a mode of biopolitics but also of distinguishing biopolitics from liberalism, with the goal of making it possible to think an affirmative biopolitics beyond liberalism (and neoliberalism).[35] Esposito's own method, which he describes as a "constructive deconstruction," does not allow him to say much about the content of an affirmative biopolitics, for he claims that it is first necessary to reveal the various "antinomies" that structure earlier political, philosophical, anthropological, and scientific thought before one can develop such a biopolitics positively.[36] While I share Esposito's goal of sketching out an affirmative biopolitics, I am less convinced by his methodological premise that one can only approach this task asymptotically, and I seek to show in this book not only that many Romantic authors sought to develop the contours of an affirmative biopolitics but also that they provide us with positive resources for our own efforts. Looking to the Romantics for a provisional orientation, then, an "affirmative" biopolitics would not set self-preservation but rather self-transformation as its goal, and it would affirm every human life (and likely also many forms of non-human life), rather than searching for those human lives that can be sacrificed for the sake of the rest.[37]

In part because I share Esposito's interest in thinking through what an affirmative biopolitics could mean, I return repeatedly to the example of inoculation in this book. From one perspective, the practice of inoculation (and later, vaccination) was just one of many examples of the results of new eighteenth-century biopolitical sciences, and to acknowledge this, I often pair inoculation with the equally new science of political economy. Yet my emphasis on inoculation is intended to have a tactical benefit that is itself bound up with Esposito's understanding of the singular importance of concepts of immunity for the modern development of biopolitics. Scholars interested in eighteenth- and nineteenth-century biopolitics have often focused on eighteenth- and nineteenth-century political economy and have adopted an understandably critical approach to the claims of authors such as Smith and Malthus. Such a critique comes easily to scholars in the humanities, in part because the class, racial, and gender inequities to which eighteenth- and nineteenth-century political economy led are hard to deny and because the persisting importance of Marxist critique within the humanities ensures that one can imagine that economic relations could and should be organized completely otherwise (even if the details of that alternative organization are a bit fuzzy). Biopolitical practices such as inoculation and vaccination lead, however, to trickier, less predictable terrains of thought, for though these practices *have* often been applied inequitably, it is less clear that they are simply ideological mystifications. Moreover, as Esposito stresses, practices such as inoculation and vaccination underscore our radical *openness* to one another, an openness that both enables infection among individuals and the immunological channeling of that openness into preemptive immunity via techniques such as inoculation and vaccination. The examples of inoculation and vaccination thus encourage us, in ways that the example of political economy generally does not, both to engage and to think through the implications of this common openness.

With that said, Esposito's approach also helps us understand better the relationships among biopolitics, political economy, liberalism, and *neo*liberalism, though the latter is not a term that Esposito uses. As both Foucault and historians of economics such as Philip Mirowski, Dieter Plehwe, Edward Nik-Khah, and Robert Van Horn have noted, neoliberalism was first formulated in the 1930s by economists such as Friedrich Hayek, who felt that nineteenth-century liberals had not recognized the extent to which governments must actively *construct* the conditions for successful markets, rather than simply adopting a hands-off, laissez-faire approach to them.[38] Combining Foucault's and Mirowski's accounts, I approach neoliberalism in this book as committed to the principle that "the market" is in its essence an information processor and, moreover, the most efficient information processor possible. However, neoliberals also believe that the market-cum–collective computer cannot fully optimize itself when simply left alone, as eighteenth- and nineteenth-century thinkers had believed. Rather, the goal of government is to force individuals in directions that will enable a full optimization of market relations. This is to be accomplished by actively encouraging individuals to become "entrepreneurs of the self," which in turn requires an aggressive reconceptualization of all human relations as market relations, the elimination of welfare provisions, and the creation of sufficient police forces and prisons to immunize society against those who either do not wish to participate or do not "succeed" in this grand vision of market optimization. Despite the hopes of many critics that the worldwide financial crisis that began in 2007 would destroy the legitimacy of neoliberalism, in fact the opposite seems to have been the case, for—in a reversal that induces rather painful intellectual whiplash—the financial crisis caused by neoliberal policies has served as a means for neoliberalism to integrate itself even more deeply, and biopolitically, into everyday life.[39]

In such a context, it is vital to keep firmly in view not only the close relationship between biopolitics and neoliberalism but also their fundamental distinction. That is, on the one hand, it is important to recognize that the neoliberal goal of establishing a fully optimized market society requires the development of ever more expansive and precise biopolitical technologies, such as the now vast range of "smart" devices and processes that make it possible to grasp and manipulate algorithmically nearly all of daily life.[40] On the other hand, it is equally important to recognize that these same biopolitical technologies exceed the neoliberal goal of a global market society, in the sense that they also enable other kinds of human relations. My hope is that, by emphasizing the difference between biopolitics and liberalism in the Romantic era, the fault lines of this distinction in our own period will also become more evident.

Liberalism, Biopolitics, and the Sciences

I follow Foucault in understanding eighteenth-century sciences as essential to the emergence of biopolitics in general and that specific form of eighteenth- and early-nineteenth-century biopolitics that we now call liberalism. Eighteenth-century sciences enabled biopolitics and liberalism by providing their advocates and proponents with putative truths—namely, those "laws" and "principles" discovered by new sciences such as political economy, physiocracy, and inoculative medicine—that could be used to reduce the power of those who believed that only sovereign commands, translated into laws that were faithfully respected by legal subjects, could enable social order. Advocates for new sciences such as political economy and inoculative medicine argued that, just as Newton had revealed immutable laws for matter, their sciences revealed immutable laws of human behavior, which determined the movements of humans, no matter what a sovereign might command.[41] Yet advocates of new sciences such as political economy and inoculative medicine also argued that knowledge of the laws that they discovered would, very much *unlike* Newton's laws of matter, enable human relationships to be altered for the better. To put this another way, those eighteenth- and early-nineteenth-century sciences especially important for the emergence of biopolitics and liberalism were what we might call "regulative sciences," in the sense that these sciences both aimed to locate self-regulatory movements within human relationships and sought to apply knowledge of those movements directly to human affairs.

The sciences that I focus on in this book thus tend to be both extraordinarily applied and—as a consequence—were often sciences that were contested, in the sense that the question of whether in fact they produce real knowledge was never a given. For many eighteenth-century commentators, for example, it was by no means clear whether smallpox inoculation actually worked. Advocates of inoculation argued that the best way to settle the question was to allow "the Experiment [to] go on" by inoculating large populations against smallpox and then checking fatality statistics; critics of inoculation responded by attacking a science that proceeded in this way as amoral at best.[42] The sciences of political economy and physiocracy were equally contested, especially when advocates such as Thomas Malthus moved demographic claims about the laws that determined the increase and decrease of population size to the center of political economy. Advocates of the sciences that I consider had to fight continually to legitimate their claims as valid contributions to natural philosophy (and, later, "science"), and that fight was never decisively determined one way or the other.[43]

As a consequence of this way that eighteenth-century biopolitics, liberalism, and the regulative sciences were linked to one another, I focus less on how literary authors explicitly engaged the work of famous scientists and more on how literature itself became part of the contests around concepts central to the regulative sciences, such as "regulation," "population," and "experiment." Eighteenth-century regulative sciences were often effective less because a particular claim of a specific author was widely accepted and more because these sciences effectively promoted or transformed the basic coordinates of a concept such as "population." Eighteenth- and early-nineteenth-century regulative sciences sought to convince readers, for example, of the importance of questions such as: What is a population? How can one best study populations? Are differences among people important or not for the study of populations? What are desirable behaviors? How can populations be manipulated to increase the incidence of desirable behaviors and decrease the incidence of undesirable ones (or people)? While the regulative sciences had a variety of answers to each of these questions, so too did authors whom we now consider to be literary, such as Wordsworth, Hazlitt, and Mary Shelley. And because the regulative sciences and their answers were contested in the Romantic era, literary authors contributed to these debates neither as tyros nor as emulators of the sciences but as advocates for different constellations of biopolitics and the sciences.

This approach helps us recognize literary texts as elements of what Foucault described as "technologies of the self." Foucault did not explicitly discuss biopolitics after the late 1970s, turning instead to historical investigations of what he called "governmentality." Foucault used this term to refer to practices by which individuals and groups employed truth claims in order to govern, or regulate, themselves. Governmentality thus includes practices by which individuals

> effect, by their own means, a certain number of operations on their own bodies, their own souls, their own thoughts, their own conduct, and this in a manner so as to transform themselves, modify themselves, and to attain [what they believe will be] a certain state of perfection, happiness, purity, supernatural power.[44]

As Thomas Lemke notes, Foucault's interest in governmentality was not a turn away from but rather an attempt to recontextualize biopolitics.[45] While from one perspective biopolitical projects such as eighteenth-century political economy and inoculation represented a lessening of juridical-political power in favor of allowing "the truth" about the nature of populations to determine political policy, these projects also required that individuals learn to look at the world in terms of truths about populations and to alter "their own souls, their own thoughts, their own conduct" accordingly. Romantic-era literature, I contend, was one of the mechanisms that enabled these biopolitical technologies of the self in the eighteenth and nineteenth centuries, and it did so by connecting otherwise abstract truth claims about populations and power with aspirations for states of "perfection, happiness, [and] purity."

As I document in each of my chapters, literature provided elements of technologies of the self that could support liberal forms of biopolitics, but literary texts often also pointed toward nonliberal forms of biopolitics. The first chapter, for example, explores how the concept of genius makes concrete otherwise abstract aspirations for maximizing qualities within populations—and makes equally concrete what will happen if a desired quality is not maximized—and this lent itself easily to the liberal project of political economy. However, as I note at the end of the chapter, the project of genius maximization also led authors such as Godwin and Wordsworth beyond liberal biopolitics, toward forms of biopolitics no longer bound to class division and ecological plunder. More briefly, the second chapter connects governmentality to the concept of population by exploring both liberal and nonliberal versions of this concept; the third chapter considers some of the technical devices (character-systems and free indirect discourse) by means of which literature could claim to articulate *truths* about populations; the fourth chapter explores liberal and beyond-liberal approaches to globalization and the natural systems of the globe; the fifth chapter connects governmentality to concepts of self and collective experimentation and to the related question of whether the "we" to which individuals are connected should be understood as a historical inheritance that must be defended or a community that can only be achieved in the future; and the final chapter explores the concept of governmentality through analysis of its key modern synonym, self-regulation.

Each chapter, in short, connects biopolitics with governmentality, with the goal of exploring both the development of liberal *and* nonliberal techniques of the self. My hope is that this approach helps us move our understanding of biopolitics and its sciences beyond the frame of liberalism, and the wager of this book is that Romantic authors introduce many of the problems, regulative concepts, and experimental stances that will be necessary for this collective work.

Part I: Romanticism, Biopolitics, and Literary Concepts

1. Biopolitics, Populations, and the Growth of Genius

The present order of society . . . is the great slaughter-house of genius and of mind.
— William Godwin, "Of the Sources of Genius"

A people is a detour of nature to get to six or seven great men. —Yes, and then to get around them. [*Ein Volk ist der Umschweif der Natur, um zu sechs, sieben großen Männer zu kommen. —Ja, und um dann um sie herumzukommen.*]
— Friedrich Nietzsche, *Beyond Good and Evil*

Michel Foucault's lectures from the 1970s on the emergence of biopolitics and liberalism in the eighteenth century emphasize the key role played by the term "population" in those developments, and this trend suggests several important implications for literary criticism that focuses on this period.[1] It did not take Foucault's work, of course, for literary critics to recognize the importance of the term "population" for literature of the eighteenth and nineteenth centuries, but earlier criticism tended to focus on the very late-eighteenth-century understanding of this term developed by Thomas Malthus.[2] For Malthus, population was always potentially a problem and always connected to the twin specters of dwindling food supplies and surging sexual desire. Foucault, by contrast, stressed earlier understandings of the term, noting that for late-seventeenth- and eighteenth-century authors, the "problem" of population was generally that of too few people rather than too many (a question of under- rather than overpopulation), that the key issue motivating discussion of populations was labor, and that emigration—which Malthus presented as simply a temporary or imaginary solution to the problem of overpopulation—was often understood as precisely what *could* resolve all kinds of population problems. Foucault also stressed the conceptual distinction between the population of a territory and the group of legal subjects within a realm. Where the latter concept presumed that the sovereign employed laws to prevent or encourage various behaviors, a population was understood as a collection of relatively recalcitrant bodies that tended to follow their own movements, no matter what laws were enacted; as a consequence, the legislator governed best when he or she found ways of channeling these existing movements.

For Foucault, the concept of population underpinned the emergence of both liberalism, which presumed that one governed best when individuals could determine, so far as possible, their own courses of action, and bio-politics, which involved gathering data about the biological aspects of a population (for example, "health, hygiene, birthrate, life expectancy, [and] race") so that one could then alter some of those dynamics.[3]

This chapter builds upon Foucault's work by considering the implications of eighteenth-century approaches to population for our understanding of literary texts. However, I focus on the link between eighteenth-century population thinking and the identification of the purportedly rare quality of *genius* to emphasize that biopolitics is not restricted solely to those clearly biological aspects of collective life, such as health, hygiene, and birthrate, that Foucault stressed. My starting point is William Petty's intriguing claim in *Another Essay in Political Arithmetick* (1683) that among the virtues of increasing the population of a territory is a potential increase in the "*Arts of Delight* and *Ornament*," for, Petty claims, "it is more likely that one Ingenious Curious Man may rather be found out amongst 4 millions than 400 persons."[4] Petty, a member of the Royal Society and founder of "political arithmetick," is an important figure in the history of economics, the history of statistics, and recent discussions of the history of population theory, primarily because of his interest in creating categories for counting and measuring people within a territory. Yet Petty's interest in population has tended to be read as a homogenizing enterprise: though political arith-metic distinguished between kinds of people, it often seemed committed to erasing differences between individual bodies in favor of abstract, shared human capacities, such as the capacity to labor. What Petty's comment about genius underscores, though, is another way that seventeenth- and eighteenth-century authors thought about both population and the tech-niques by which legislators could know and channel its resources. Rather than restricting political arithmeticians to broad categories (for example, kinds of laborers), the concept of population also encouraged Petty to consider rare instances of something virtuous—not simply the Ingenious Man, not simply the Curious Man, but the Ingenious Curious Man—and to consider how, through regulation, one might increase the number of such individuals.

Petty's comment illuminates an important series of links between three eighteenth-century topics: genius, population, and "literature" (a term that, by around the end of the eighteenth century, had come to refer solely to "elevated" forms of fictional writing). In schematic form, my argument is the following. The tentative link that Petty established between genius and population was consolidated neither in political arithmetic nor in its successor, political economy, but rather in two more "literary" arenas: mid-eighteenth-century debates about the nature of genius and mid- to late-eighteenth-century poetry that reflected on how geniuses might be overlooked or lost. These discourses helped transform the link between genius and population from an abstract connection to a topic of vital concern. Mid-eighteenth-century discourses on genius fleshed out Petty's aspiration for a political arithmetical approach to genius by emphasizing forms of regulation purportedly capable of increasing the number of geniuses in British territory. Poetry, for its part, illuminated Petty's aspiration of maximizing genius by surrounding this positive goal with two "negative" possibilities, or what I will call biopolitical forms of concern about what might result were genius not to be regulated. First was the worry, exemplified in Thomas Gray's *Elegy Written in a Country Churchyard* (1751), that, in a large population without sufficient forms of institutional discovery and nurturing, potential geniuses would be lost. The second kind of worry, exemplified by William Wordsworth's 1800 Preface to *Lyrical Ballads* and the "Arab Dream" section of *The Prelude*, was that, in a large population with its own dynamics of interest and entertainment, a genius once glorious, such as Milton, could be forgotten. Both kinds of worry, I suggest, were key to the emergence of population as an effective biopolitical concept and provided means for transforming the abstract possibility of maximizing the incidence of a quality (genius) into a pressing concern for individuals. At the same time, both worries—that of unseen literary potential and that of misplaced literary value—can only be problems (that is, can only become legible and insistent sources of concern) from the viewpoint of population thinking. In the final part of my argument, I contend that the modern, restricted notion of "literature"—literature understood as especially valuable instances of imaginative writing, primarily in the genres of poetry and drama—emerged as the concept and institution that fully sutures Petty's biopolitical hope of maximizing genius with Gray's and Wordsworth's (equally biopolitical) worries about losing genius.

As some readers may already have noted, my literary examples parallel those employed by John Guillory in his well-known sociological account of the emergence of the modern concept of literature.[5] This is not a coincidence, for in addition to expanding our understanding of biopolitics beyond those clearly biological elements upon which Foucault focused, I also hope to clarify important differences between sociological and biopolitical readings of eighteenth-century literary culture. Guillory focused on

Gray and Wordsworth precisely because these two authors had become part of the "canon" within the schooling systems of the United Kingdom (and its colonial possessions) and the United States. His powerful and nuanced reading of the reasons why Gray and Wordsworth became part of the canon (which depended in part on their contributions to the development of our modern concept of literature) provides a helpful contrast to my biopolitical reading of this same development, for I stress a different understanding of the affective work of literary texts than that outlined by Guillory: Rather than functioning solely as what Guillory describes as points of "rest" in the midst of general social competition, literary texts also functioned as instigators of those emotions and affects essential to the operation of biopolitics.

Political Arithmetic and the Increase of Genius

"Political Arithmetick" was the name William Petty gave, in the late seventeenth century, to what he hoped would be a new science that, by counting and measuring people and things within a country—that is, by the use of "Number, Weight, or Measure"—would increase the wealth and military strength of the nation. This new science was arithmetical in the sense that it would allow legislators to answer conclusively, in the manner of arithmetical demonstrations, quantitative questions, such as whether "the Rents of lands [in the kingdom] are generally fall'n," whether "there is a great Scarcity both of Gold and Silver," and whether "the land is under-peopled."[6] Unequivocal demonstrations of the answers to such questions would not only displace unfounded opinions upon which legislators might otherwise base political decisions but would also encourage legislators to consider how best to maximize desired aspects of a specific country. Petty's interest in counting people and things was fundamentally oriented toward identifying limits—that is, maxima and minima of qualities such as national wealth or the amount of daily food a laborer needed to survive—and locating those points in the social field at which pressure could be applied in order to encourage movement of some quality toward the desired maximum or minimum. Petty insisted that maximizing qualities such as wealth or military strength was not simply a matter of having more people or land, for even "a small Country [with] few People" can, if it sufficiently exploits "its Situation, Trade, and Policy . . . be equivalent in Wealth and Strength, to a far greater People and Territory."[7]

As twentieth-century commentators have noted, though Petty's new science of political arithmetic attracted followers and commentators, it remained more of an aspirational than actual science.[8] Petty, and subsequent advocates of political arithmetic such as John Graunt, provided legislators with numbers about all kinds of things, such as the historical and projected future population growth for England, London, and the entire globe and the basic cost of keeping alive a laborer (seven pounds per annum, according to Petty).[9] Yet contemporary legislators seemed neither particularly interested in basing policy on this data nor in establishing the kinds of data collection systems that Petty and his followers stressed were necessary for the development of this new science. Political arithmetic thus remained during the eighteenth century a well-known potential science, but its approach to governing by numbers was not put into practice until the nineteenth and twentieth centuries.

Yet if political arithmetic failed to materialize as a science of governing in the way that Petty and his followers had hoped, it nevertheless fundamentally transformed the intellectual landscape around several key concepts, especially the connection between "population" and "people."[10] In his seventeenth-century essay "Of Seditions and Troubles," Francis Bacon had urged legislators to see the state of "the population" as a key factor in encouraging or hindering political upheaval, and he had stressed that "the Population" is not "to be reckoned only by number; for a smaller number that spend more, and earn less, do wear out an Estate sooner than a greater number that live lower, and gather more."[11] Though Petty and Graunt tended to use terms such as "the People" or "the Nation" rather than "population," political arithmetic, even just as aspiration, helped eighteenth-century authors begin to understand what it would mean to reckon population not as a single number but as a complex collection of bodies and desires that contained numerous potential maxima and minima of qualities such as wealth, subsistence, and military power.

It is in this context that Petty's claims about genius take on their real significance. Political arithmetic allowed eighteenth-century authors to understand population as, among other things, that within which legislators could maximize desired qualities and minimize undesirable qualities. In some of Petty's examples, maximizing a value such as wealth or military power depended on determining the absolute minimum of something that everyone needs, such as food, or something that the vast majority of people should do, such as labor. But Petty's example of the "Ingenious Curious Man" underscores that political arithmetic was also interested in qualities possessed by only a few individuals. Petty discussed the maximization of geniuses in the context of a political arithmetical analysis of London, in which he proposed "two Imaginary states" of London, one in which the city was seven times its current size and one in which it was one-seventh its current size.[12] Petty analyzed these two states according to a number of criteria, such as defensibility of the city from foreign attackers; prevention of "*intestine Commotions of Parties* and *Factions*; "*Gain* by Foraign [*sic*] Commerce"; and "*Husbandry, Manufacture,* and . . . *Arts* of Delight and Ornament."[13] Petty concluded that since the "*Arts* of Delight and Ornament" are "best promoted by the greatest number of Emulators," and since "it is more likely that one *Ingenious Curious Man* may be found out amongst 4 millions than 400 persons," the more populous of London's imaginary states would best serve that goal. Though the maximization of genius was not a capacity to which Petty returned in his texts on political arithmetic, this aspiration was nevertheless fully consonant with—and arguably a necessary implication of—his understanding of population as a collection of individuals within which qualities can be maximized and minimized.

Though Petty's assumption that "it is more likely that one Ingenious Curious Man may be found out amongst 4 millions than 400 persons" may seem commonsensical, not all eighteenth-century authors agreed. In his *Letter to D'Alembert on the Theater* (1758), for example, Jean-Jacques Rousseau—an author obsessed with the economic dimensions of modern life and who likely read Petty—claimed something quite different.[14] Rousseau's open letter was intended to contest Jean d'Alembert's glowing description of the institution of the theater and his suggestion that a theater should be established in Geneva. In earlier publications such as *Discourse on the Sciences and Arts* (1750) and *Discourse on the Origin and Basis of Inequality among Men* (1754), Rousseau had already written critically about the modern increase of genius in the arts and sciences and established his antagonism to both theater and spectacle-based relationships of modern urban life. In his *Letter to D'Alembert*, he sought to convince his readers that the institutions of small-town life were not stultifying but instead would, if fully embraced, result in an increase in genius:

> In a little town, proportionately less activity is unquestionably to be found than in a capital, because the passions are less intense and the needs less pressing, but more original spirits, more inventive industry, more really new things are found there because the People are less imitative; having fewer models, each draws more from himself and puts more of his own in everything he does; because the human mind, less spread out, less drowned in vulgar opinions, elaborates itself and ferments better in tranquil solitude; because, in seeing less, more is imagined; finally, because less pressed for time, there is more leisure to extend and digest one's ideas.[15]

While Rousseau also dealt with the question of genius and innovation from the perspective of population, his understanding of the connection between genius and imitation led him to conclude that the relevant "incubator" population for genius could be a small town, rather than a very large city. For Rousseau, what allowed genius to emerge from a population was not the size of the latter but rather the kinds of social relations it encouraged.

Though Petty's reflections on political arithmetic isolated the maximization of genius as a potential goal for future legislators, they also raised several questions. For example, was population increase the key to producing more genius, or were social institutions of the sort described by Rousseau necessary both to identify and cultivate genius? If population expansion was indeed necessary for an increase in the number of geniuses, what other forces, social or otherwise, might hinder its actual increase? Though these questions might be addressed in part by counting and numbers, they were more fundamentally about models of populations and the relationship between populations and social institutions such as those described by Rousseau. As I note in the next two sections, though some of these considerations were taken up obliquely in late-eighteenth-century sciences such as political economy, the implications of population thinking for the maximization of genius were more thoroughly investigated in philosophical and literary reflections on genius itself.

Political Arithmetic, Political Economy, and Genius

Though most elements of Petty's political arithmetic, such as national censuses, remained aspirations rather than implemented technologies during the eighteenth century, his "liberal" approach to social order nevertheless established a template that was subsequently taken up more successfully by the new mid-eighteenth-century science of political economy.[16] Petty's approach was liberal in that he understood the legislator not as a sovereign who possessed the capacity to control the behavior of his subjects via legislation but rather as an observer-regulator who could, through the new science of political arithmetic, locate and then harness existing "natural" forms of behavior that evade the reach of law. Foucault captures this in his description of liberalism as fundamentally oriented toward truth rather than justice: for liberals, legislators must first focus on the truth of how an institution, such as "the market," functions naturally, rather than seeking to intervene in the operation of this natural institution in the name of justice (by, for example, establishing a "just" price for wheat).[17] Foucault's account explains Petty's desire for political *arithmetic*; this numerical dimension indexed Petty's effort to shift the activity of the legislator from justice and right and toward truth and knowledge. Foucault's account also helps us understand better Petty's understanding of population as a collective characterized by autonomous forms of natural movement, rather than as simply a collection of legal subjects. As authors such as David Hume, James Steuart, and Adam Smith subsequently created what would become known as political economy, they fully embraced this liberal approach pioneered by Petty, even if they did not always adopt Petty's particular techniques.[18]

Perhaps counterintuitively, Petty's liberalism also established the basic framework for the mid-eighteenth-century explosion of texts on genius. In precisely the same period that Hume, Smith, and Steuart published their texts on the wealth of nations and political economy, authors such as William Sharpe, Alexander Gerard, William Duff, and Edward Young published extended "dissertations," "essays," and "conjectures" on genius. This coincidence has been the subject of an important debate among literary critics who have sought to shift discussion of mid-century texts on genius from an exclusively aesthetic discourse to one that acknowledges the dependence of aesthetic categories, such as genius, on class-based ideologies. The key point of debate has been whether, as Martha Woodmansee and Mark Rose have argued, the discourse of genius was part of political economy or whether, as Zeynep Tenger and Paul Trolander have argued, the discourse of genius was instead a rival to political economy: that is, an alternative understanding of how to generate the "wealth of nations."[19] Yet arguably lost in this focus on the ideological dimensions of the genius debate are its liberal and biopolitical dimensions. Authors such as Sharpe, Gerard, Duff, and Young sought to determine the truth of genius—its various modes, its relationship to our mental faculties, and its relationship to imitation—*so that* the incidence of genius could be increased. From this perspective, the ideological convergence or tension between political economy and debates about genius is less important than the fact that both develop within the liberal, biopolitical framework initiated by Petty.

The desire to maximize genius within a national territory was explicit in most mid-century British texts on "original genius." In *A Dissertation upon Genius* (1755), Sharpe contended that genius was not implanted by nature but was the result of circumstance and education; as a consequence, "multitudes of Geniuses are scatter'd" not randomly, as would be the case if nature implanted genius in individuals, but instead grouped in areas where "opportunity, example, and encouragement concur."[20]

Sharpe contended that understanding how genius emerged enabled a polity to increase its number of geniuses; by the same token, failing to take these forces into account would be a loss not just for certain individuals but for the country as a whole.[21] Though most other commentators argued that genius was a natural endowment, rather than the result of education, they nevertheless stressed the possibility of regulating its emergence and expression. Gerard, for example, began *An Essay on Genius* (1774) with the claim that genius is "the grand instrument of all investigation," whether scientific or artistic, and that understanding its nature would enable a "regular method of invention" (and that without such a method, "useful discoveries must continue to be made, as they have generally been made hitherto, merely by chance").[22] In similar fashion, Edward Young suggested in *Conjectures on Original Composition* (1759) that though the "modern powers [of genius] are equal" to those in earlier times, "modern performance in general is deplorably short" as a consequence of the fact that earlier ages provided more effective governmental "encouragement" than did the present age.[23] For these authors, genius was a topic directly related to the national capacity for scientific and artistic innovation and was hence something that could, and should, be regulated to whatever extent possible.

Understanding mid-century discussions of genius as oriented toward the maximization of this capacity emphasizes that debates about whether genius was natural or acquired were first and foremost efforts to locate those points at which regulatory technologies could be most efficaciously introduced. For example, Sharpe's explicitly Lockean claim that each of us is born as an equivalently "blank paper"[24] and that genius must therefore be the result of education and circumstance suggested that educational institutions were key sites at which genius could be increased; it further implied that the educational formula for enabling genius, if discovered, should be applied universally. Young, by contrast, claimed that nature "brings us into the world all *Originals*: No two faces, no two minds, are just alike; but all bear Nature's evident mark of Separation on them."[25] However, "that medling Ape Imitation . . . blots out nature's mark of Separation, cancels her kind intention, destroys all Individuality" (42), which would likely militate against a uniform system of schooling. Sharpe's and Young's differences with respect to the origin of genius—is it innate or acquired?—are more fundamentally differences about how and where best to regulate those educational institutions through which children pass, so as to increase the number of geniuses.[26]

Focusing on regulation as a key desideratum of discussions of genius allows us to reconsider the plant metaphor so central to eighteenth-century discussions of genius. Joseph Addison had established the topos of "vegetable" genius in his famous *Spectator* essay (#160, September 3, 1711), in which he distinguished between two kinds of "great geniuses," those who write without following rules and those who write within the constraints of rules. "The genius in both these classes of authors may be equally great, but shows itself after a different manner," Addison wrote.

> In the first [the genius who does not follow rules] it is like a rich soil in a happy climate, that produces a whole wilderness of noble plants rising in a thousand beautiful landscapes without any certain order or regularity; in the other [the genius who follows rules] it is the same rich soil, under the same happy climate, that has been laid out in walks and parterres, and cut into shape and beauty by the skill of the gardener.[27]

For Addison, genius is simultaneously a medium (a rich soil) and an agent (a gardener) who can decide whether to manipulate that soil and its productions; this manipulation or its lack then alters the distribution and nature of the productions of genius. Though Addison suggested that one kind of great genius arises without the genius-gardener having to cultivate his rich soil, the second kind of genius—which includes authors such as Plato, Aristotle, Virgil, Tully, Milton, and Bacon—requires the genius-gardener to self-regulate his own genius-soil. Addison's proposal that one kind of genius requires self-regulation hints at the question of what more general regulatory measures, of the sort considered by Sharpe and Young, might encourage such self-regulation.

Addison's vegetable image was taken up by Young, who contended that an "original" composition of genius "may be said to be of a *vegetable* nature; it rises spontaneously from the vital root of Genius; it *grows*, it is not *made*."[28] In his classic account of Young's metaphor, M. H. Abrams emphasized autonomy, reading Young as stressing that just as plants grow on their own, so too does the product of genius.[29] Yet one senses that Young was far more interested in plants that can be cultivated than those that grow wild, since Young asserted that

> an Evocation of vegetable fruits depends on rain, air, and sun; and Evocation of the fruits of Genius no less depends on Externals. What a marvellous crop bore it in *Greece*, and *Rome*? And what a marvellous sunshine did it there enjoy? What encouragement from the nature of their governments, and the spirit of their people?[30]

Pace Abrams, Young stressed the vegetable nature of genius not primarily to depict its freedom from human meddling but to underscore its capacity for management and regulation, that is, to be encouraged and directed by the "sunshine" of government and the spirit of a people. Humans cannot produce plants ex nihilo, but they can facilitate the vitality of plants by managing the media, such as soil, warmth, and exposure to light, that enable plants to grow. In the same way, Young suggested that, though genius emerges from sources unknown, once it has come into being, at least some kinds of genius can be facilitated by managing the cultural "media" within which humans grow and thrive.

Though Tenger and Trolander are in one sense correct that the discourses of genius and political economy provided competing explanations for the phenomena of national wealth increase and technological progress, from a wider perspective both discourses cooperated in the sense that both encouraged a liberal imagination. This liberal imagination, or perspective, sought to locate and demarcate the realms in which "nature" expressed itself in human relations, to determine the truth of nature's operations in those realms, and to devise forms of regulation that could more successfully exploit those natural dynamics. For both discourses, the operations of nature in social relations were intimately linked to differences among individuals, though each discourse understood this link differently. In the case of political economy, the diversity of individuals' "interests" served as the natural force that enabled the market, the division of labor, and wealth, and so the legislator should orient his minimal regulatory activities toward a careful balancing of interests.[31] In the discourse of genius, the diversity of "potential" and the diversity of different kinds of genius were instead the concepts around which one could locate nature's entry into the field of social relations. Commentators focused their attention on those institutions such as schools, social groupings, or government patronage that purportedly translated potential genius into actual expression and hence increased the amount and diversity of genius.

Poetry, Genius, and Worry: Gray's *Elegy*

Interest in maximizing genius by harnessing differences among individuals was pursued not only in texts about genius but also in works *of* poetic genius (or, at any rate, works by authors hoping to be perceived as geniuses by peers and subsequent generations). Where essays, dissertations, and conjectures sought to produce knowledge about genius in order to locate possible sites of regulation, poetic texts strengthened the biopolitical project of maximizing genius by linking Petty's positive ideal, the increase of genius, to negative alternatives, bodied forth in figures of unharnessed value. These negative possibilities clarified both that genius maximization would not necessarily happen of its own accord and that the consequences of allowing geniuses to remain undiscovered were dire. Thomas Gray's *Elegy Written in a Country Churchyard* (1751) expressed one of these negative alternatives in the worry that a polity might fail to locate potential geniuses—that is, that would-be Miltons would remain mute and inglorious—and then surrounded that image of lost potential with other images of undiscovered or unappreciated value.

Among the many peculiarities of Gray's *Elegy* is its ambivalence concerning the condition of poverty that it presents as that which mutes potential Miltons. The *Elegy*'s narrator, viewing a humble country churchyard, rues the fact that only "Chill Penury" prevented those rustic poor whose hearts were "once pregnant with celestial Fire," or whose "Hands . . . the Reins of Empire might have sway'd," or who might have "wak'd to Extacy the living Lyre," from attaining the "Knowledge" that would have turned potential into actual achievements.[32] Yet as William Empson famously noted, the *Elegy* also suggests that poverty is natural and cannot be ameliorated, which makes this loss of genius seem irremediable. Figuring the loss of potential among the poor by means of natural objects—a "Gem" hidden in the "dark, unfathom'd Caves of Ocean" or a flower born to "blush unseen" in the desert (8; ll. 53–56)—the social causes of poverty become part of the natural order. "By comparing the social arrangement to Nature," Empson wrote, Gray "makes it seem inevitable, which it was not, and gives it a dignity which was undeserved."[33] This sense of inevitability is further consolidated in both the elegiac mood of the poem and its setting in a churchyard: "The tone of melancholy claims that the poet understands the considerations opposed to aristocracy, though he judges against them; the truism of the reflections in the churchyard, the universality and impersonality this gives to the style, claims as if by comparison that we ought to accept the injustice of society as we do the inevitability of death" (4). Empson makes these points in support of his more general claim that the poem documents an England that has no "scholarship system or *carrière ouverte aux talents*" capable of locating and unearthing these hidden gems of potential poetry among the poor (4).

Where Empson interprets the *Elegy*'s ambivalence about poetry and the loss of potential genius through a moral lens—he suggests that this ambivalence is the source of many readers' irritation with the poem's "complacence" and what they feel is a "cheat in the implied politics"

(5)—John Guillory develops a more ambitious account, finding in the poem's ambivalence one of several instances of "rest," which, in his reading, enabled the poem to ensure its central place in an emergent canon of vernacular literature.[34] Guillory's reading of the *Elegy* is extraordinary and compelling in large part because he illuminates the extent to which the poem carefully suspended itself between numerous eighteenth-century cultural developments: For example, the poem alludes to the refinements of the classical poetic heritage while remaining firmly vernacular in its own language, and it serves as a sort of commonplace of poetic allusions while at the same time appealing to the private retreat of the pastoral. For Guillory, then, the poem's ambivalence about poverty is just one of the many ways—or rather, it is the synthesis of the multiple ways—that the *Elegy* offers its readers a "unique place of *rest*," or suspension, in a rapidly transforming society.[35] For Guillory, what may seem like the poem's equivocating refusal to make a strong claim about the relationship between poverty and lost genius—is poverty a condition one should seek to ameliorate in order to maximize genius, or should one instead simply accept poverty, and its elimination of would-be geniuses, as natural?—is better understood as a canny lessening of tension, captured in Gray's own deflationary image of a "Tribute of a Sigh" (9; l. 80), that allowed readers to claim the cultural capital of appreciating vernacular poetry without having to choose between these disjunctive positions on poverty.

Yet one wonders if *rest* is the best term to describe the effects of the poem's lament over the loss of potential. Empson and Guillory are no doubt correct that Gray's poem calls forth no more than virtual readerly tears (and certainly no kind of concrete ameliorative action) for those "mute, inglorious Miltons" who perish because of poverty. However, the indefinite nature of the subject of the *Elegy*—the poem is set in an unspecified country churchyard and is about an equally abstract category of gifted poor—encourages a quite modern kind of imaginative activity, namely, the imagination of populations and the ways that mute inglorious Miltons among them *could* be given voice and glory. From this perspective, the poem does not ask its readers to determine whether poverty is remediable but rather employs chill penury as a vector that allows readers to imagine the population of a national territory and how the members of that population could each be assayed for a given quality. Precisely because poverty is widespread—because its victims cannot be restricted to those buried in *this* graveyard, wherever this graveyard might be—the poem authorizes and encourages the reader to imagine this untapped potential of possible Miltons as distributed throughout the geographic and linguistic territory of

Britain. This in turn suggests that identifying and bringing to voice otherwise mute Miltons will require the discovery or creation of a "surface" through which all members of the population would pass and that can be used to test for a given capacity (in this case, potential genius); this surface is the key element of a system that can then locate those few (Petty's one in 4 million) who are indeed potential rivals of Milton.[36] The unintended irony of Empson's reading is that his almost dismissive suggestion that the *Elegy* reveals the pathos of a polity that lacks a "scholarship system or *carrière ouverte aux talents*" exemplifies precisely this kind of imaginative activity. In proposing a system that *could* identify talent wherever it might arise, Empson responds to the form of worry that the poem encourages by imagining populations and the institutions that might identify differences among the members of a population.

A surface capable of encompassing a population and identifying talent among individuals could take forms other than Empson's proposed scholarship system. Such a surface could take the form of aristocratic patronage of "peasant poets" or widespread literacy plus access to a literary market. In the early twentieth century, W. E. B. Du Bois imagined historically black colleges and universities as such a surface capable of locating the "talented tenth" who could "guide the Mass away from the contamination and death of the Worst, in their own and other races," while Virginia Woolf suggested that the means for giving voice to an otherwise "mute and inglorious Jane Austen" was five hundred pounds a year and a room of her own.[37] For other authors, only the complete transformation of social relations would enable hidden value to be uncovered. P. B. Shelley, for example, imagined in the first scene of act IV of his drama *Prometheus Unbound* (1820) "vast beams" of light that "pierce the dark soil, and as they pierce and pass, / Make bare the secrets of the earth's deep heart," secrets that include not only "valueless stones, and unimagined gems," but also past human achievements now buried and otherwise lost.[38]

Empson's solution to the problem posed by the poem emphasizes that the transcendence implicit in the religious aspects of the *Elegy* operates in service of a more secular, biopolitical form of longing and redemption. The poem's narrator echoes the traditional claim that death is the great equalizer—the "Paths of Glory," just like the paths of the unknown villagers buried in the country churchyard, all "lead but to the Grave" (7; l. 36)—and stresses that the memorials left by the rich, such as a "storied Urn, or animated Bust" (7; l. 41) cannot restore life to those who were once famous but are now dead. That is, the social differences that seem so important to the living are of no consequence in our final destination, for we all pass into the afterlife through the common surface of "silent Dust" (7; l. 43). Yet the *Elegy* is not satisfied with this claim of afterlife equality, for it follows these accounts of converging paths and useless memorials with the poem's key images of unrealized potential (the heart pregnant with celestial fire, the submerged gem, the unseen desert flower). Rather than orienting readers toward imagination of a world beyond ours, the *Elegy*'s emphasis on our common condition instead turns us toward an earthly future in which potential that went unseen in the past can be redeemed, and in a sense even resurrected, by locating ways, such as Empson's scholarship system, that can give voice to otherwise mute and inglorious Miltons. To put this another way, the *Elegy*'s images of individual death lead us to discern a more primal vital body—what Cleanth Brooks described in his reading of the poem as our "common humanity" but that, following Foucault, I am more tempted to call the multiple body of the population—that continues to generate genius in every generation.[39]

While Gray's *Elegy* may indeed have provided readers with a point of rest within what Guillory calls "the agon of social mobility," it also encouraged a form of intense biopolitical imaginative activity oriented toward the discovery or invention of surfaces that could subtend entire populations in order to identify and develop qualities of interest, such as Milton-like genius. The *Elegy* did not simply provide its bourgeois readers with cultural capital by letting them off the hook (that is, facilitating a form of social advancement that feels like rest); it also, and arguably more fundamentally, encouraged biopolitical forms of worry (how to locate would-be geniuses) that in turn encouraged the imagination and eventual implementation of solutions such as the scholarship system that Empson extolled.

Biopolitics, Worries, and Smallpox Inoculation

I will return below to the second form of biopolitical worry that emerged, a few decades later, from the crucible of these debates on genius. First, though, I want to underscore the extent to which forms of worry analogous to those that I have analyzed in the context of debates about genius were also central to eighteenth-century projects such as smallpox inoculation, which conform more closely to those campaigns and apparatuses that Foucault had in mind with (and many readers will likely understand by) the term "biopolitics." While smallpox inoculation in Britain in the early part of the eighteenth century was restricted primarily to the aristocratic and middle classes, mid-eighteenth-century advocates of this medical practice imagined and began to develop institutions, such as the Foundling Hospital and the Smallpox and Inoculation Hospital, intended to inoculate those who lacked the means to pay for treatment (that is, the poor), with the eventual goal of inoculating the entire population.[40] Eighteenth-century advocates of smallpox inoculation did not support this goal of population-wide inoculation with our contemporary concept of "herd immunity" but instead sought to garner support for their efforts by means of negative images of stalled commerce and difficult daily life for the middling and upper classes.

An especially important site for the generation of biopolitical images of worry were the annual sermons commemorating the 1746 founding of Middlesex Hospital for the Small-Pox and Inoculation.[41] In addition to providing yearly accounts of the number of successful inoculations (and the increasingly minuscule percentage of deaths from inoculation), the published versions of these sermons clarified the aspirations behind the desire to expand smallpox inoculation. These aspirations often had to be inferred as the inverse of worries about what would happen should the poor *not* be inoculated against smallpox. In 1753, for example, Isaac Maddox asserted that the key problem with smallpox was that it caused "*Cessation* of Trade and Business."[42] This claim was repeated by John Green a decade later, and he expanded on why this was the case, contending that since "men cannot flee from place to place, to avoid the danger of infection," epidemics ensure that "multitudes will soon be reduced to poverty, manufactures will be stopt, [and] commerce will stand still," among other evils.[43] In his 1760 sermon, Samuel Squire commanded his listeners to "See that multitude of *industrious poor* thronging in every quarter of this immense theatre of commerce, business, and action! See them distributing themselves through all the laborious offices of society!" but also encouraged his audience to imagine this scene of pacific commerce disrupted by the outbreak of smallpox.[44] Five years later, Richard Eyre suggested that, without the smallpox hospital and inoculation, his listeners' "Prosperity" and

> success . . . might not otherwise have been obtained, with so much ease,
> and readiness, had they been debarred, or affrighted, from a Due and
> Regular pursuit, of the business, of their several callings, either, from their
> own, or their friends apprehension, (for them,) of concurring the danger,
> of so Precarious a distemper; in consequence of a diligent, and a necessary

attendance, upon each respective branch, of publick trade, or commerce, which could not but expose them to such hazard![45]

This same theme was again stressed by Brownlow North in his 1773 sermon, in which he asserted that smallpox "impends on every village, on every seat of manufacture and trade, on every useful assemblage of men whatever, whose extreme poverty, or unremitted industry, has prevented their timely preparation, by the easy method established in this hospital."[46] These advocates saw smallpox inoculation as a practice that would, ideally, be extended to all individuals within the national territory, thereby guaranteeing security of health and hence ensuring the uninterrupted commerce that many eighteenth-century authors seem to have understood as the final cause of health.

Though smallpox and genius were understood by eighteenth-century commentators as quite different kinds of qualities—for most authors, only a few individuals could be geniuses, while smallpox could affect everyone—both inspired reflections about how populations might be regulated in order to maximize or minimize their incidence. These reflections were enabled in both discourses by negative images of worry and concern: in the case of the genius debates, what would happen if too many geniuses were left undiscovered; in the case of debates about smallpox inoculation, what would occur if smallpox were to affect too many people. This correspondence between reflections on genius and more obviously biopolitical measures such as the eighteenth-century British smallpox campaigns thus sheds light on an aspect of biopolitics not often stressed in accounts of its development, namely, the extent to which biopolitical technologies require the development of images of negative and positive population possibilities. In the case of health-oriented measures such as smallpox campaigns, the positive pole (embodied in figures of health, political stability, and uninhibited commerce) and the negative pole (figured in images of death, political instability, and commercial stasis) were developed in medical pamphlets and religious sermons. In the case of genius, the positive consequence of maximizing this quality—namely, the progress of commerce and the arts—was developed early in political arithmetic and subsequently in a sui generis discourse on the nature of genius. Yet this positive biopolitical pole could not exist independently; for its positivity to shine forth, it had to be counterposed to what would feel like an equivalent form of loss. The consequences of losing geniuses remained implicit in Petty's text and in subsequent treatises on genius but were developed explicitly in poetry, such as the figures of unrealized potential in Gray's *Elegy* (mute inglorious Miltons, submerged gems, unseen flowers) and the effort of the poem to make these feel like British losses.

Worry, Preservation, and the Emergence of "Literature"

Near the end of the eighteenth century, commentators on genius also began to worry that, in addition to the form of loss lamented in Gray's *Elegy*—potential genius that did not become actual—geniuses that *had* been properly valued might also be lost. Young had touched briefly on this possibility in the mid-eighteenth century, noting that many ancient authors whom we value as geniuses were simply imitators of earlier geniuses now lost to us: "It is said, that most of the *Latin* classics, and all the *Greek*, except, perhaps, *Homer, Pindar,* and *Anacreon,* are in the number of Imitators, yet receive our highest applause . . . [because] the works they imitated, few excepted, are lost."[47] Young concluded, however, that the "perpetuating power" of the printing press rendered such losses less likely in the present, and there was thus relatively little danger that the works of a modern genius such as Milton could disappear.

For late-eighteenth-century commentators such as William Wordsworth, though, the printing press was not necessarily Milton's salvation but rather that which threatened his works. Wordsworth saw a conflict between cultural attention and memory, for the printing press's perpetuating power multiplied not only copies of Milton's works of genius but also mediocre novels, which successfully competed with Milton for public attention. Wordsworth worried in the 1800 Preface to *Lyrical Ballads* that, as a consequence, the capacity for appreciating the works of past geniuses was disappearing:

> For a multitude of causes unknown to former times are now acting with a combined force to blunt the discriminating powers of the mind, and unfitting it for all voluntary exertion to reduce it to a state of almost savage torpor. The most effective of these causes are the great national events which are daily taking place, and the encreasing accumulation of men in cities, where the uniformity of their occupations produces a craving for extraordinary incident which the rapid communication of intelligence hourly gratifies. To this tendency of life and manners the literature and theatrical exhibitions of the country have conformed themselves. The invaluable works of our elder writers, I had almost said the works of Shakespeare and Milton, are

driven into neglect by frantic novels, sickly and stupid German Tragedies, and deluges of idle and extravagant stories in verse.[48]

For Wordsworth, the printing press was as much a curse as a saving power, for its indifference to the quality of its productions enabled works of genius to be drowned in a deluge of popular literature.

Guillory seems to me correct when he suggests that the modern concept of "literature" requires Wordsworth's apocalyptic image of culturally created deluge and loss. The well-researched narrowing of the concept of literature at the end of the eighteenth century involved both the restriction of the term to include only imaginative works (rather than, as had been the case earlier, also historical and philosophical works) and a subsequent further belt tightening that left only poetry, novels, and plays.[49] Guillory argues that this restricted concept of literature encouraged belief in a purported distinction of quality between "the works of Shakespeare and Milton," on the one hand, and a deluge of popular literary works written in the vernacular, on the other. Yet as Guillory notes, this purported difference in quality could no longer be *marked* in any clear way, since the difference was not that of, for example, works composed in classical languages and those composed in the vernacular. In this sense, "Wordsworth can conjure up an apocalyptic scenario in which the words of Milton and Shakespeare are swallowed up in the sea of popular writing . . . [only] because the distinction between serious and popular genres produces no corresponding linguistic differentiation within the reading public."[50] Guillory's larger point is that this new concept of literature, like its canonized exemplar, Gray's *Elegy*, served the interests of a rising bourgeoisie, who desired the cultural capital associated with the lettered aristocracy but were suspicious of the classical education of the latter. Or, as Guillory puts it, "the fact of increased upward mobility is at once the premise of 'bourgeois ideology'—that anyone can succeed—and its prime source of social anxiety. Hence the continuous appropriation by the bourgeoisie of aristocratic caste traits, precisely in order to reinforce and stabilize a class structure founded upon a necessary degree of instability or fluidity" (93). The new concept of literature promoted the aspirations of the bourgeoisie for upward mobility by suggesting that they too now possessed what had previously been an exclusively aristocratic canon of works. At the same time, though, this new concept of literature ensured, via an (invisible) line separating works of genius and the "sea of popular writing," that there was not too *much* social mobility to threaten bourgeois distinction.

Even as literature undoubtedly performed this cultural work for a rising bourgeoisie, this concept emerged from the biopolitical discourse on genius that I have traced in this chapter, and this discourse is different than, and not reducible to, the logic of class conflict stressed by Guillory.[51] Or, to put this another way, the bourgeoisie drew upon a biopolitical logic of genius more expansive than bourgeois class interests. The difficulty of collapsing the biopolitical logic of genius into bourgeois class interests becomes evident in both William Godwin's reflections on genius and literature in *The Enquirer* (1797) and Wordsworth's further reflections on deluges and genius in the 1805 *Prelude*. Both accounts contested the bourgeois cultural monopoly on literature central to Guillory's account: Godwin, by presenting literature as that which produces geniuses able to destroy the bourgeois order, and Wordsworth, by connecting genius and literature to what we would now describe as "Anthropocene" ecological concerns.

Godwin's attacks in *Enquiry Concerning Political Justice* (1793) on all institutions, especially those of property, marriage, and law, make it difficult to describe him as an advocate of bourgeois interests. *The Enquirer* continued some of those reflections via a shift in both genre and assumptions about reading. Where *Political Justice* employed the genre of the philosophical system, Godwin described *The Enquirer* as a paratactic series of essays based on "experiment and observation" and on conversations, rather than a "system" resulting from principles and deductive argument.[52] This new literary form encouraged Godwin to focus on a problem he had not engaged in *Political Justice*, namely, how to educate necessarily dependent individuals— that is, children—such that they could become the rational, free subjects at which *Political Justice* had aimed. To solve this kind of problem, Godwin turned, in the opening essays of *The Enquirer*, to the topic of genius.

The first three essays of *The Enquirer* represent a precise, albeit largely implicit, engagement with the mid-eighteenth-century genius debate. In the book's first essay, entitled "Of Awakening the Mind," Godwin considered the importance of innate differences among humans, contending that though "children bring some qualities . . . into the world with them," education is nevertheless more important than these innate qualities in establishing the capacities of the individual.[53] Godwin also assumed, like earlier contributors to the genius debate, that the point of analyzing the nature and causes of genius was to determine how to maximize the number of individuals with great talent in different fields and that increasing the number of geniuses facilitated social progress. However, where earlier authors frequently linked genius to progress through the medium of commerce, Godwin argued that geniuses directly reformed society. He wrote that "the affairs of man in society are not of so simple a texture, that they require only common talents to guide them," for "tyranny grows up by a kind of necessity of nature." The complicated affairs of men in society also required that "men of genius . . . rise up, to show their brethren that these evils [of tyranny], though familiar, are not therefore the less dreadful" (10) and to reveal how to reform social institutions (10–11). The genius can do this because genius itself is characterized by that same capacity—namely, "a spirit of prying observation and incessant curiosity" (16)—that enables social reform. Godwin's geniuses do not require the invisible hand of commerce to coordinate their activities into a progressive unitary movement but are instead themselves the points at which potential coordination becomes visible and can be directed.

This redefinition of genius leads into Godwin's third and fourth essays, both entitled "Of the Sources of Genius." In the first of these paired essays, Godwin revived the botanical-agricultural image of the earlier genius debate, contending that the "talents of the mind, like the herbs of the ground, seem to distribute themselves at random" (29). Part of the task of the genius was to subject the emergence of talent in society to "rules" and "system" (30), so that the incidence of genius could be maximized.[54] This meant, in part, reforming educational institutions, so that more individuals developed that "spirit of prying observation and incessant curiosity" that characterized geniuses. (And for this reason, the topic of genius is not simply a theme treated in *The Enquirer* but is the primary topic and the telos of the book as a whole.) The key to producing geniuses through education is to keep the forming mind "ductile," rather than producing that mental transformation that characterizes most current forms of education, in which "what was at first cartilage, gradually becomes bone"—that is "stiff, unmanageable and unimpressible" (17).

Godwin claimed that current modes of education—and, more generally, the current order of society and especially its class divisions—were directed against mental plasticity. As a consequence, Godwin contended, "the present order of society . . . is the great slaughter-house of genius and of mind" (17). Where Thomas Gray's earlier image of poverty implicitly encouraged readers to make an imaginary survey of England to consider how many would-be Miltons had been muted, Godwin made such a survey explicit. "If a man could go through the island of Great Britain," Godwin wrote,

> and discover the secrets of every heart . . . how much genius, what a profusion of talent, would offer themselves to his observation? In one place he would discover an embryo politician, in another a philosopher, in a third a

poet. There is no benefit that can be conferred upon the human race, the seeds and materials of which would not present themselves to his view. Yet the infinite majority of these are destined to be swept away by the remorseless hand of oblivion, and to remain to all future ages as if they had never been. . . . Centuries perhaps will glide away, and pine in want of those benefits, which seemed ready to burst from their bud and gladden the human race. (286)

As in the case of the earlier genius debate, Godwin's imaginary survey encouraged readers to imagine technologies or techniques that could make a wide and deep survey throughout the polity and hence capture would-be geniuses on a surface before they were slaughtered by an oppressive class system.[55]

For Godwin, what he called "literature" provided both the image for and a key mechanism of this surface. This is in part because, for Godwin, literature provided the template for the genius him- or herself. For Godwin, the "prying observation and incessant curiosity" of the genius enables what we might call, somewhat anachronistically, an "optimized" use of mental resources. Thus, Godwin wrote, "the chief point of difference between the man of talent and the man without, consists in the different ways in which their minds are employed during the same interval. They are obliged, let us suppose, to walk from Temple-Bar to Hyde-Park-Corner." The dull man goes straight from point A to point B, has few thoughts along the way, and does not look around him. The man of talent, by contrast, "gives full scope to his imagination. . . . He enters into nice calculations. . . . He makes a thousand new and admirable combinations" (32). Literature provides the paradigm for this optimizing mental activity of the man of talent, for books "gratify and excite our curiosity in innumerable ways. They force us to reflect. They hurry us from point to point. They present direct ideas of various kinds, and they suggest indirect ones." Literature—exemplified by Godwin with authors such as Thomson, Milton, Gray, Pope, and classical Greek and Latin authors—enables a mind to "becom[e] ductile, susceptible to every impression" (33).[56] Reading literature transforms what Godwin describes as an uncultivated mental wilderness into a "regulat[ed] mind" (49).[57]

Literature is also a key element of a surface, or medium, that extends the effects of genius throughout a single polity and the world more generally, though more in the way of invisible gas than an invisible hand. Godwin contended that he "can guess very nearly what [he] should have been, if

Epictetus had not bequested to us his Morals, or Seneca his Consolations."
But, he continues,

> I cannot tell what I should have been, if Shakespear or Milton had not writ-
> ten. The poorest peasant in the remotest corner of England, is probably a dif-
> ferent man from what he would have been but for these authors. Every man
> who is changed from what he was by the perusal of their works, communi-
> cates a portion of the inspiration all around him. It passes from man to man,
> till it influences the whole mass. I cannot tell that the wisest mandarin now
> living in China, is not indebted for part of his energy and sagacity to the
> writings of Milton and Shakespear, even though it should happen that he
> never heard of their names. (140)

Because literature is, for Godwin, a technology of mental optimization—
that is, a means by which the mind indirectly learns how to regulate itself
more generally—its effects are spread by all the actions of those who read
literature and allow it to perform its work of mental optimization on
themselves.

Precisely because Godwin developed his image of literature by reconfiguring the earlier genius debate, literature emerged in his text as something other than a form of cultural capital that advanced the class interests
of the bourgeoisie, as in Guillory's account. For Godwin, literature was a
sediment of human collective intelligence that, in reflexive Romantic fashion, deepened human collective intelligence because of its indirectness,
that is, its capacity to enable unexpected connections between places.
Hence, rather than confirming bourgeois readers in a sense of aristocrat-
like exclusivity, literature facilitated the emergence of "men of genius"
who "show[ed] their brethren" that the evils of tyranny, "though familiar,
are not therefore the less dreadful."[58]

The connection that Wordsworth made between literature and the
genius debate also drove him beyond the narrow realm of class interests,
in this case toward an ecocritical horizon. From the perspective of his later
Prelude, Wordsworth's worries in the 1800 Preface to *Lyrical Ballads* about
the potential loss of works of literary geniuses turned out to be simply a
special case of a more general worry about the loss of works of genius in
all fields (that is, the arts and sciences). Moreover, Wordsworth's fear of
metaphorical apocalypse in the Preface to *Lyrical Ballads*—the "deluge" of
novels, sickly and stupid German tragedies, and stories in verse—turned
out to be simply a pale reflection of his fear of real deluges, which threatened not only works of genius in all fields but all human works. In the

"Arab Dream" section of the 1805 *Prelude*, for example, Wordsworth's narrator suggested that neither the printing press nor the category of literature could guarantee the survival of works of genius in the face of a worldwide natural disaster, such as a flood, which contemporary geologists suggested had occurred in the past and might again occur in the future.[59] While the narrator of *The Prelude* suggested that the human species would likely survive such a natural disaster, he was convinced that works of genius, and human works in general, would not:

> But all the meditations of mankind,
> Yea, all the adamantine holds of truth
> By reason built, or passion . . .
> The consecrated works of bard and sage,
>
> . . .
>
> Where would they be?[60]

In the face of worldwide natural disaster, works of genius are no more durable than any other kinds of works, leading the narrator of *The Prelude* to despair that the mind had no "element"—that is, nothing with the eternal persistence of air or light—"to stamp her image on" (154; l. 45). Wordsworth's worry about the loss of human works, in other words, was restricted neither to literary works, nor to national context, nor even to the valued category of genius. Instead, Wordsworth's worry encompassed the global population of humans and their works and focused on the embodied relationship among humans, their natural environments, and the media by means of which they preserved their works.

Just as Malthus's reflections on population depended for their force on the imagination of an expanding swell of humans that encircled the globe, Wordsworth's worries about the loss of all human works depended upon the imagination of disasters that would engulf not simply this or that national population or literature but the entire global population of the human species. Since literature depends upon the medium of print, humans would be as powerless to preserve its canon in the event of such global catastrophe as they would artistic productions in any other medium. Yet what literature could do—or, at least what Wordsworth sought to achieve within that instance of literature which was *The Prelude*—was to make such global loss thinkable and affectively pressing. Wordsworth accomplished this by treating his literary text as a sort of gathering place, which linked the biblical Flood and its tremendous importance for the European literary tradition with those geological sciences of Wordsworth's day that proposed the possibility of a new deluge or other form of ecological disaster.

Such a gathering place enabled Wordsworth, in Godwin's terms, to present "direct ideas of various kinds" and to suggest "indirect ones." This in turn enabled Wordsworth to formulate worries about the fate of human works in the face of a global ecological disaster that would destroy the works produced by past populations, impact all current populations of the world, and, as a consequence, affect future populations as well. From this perspective, both Godwin and Wordsworth indeed suggested a narrowing of the concept of literature from its earlier capacious inclusion of most kinds of printed texts to solely those texts that employ the combination of directness and indirectness described by Godwin. Yet both authors also linked this narrowed concept of literature to future states—the rational, classless society of the future, in Godwin's case, and global ecological devastation, which respects no class boundaries, in Wordsworth's—that cannot be aligned with the functional role of literature in class conflict stressed by Guillory.

Both Godwin's reflections on genius and literature in *The Enquirer* and Wordsworth's worry in *The Prelude* about the durability of the works of humans emphasize that though the modern, restricted concept of literature may have been co-opted for the purposes of class conflict, the basic concepts of genius and literature emerged from a crucible of aspirations and worries focused on *populations* and their embodied dynamics, rather than classes and their social dynamics. This intrinsic link between genius and population was evident in Petty's political arithmetic, but it is equally present in Wordsworth's earlier discussion of works of genius in the Preface to *Lyrical Ballads*. As Guillory notes, Wordsworth worries in the Preface about how to save works of genius from a literary deluge of inferior works. Yet that literary deluge is itself enabled by what Wordsworth describes as "the encreasing accumulation of men"—that is, an increasing population. This in turn underscores the extent to which the double worry of "overlooking" associated with genius in the eighteenth century—the *Elegy*'s worry of overlooking would-be geniuses and Wordsworth's worry of overlooking what had earlier been recognized as genius—only became possible through the imagination of populations characterized by complex internal dynamics. Petty had seen the increase of population, combined with the creation of surfaces capable of tracking and cultivating populations, primarily as grounds for hope: the possibility of more labor, more wealth, more genius. Gray and Wordsworth—and Thomas Malthus, in a different register—understood that these positive aspirations of regulation and maximization were not thinkable, or at least not affectively moving, unless they were surrounded and illuminated by negative possibilities, that is, images of what might be lost.

Conclusion

As I hope has been clear, my argument is not that an autonomous eighteenth-century form of biopolitics that first emerged around Petty's new science of political arithmetic provides the explanatory context for discussions of genius in literary texts. Rather, I have argued that these subsequent discussions of genius were essential to the more general development of the logic of biopolitics itself. Discussions of genius were important to the development of this logic in several ways. The genius debate, for example, linked biopolitical regulation to a principle of individual uniqueness, and it connected positive aspirations for regulation, such as Petty's desire to maximize genius, to negative, even apocalyptic, alternatives. Understanding these discussions of genius as part of biopolitical discourse also helps us see the latter as encompassing more than simply those clearly biological aspects of collective living stressed by Michel Foucault ("health, hygiene, birthrate, life expectancy, race . . ."). Finally, understanding discussions of maximization and loss of genius as biopolitical presents us with an image of the effects of literary texts, such as Gray's *Elegy*, different from that described by Guillory: Instead of functioning just, or primarily, as a place of "rest" in the midst of a field of social competition, the *Elegy* encouraged intense imaginative surveys of population and national territory.

Though there is not necessarily a disjunctive relationship between Guillory's sociological perspective and the biopolitical perspective that I have outlined here, these two perspectives also cannot be simply correlated or combined. The difficulty of coordinating or combining biopolitical and sociological perspectives is a function of the different object of each: Biopolitics focuses on populations, while a sociological perspective focuses on social conflicts, tensions, and forces. The latter are important for population technologies because social conflicts, tensions, and forces in essence constitute key elements of the milieu within which population technologies are employed. There is also no doubt that the appreciation of genius (that is, "taste") functioned in the eighteenth century as a marker of cultural capital and hence served the interests of a bourgeoisie that sought to establish forms of cultural value in place of the hereditary capital provided by aristocratic birth. Yet the discourse on genius, in both its prose and poetic forms, also encouraged the imagination of populations and institutions that could locate and harness differences among individuals in a population. Though this logic can at times be exploited by a social logic of distinction, it fundamentally differs from this latter. Gray's *Elegy*, for example, indeed likely served the purposes of upward social mobility for a specific form of bourgeoisie. However, by means of its abstract image of gifted noble poor, it also facilitated the logic of biopolitics by encouraging the imagination of population, and this latter operates on terrain quite different from the social space within which the dynamics of class play out. As the *Elegy* highlights through its images of mute Miltons, submerged gems, and inaccessible blooming flowers, the imagination of population meant the thought of an internally differentiated multitude and illuminated possible futures of that multitude through images of radiant gain and light-engulfing loss. As Godwin illustrated in his image of status quo–destroying "men of genius," and Wordsworth underscored in his yearning for a fundamental "element" capable of withstanding global natural catastrophes, the biopolitical logic that connects the discourses of genius and literature could lead authors beyond, or below, the class logic of society and toward concepts of nature that were not simply ideological fronts for "naturalizing" social hierarchies but were rather the thought of a quasi-elemental source of difference from which variation, new qualities, and transformation perpetually emerge.

As I noted in my introduction to this book, literary critics have tended to approach biopolitics as a form of politics even more nefarious and pernicious than class conflict. My suggestion that eighteenth- and early-nineteenth-century concepts of genius and literature emerged from a biopolitical matrix may thus seem like even more bad news, in the sense that this account would provide even more reason to remain ambivalent about or perhaps outright dismissive of these concepts. However, this is not the conclusion that I draw from the account I have developed here. Rather, I see in this account—and especially in Godwin's and Wordsworth's reformulations of the connection between genius and literature—grounds for both hope and for a rethinking of the redemptive potential of literature. Both Godwin and Wordsworth suggest that it is in and through literature that one can approach populations as entities that have capacities for creating new norms, and the goal of the chapters that follow is to describe some of the means by which Romantic literature sought to accomplish this task.

2. Imagining Population in the Romantic Era

Frankenstein, Books, and Readers

Though the Romantic-era debate between William Godwin and Thomas Malthus about the limits of social progress seemed to have concluded in the 1820s in something of a stalemate, this conversation has recently been revived, though with a rather peculiar twist.[1] The Romantic-era version of this debate pitted Godwin's principle of perfectibility against Malthus's principle of population, with Godwin arguing that social relations could be slowly perfected as legal and political institutions were eliminated and Malthus countering that a key determinant of collective behavior was located in the biological register of "population." Malthus contended that the register of population was inaccessible to human control or intervention and thus concluded that strong social institutions were, *pace* Godwin's claims, necessary in order to reduce human suffering. Malthus's account of population infuriated many Romantic-era authors. William Hazlitt charged that in *An Essay on the Principle of Population*, Malthus "vibrat[ed] backwards and forwards with a dexterity of self-contradiction which it is wonderful to behold," and P. B. Shelley was even more direct, writing that he would "rather be damned with Plato and Lord Bacon, than go to Heaven with [William] Paley and Malthus."[2] The debate between Malthus and Godwin helped establish a stark division, one that would persist into the twentieth century, between progressives on the left who argued for a malleable social subject capable of self-improvement and those on the right who argued for biological limits on perfectibility. Karl Marx's claim in *Capital* that "the great sensation of [Malthus's] pamphlet . . . was due solely to the fact that it corresponded to the interests of a particular party" also encouraged the left to see appeals to purportedly biological facts as ideological illusions that defused efforts to improve social relations.[3]

Yet beginning in the late 1960s and continuing into the present, the political valences associated with the Godwin-Malthus debate underwent an extraordinary doubling and reversal. On the one hand, while "Malthus" continues to serve some on the left as a shorthand for attempts to naturalize class relations, ecologically oriented left-leaning groups discovered in the principle of population a resource for critiquing the institutions of capital. The famous 1972 Club of Rome report on *The Limits of Growth*, for example, argued on Malthusian grounds that the dominant Fordist model of manufacture produced ecological and social crises, and the ecologist Garrett Hardin argued in "The Tragedy of the Commons" that the threat of global human population could be combated only by "relinquishing the freedom to breed."[4] This leftist neo-Malthusian emphasis on the natural limits of economic growth encouraged neoliberal economists and journalists to promote even more aggressively "the market" as a mechanism capable of overcoming all apparent limits and—perhaps counterintuitively—to link this neoliberal vision of infinite economic expansion with Godwin's claims about the possibility of perpetual social improvement.[5] While Godwin's *Of Political Justice* continues to be seen as an angry attack on class-based privilege, the right has embraced a neo-Godwinian form of institutional critique in order to cut the purse strings of (for example) public funding for the natural sciences, arguing that academic science is simply one more self-interested institution that ought to be opened up to the market.[6] A left that grounds its program for human improvement in the biological register of population and a right that appropriates Godwin's emphasis on institutional critique: We find ourselves in a strange neo-Romantic era, in which the ghosts of Malthus and Godwin have doubled, with the result that each can serve as a tutelary spirit for both the left *and* the right.

This uncanny resurrection, splitting, and reconfiguration of the debate between Godwin and Malthus presents us with an opportunity to reconsider and reconfigure the role of literary theory and its relationship to social progress. A key development in literary theory in the 1970s was the reevaluation of the institutional status of "literature." Where earlier humanist critics had presented literature as an institution that provided readers with eternal truths, positive normative models, or occasions for the healthy exercise of the powers of reason and feeling, 1970s critics inspired more by Freud, Marx, and Foucault saw instead a problematic technology of normativity that socialized readers by encouraging them to adopt social norms that served ideological, rather than rational, ends. These new forms of institutional critique were invariably aligned with the rejection of appeals to a fixed biological nature, neo-Malthusian or otherwise; for these neo-Godwinians, the reader-subject is a malleable substance upon which the institutions of literature inscribed ideological contents.[7] However, when neoliberals have now added their voices to the chorus of critiques of the institutions of literature and the humanities, it is perhaps a good time to revisit the other pole of the Godwin-Malthus debate—the concept of population—for tools that might help us understand better the nature of creative literature and to redeem its critical potentials.

This chapter pursues this task across six sections. These are collectively structured as a narrative of rivalries and romances, and they tell the story of two hidden trysts and their multiple monstrous offspring. I begin by noting that Malthus's and Godwin's public antagonism masked a more fundamental compatibility, for both believed that explaining social phenomena meant assuming that individuals are, for all intents and purposes, the same. More specifically, both assumed that population-level analyses could disregard individual differences; for Godwin specifically, this meant assuming that social institutions produced the same effect in many individuals. The second section clarifies that the true rivals of the Malthus-Godwin couple were theorists committed to the principle that the individuals who made up populations *differed* from one another in innumerable ways and that population-level analyses required a recognition of such differences. The third section emphasizes the implications of these hidden Romantic-era affinities and rivalries for our understanding of twentieth-century literary interpretation, suggesting that accounts of literature as a technology for encouraging normative behavior are direct descendants of the Malthus-Godwin pair. The fourth and fifth sections then consider another hidden, and even more unconventional, coupling, one that brought into intimate proximity the Malthus-Godwin pair and their populationist rivals. The site of this tryst was Mary Shelley's *Frankenstein*, a text that, like Malthus's and Godwin's, continues to have important resonance in our own moment.[8] I account for the continuing relevance of *Frankenstein* in part as a consequence of its interest in helping its readers see the world in terms of a difference-oriented concept of population. Seeing the world in terms of the differential aspects of populations could mean searching for biological explanations of social relations, but it could also mean looking for cultural phenomena in which unlikely and improbable events or behaviors were as important as those closer to the normative center. As I note in the fifth section, some of Shelley's first readers—namely, periodical reviewers—demonstrated that this view of the world could be spread even in the form of criticisms of her novel. The sixth section connects these earlier attempts to understand literary dynamics in terms of populations to Franco Moretti's interest in describing the publishing dynamics of short stories and novels in terms of the "culling" performed by literary markets on literary populations, and I conclude with a discussion (and critique) of recent neoliberal efforts to identify population logic completely with the logic of the market.

Society and Population in the Romantic Era

The political philosophy Godwin developed in the 1793, 1796, and 1798 editions of his *Enquiry Concerning Political Justice* pursued to its logical conclusion the Enlightenment project of identifying and criticizing those social structures that had cast long shadows of illusion and error. Earlier eighteenth-century Enlightenment authors focused their critique on specific institutions, usually those of "kings and priests" (that is, bad government and false religion). Godwin went further, arguing that the real impediments to enlightenment were not *specific* institutions but institution itself. Godwin argued that institutions, by their nature, forced individuals to adopt the opinions of others, rather than allowing each to employ his or her own reason. Taking on the opinion of another was the real obstacle to enlightenment and social perfectibility, and Godwin thus opposed all institutions, including those of politics, religion, economy (for example, property), and private life (for example, marriage). However, because he understood himself to be living in an era in which institutions did most of the individual's thinking for him or her, Godwin did not support the immediate overthrow of institutions. Such a step would lead to chaos, as people sought to grasp the new situation by means of habits of thinking formed not by reason but by now-absent institutions. He advocated instead for the gradual elimination of institutions, a process that would slowly and safely increase the occasions for the exercise of individual reason.

Malthus's "principle of population" was intended to trump Godwin's principle of perfectibility not by denying Godwin's claims about institutions but rather by locating a noninstitutional register of darkness—namely, the dynamics of population—that was inaccessible to the enlightening exercise of reason. Malthus's concept of population thus emphasized the relative sterility of Godwin's version of materialism. Godwin argued in the *Enquiry Concerning Political Justice* that human beings and their social relations are complicated constellations of the same matter and movement that make up the rest of the universe. Yet this kind of materialism focused on a register of reality so far below that of individual decision making or institutional dynamics that discussions of matter and movement occupied very limited space in Godwin's long text. Malthus, by contrast, presented a more complex materialism, one that focused not, like Godwin, on the physics of material bodies but rather on an intermediate realm—the realm of population—which lay between Godwin's realms of bare matter and movement, on the one hand, and individual and institutional dynamics, on the other. Human reason could illuminate facts about population dynamics, such as rates and causes of population increase or decrease. However, these were not properly "human" dynamics, for they applied to all living beings and were, as a consequence, largely inaccessible to human control. Malthus asserted that though population dynamics resulted from individual decisions about when and where to reproduce, one could only make sense of these facts of population by abstracting from individual decisions. In place of Godwin's materialism, upon which little of the argument in his *Enquiry* depended, Malthus introduced a more complex materialism that impinged directly upon human affairs and institutions.

Yet Malthus's introduction of this intermediate materialist realm drew on a *pre*-Romantic sense of population, one that was already being partially displaced—or at least questioned—at the time *An Essay on the Principle of Population* was published. As Michel Foucault noted, seventeenth- and early-eighteenth-century authors had used population as the opposite of *depopulation*; that is, population referred to processes by which a "deserted territory was repopulated after a great disaster, be it an epidemic, war, or food shortage."[9] Since populousness was associated with the polity's strength and health, population was invariably understood as good. Malthus drew upon this concept of population but simply reversed the valence of increasing population from positive to negative, in the sense that, as Frances Ferguson notes, where earlier authors saw increasing population as intrinsically good, Malthus saw it as a threat.[10]

By the mid-to-late eighteenth century, though, population also denoted something quite different, namely, a conceptual framework for discovering new facts about large collections of people, facts that were in turn used to determine where, when, and how to apply regulatory measures such as disease inoculation or fiscal policies. For the French physiocrats, and also for physicians and mathematicians in Britain and France interested in questions of disease management, population—along with related terms such as "generation" (*génération*) and the "human species" (*genre humain*)—denoted not a homogenous mass of individuals that increased or decreased in size but rather a heterogeneous collection of individuals, subgroups of which differed in key respects from one another. The Swiss mathematician Daniel Bernoulli, for example, argued that, given a population (or "generation") of 13,000 infants, a specific percentage would contract smallpox; a percentage of that subgroup would recover and a percentage would die—and, most significant, these percentages could be changed by means of inoculation.[11] Population thus denoted a heterogeneous object of analysis that changed in accordance with its own natural logic—that is, changed largely of its own accord, whether or not laws and institutions forbade these changes—but that could be nudged in certain directions provided that one located the proper pressure points and thresholds. For example, inoculation policies could be justified by calculating and comparing the percentage of deaths that occurred in a population both with and without smallpox inoculation.[12] As Foucault noted, a population was thus for many late-eighteenth-century authors "a set of elements that, on one side, are immersed within the general regime of living beings and that, on another side, offer a surface on which authoritarian, but reflected and calculated transformations can get a hold" (75). Determining where, precisely, authoritarian state measures could gain purchase was a matter of determining the "constants and regularities even in accidents" and the "modifiable variables" on which these constants and regularities of the population depend (74).

From this perspective, the debate between Godwin and Malthus looks less like a conflict between modern principles of socialization and population and more like a conflict between a modern principle of socialization and a premodern approach to population. Though Malthus, like his contemporaries, emphasized a biological register of reality amenable to quantification, his approach to this register was extraordinarily coarse, for the only number about population that interested Malthus was its rate of increase. His approach was also necessarily coarse, for he focused attention on this biological register primarily in order to produce fear about a population that perpetually threatened to increase beyond bounds. His account of population thus stood in stark contrast to those of his contemporaries who deployed this term as a means for generating new facts intended to assist in the *transformation* of the biological realities of populations (for example, suggesting measures that would push the current normal curve of smallpox mortality in a specific population toward a better normal curve of smallpox mortality).

Godwin did not seem to recognize this point in *Of Population* (1820), his rather delayed response to Malthus's *Essay*. Instead, Godwin implicitly accepted Malthus's concept of population but claimed that the biological register of population did not have the significance that Malthus claimed. Like Malthus, Godwin focused solely on the rate of population increase but contended that Malthus's claim that populations exponentially increase unless otherwise checked bore no correspondence to the actual facts of population increase and decline.[13] This line of argument seems to grant the importance of determining correctly facts about populations. In fact, though, it brackets progressive materialism from questions of social amelioration. Godwin and Malthus agreed that the only fact of interest about a population was its rate of increase, but Godwin implied that one could simply disregard the entire problematic of population if its rate of increase did not threaten in the way that Malthus had suggested. Godwin's response thus helped solidify what eventually came to seem like an unbridgeable and politically inflected methodological division between "conservatives" who grounded their arguments in the purportedly fixed biological characteristics of populations and "progressives" who placed their bets on the malleable and perfectible socialization technologies of society.

The Metaphysics of Population

Not only did Malthus's approach to the concept of population differ from that of contemporaries who understood populations as collections of differences, but so did his goals. Where Malthus sought to ground normative claims about social institutions in biological invariants, his contemporaries employed concepts of population to relativize norms. Foucault stressed that the new approach to population was *not* disciplinary, if by discipline one understands a socialization technique of the sort that Godwin criticized. Foucault notes that in

> the disciplines one started from a norm, and it was in relation to that training carried out with reference to the norm that the normal could be distinguished from the abnormal. Here [i.e., the new sciences of population], instead we have a plotting of the normal and the abnormal, of different curves of normality, and the operation of normalization consists in establishing an interplay between these different distributions of normality and [in] acting to bring the most unfavorable in line with the more favorable.[14]

To return to the example of inoculation, eighteenth-century authors tracked many different normal curves of smallpox, parsed by age, region, town, and occupation, but sought, by means of decisions about which people to inoculate, to nudge some of these normal curves toward other normal curves judged to be more favorable. This was thus not a matter of socializing or disciplining each individual but rather of identifying and intervening *only* at those points that enabled one to shift one curve toward another.

Foucault's account helps us think further about what we might call the metaphysical assumptions of the modern concept of population. A "population" in the modern (that is, non–Malthusian) sense was premised on the existence of:

(1) a *source* of constant variation;

(2) a *malleable collective body* within which those variations emerge, that presents a surface by means of which observers can locate regularities, and that is itself the point of application for human initiatives designed to change those regularities; and

(3) *forces of selection* that traverse the surface and destroy some, but not all, of those variations.

The source of variations can be labeled "nature" or "chance," or (in the case of cultural phenomena) "desire" or "preferences." However the source is understood, it must produce multiple variations, which observers can group into different frequencies of occurrence. The malleable collective body, made of the individuals who live within a given geographic region, is what holds these shifting distributions of variations. However, one can only speak of this malleable collective body as a "population" when scientific observers can locate (or create) within it a surface that both allows them to document distributions of variations and to modify those distributions by means of different methods.[15] Finally, forces of selection are responsible for changes of distributions of variations over time.

We can flesh out this abstract description through the example of smallpox and smallpox inoculation. Late-eighteenth-century observers noted that of those adults who contract this disease, roughly one person in eight will die. The collection of individual living bodies in a given geographic region is the surface that holds variations—in this case, the tendency of each body to succumb or not to the smallpox virus—and the smallpox virus itself is a force of selection that destroys some of those variations (by killing some of these individual bodies) and leaving others unaffected. The similarities of smallpox symptoms across the bodies of individuals present observers with a surface that allows them to identify instances of smallpox. Smallpox inoculation can be introduced into that same surface (that is, individual bodies), which alters the distribution of the force of selection represented by the smallpox disease.

This particular example identifies only two variations—susceptibility or resistance to smallpox—which may suggest that most members of a population are in fact "the same." However, the key to the modern concept of the population is that one can locate in the same population constants and regularities that bear on *many* different qualities—responses to other diseases, suicide rates, height distributions, and so on—and each additional survey of the same population renders each individual increasingly unique. I am like roughly 90 percent of the adult population with respect to my response to the smallpox virus but like only 40 percent of the adult population with respect to *both* my response to the smallpox virus and my response to disease B; like only 20 percent of the population with respect to my response to the smallpox virus, disease B, and my eye color; etc. The deep premise of the modern concept of population, in other words, is that each individual is a *unique* collection of variations.

The twentieth-century geneticist Ernst Mayr captured this point in a contrast that he drew between "typological" and "population" thinking:

> The assumptions of population thinking are diametrically opposed to those of the typologist. The populationist stresses the uniqueness of everything in the organic world. . . . All organisms and organic phenomena are composed of unique features and can be described collectively only in statistical terms. Individuals, or any kind of organic entities, form populations of which we can determine the arithmetic mean and the statistics of variation. Averages are merely statistical abstractions; only the individuals of which the populations are composed have reality. The ultimate conclusions of the population thinker and the typologist are precisely the opposite. For the typologist, the type (*eidos*) is real and the variation an illusion, while for the populationist the type (average) is an abstraction and only the variation is real. No two ways of looking at nature could be more different.[16]

For Mayr, this understanding of populations as a collection of unique individuals was the only way to make sense of the emergence of *new* species as a consequence of geographic difference.[17] The fact of individual uniqueness means that a population functions as a kind of reservoir of both visible (phenotypic) and genetic differences. If a subpopulation of a bird species located on one island migrates to a different island, differences among individuals of that migrating population of birds "permit the rapid adaptation of [that] population to [the new] local environment."[18] If this subpopulation of birds remains geographically isolated from the original bird population, it can eventually become a new species, which is unable to breed with the original species from which it has now diverged (see Figure 1).

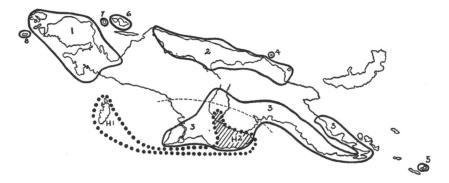

Figure 1. Illustration of speciation by means of geographic isolation. From Ernst Mayr, *Systematics and the Origin of Species from the Viewpoint of a Zoologist* (Cambridge, MA: Harvard University Press, 1942). Copyright © 1942 by Ernst Mayr. Copyright © renewed 1970 by Ernst Mayr.

Though Mayr is likely correct in his contention that population think-ing first became an *explicit* theoretical approach only after Charles Darwin's. work on evolution, Foucault's work on both biopolitics and liberalism suggests that something like a "practical" mode of population thinking emerged in the eighteenth century in the context of problems such as smallpox inoculation, interest in how the number of geniuses in a nation might be increased, debates about the limits of government control over economic phenomena, and concerns about how to price life insurance.[19] Advocates and critics of smallpox inoculation, for example, had no interest in a natural state or "type" for the human species against which individual variations would be judged but were instead interested in how regularities of smallpox infection within a population could be altered by the practice of inoculation. As I noted in the first chapter, political arithmeticians such as Petty were equally uninterested in a natural state or type of humans but rather in how to increase the incidence of a rare, anomalous variation, the genius, within national populations. Mid-eighteenth-century political economists and physiocrats also tended to valorize differences among indi-viduals with respect to economic decisions, treating these not as deviations from a natural state but simply as givens, in the sense that each individual's choices were understood to be just as "natural" as those of every other indi-vidual. In other words, the late-eighteenth- and nineteenth-century dis-cursive explosion of facts about populations was often predicated, in practice if not in theory, on assumptions more or less identical to those later articulated by Mayr.[20]

This new approach to population implied, in ways that the Malthusian model decidedly did not, that unusual, anomalous variations could serve as the motor of qualitative population transformation. For all its pressure and dynamics, the Malthusian population did not actually change *qualitatively* but only changed in size. The new sciences of population, by contrast, presumed that the distribution of qualities and potentials in a population *could* change over time and that anomalous qualities could be the means by which such changes occurred. What Foucault called "authoritarian measures," relied upon the premise that the distribution of qualities in a population could change and might be directed *toward* an unusual trait. In the case of smallpox, for example, a small population with unusually high resistance to smallpox could serve as the "norm" that policy makers aspired to replicate in the more general population.[21]

Society, Normalization, and Literature

The "metaphysics of populations" may seem rather distant from the concerns of literary criticism, and indeed literary critics of essentially every stripe have followed Godwin's lead by assuming that the register of population has no significance for our understanding of institutions, literary or otherwise. Formalist literary methodologies are, of course, no more interested in populations than in any other extratextual entities, institutions, or concerns. However, even methodological approaches that explicitly theorize the effects of the world upon literary texts (and vice versa) focus more or less exclusively on the ways that literature serves the institutional function of inculcating normative behaviors.

Consider, for example, Franco Moretti's account of how the nineteenth-century *Bildungsroman* sought to resolve the task of socialization that previously had been assured by religious rituals. Moretti contends that traditional societies divide social life into "two parts that have nothing in common," and the purpose of an initiation ritual is to "die" in one social role (say, "boy") so as to become reborn into another ("man").[22] The initiation ritual is thus a period of suspension between two distinct and discontinuous social roles. The *Bildungsroman*, by contrast, was committed to convincing its readers that each moment in life was continuous with everything that precedes and follows it. In Goethe's *Wilhelm Meister*, there is "no irreversible moment in which everything, in one fell swoop, is decided"; one must instead "be able to dispose of one's energies *at every moment* and to employ them for the countless occasions or opportunities that life, little by little, takes upon itself to offer."[23] In this way, Moretti suggests, the early

Bildungsroman reflects and reveals the dilemma of modern socialization: In place of those institutions of ritual by means of which traditional societies enable transitions between discontinuous social roles, modern culture encourages subjects to engage in perpetual, continuous, and apparently self-directed processes of language-oriented "socialization" (and its corollary, "normativity"). The *Bildungsroman*—as well as literature more generally—thus becomes, in Moretti's account, an attempt to resolve symbolically "a dilemma conterminous with modern bourgeois civilization: the conflict between the ideal of *self-determination* and the equally imperious demands of *socialization*" (15).

As an explanation of the assumptions that underwrite the specific genre of the *Bildungsroman*, Moretti's account is compelling, as are his illuminations of the logical and affective double binds that traverse the modern project of socialization. It is worth stressing, though, how emphatically his account brackets not only the fact of biological variation but variation more generally. Given his role as a literary critic, Moretti is perhaps justifiably uninterested in questions of biological variation (for example, the degree of genetic variability that would be necessary for either a traditional or a contemporary society to persist in time). However, this disinterest in variation reoccurs at the level of culture.[24] For Moretti, the "problem" of cultural reproduction is entirely that of reproducing the same. Whether in the form of traditional rituals that assign the same role (for example, "man") to all who successfully endure its trials or in the form of those modern socialization rituals by means of which individuals engage in normative "self-determination," what is at stake is how the many become the same.

Moretti's approach to variation is not an anomaly within literary criticism; in fact, it expresses in especially clear fashion an understanding of literature as a technology of normativity that underwrites most important accounts of the modern novel. In the introduction to this book, I noted a number of now classic literary critical accounts of novels as norm-enforcing technologies, such as the "monitory image" that Ian Watt locates in *Robinson Crusoe*; the limiting symbolic acts, ideologemes, and assumptions about genre that Fredric Jameson diagnoses in nineteenth- and twentieth-century literature; and the domestic novel's representation of the household, which Nancy Armstrong describes as establishing "the context for representing normal behavior."[25] For these well-known critics, novels inculcate *normative* beliefs and practices. Though they employ several quite different theoretical methodologies, these literary critics agree that the novel functions as a modern institution that produces sameness out of difference and does so by naturalizing normative beliefs and behaviors.

Frankenstein's Populations

Without contesting that the novel has played this role, we can nevertheless ask what it might mean to return to early-nineteenth-century literature and see in it not only an institution of socialization but also a technology that emerged in tandem with the new, non-Malthusian sciences of population. If socialization techniques are premised on an essential malleability of the individual, which allows many individuals to internalize the same common norms, but population technologies are premised on the importance of individual differences, what implications might this latter premise suggest for our understanding of the roles of creative literature?

Mary Shelley's *Frankenstein* allows us to explore these implications. Shelley does not explicitly use the term "population" in *Frankenstein*, but Maureen N. McLane has established the centrality of this topic to the novel's plot. McLane notes that population comes to the fore in Victor's two key experiments: his initial creation of a creature and his subsequent partial creation, then destruction, of a mate for his creature. McLane stresses that the former is not "an experiment to create a human being but rather an experiment in speciation"—that is, an attempt to create a new population.[26] For McLane, Victor is a sort of closeted Malthusian, one who shows his true colors when his creature demands that Victor allow this experiment in speciation to continue. At this point

> Victor shows his Malthusian hand and gropes his way toward the principle of population, a principle through which he finally excuses his frenzied dismemberment of the half-finished female "thing." . . . What the monster proposes as a solution—a species companion—becomes in Victor's prospectus the route to a further and more horrifying problem, that of species competition [between humans and what Victor fears would be a new "race of devils"].

Tearing up the would-be mate of his creature, Victor "shows himself to be an adept not of Paracelsus nor even of Humphry Davy but rather of Malthus, who wrote, regarding progress in human society, that 'in reasoning upon this subject, it is evident that we ought to consider chiefly the mass of mankind and not individual instances'" (103–4). Contrasting Victor's commitment to Malthusianism to the creature's commitment to *Bildung*— that is, the creature's belief that if he internalized proper social norms through literature, he would be accepted by the human community— McLane concludes that the novel reveals the failure of *Bildung* when it comes into conflict with the discourse of population.

McLane argues compellingly that the topic of population is essential to *Frankenstein* and reminds us of the ways that conservative commentators appeal to "natural laws" to trump progressive appeals to the power of nurture. Yet is Malthusianism really the key to the role of population in *Frankenstein*? To describe Victor as Malthusian is to suggest that he understands population as a homogenous mass characterized by one dynamic, its reproductive rate. In Victor's two experiments, though, we find two different conceptions of population, neither of which is precisely Malthusian. As McLane notes, Victor investigates the principle of life in part so that he can create a "new species." For Victor, creating a new species would give him a claim on their "gratitude": "A new species would bless me as its creator and source; many happy and excellent natures would owe their being to me. No father could claim the gratitude of his child so completely as I should deserve theirs."[27] Like Malthus, Victor understands this virtual population primarily as a homogeneous aggregate. However, *pace* Malthus, Victor sees its increase as good, for he presumes that the many individuals of this new species will each feel gratitude toward him. At this point in the novel, Victor adopts an early-eighteenth-century approach to population, for he understands the latter as something under the control of a sovereign authority and that enables the maximization of a desired good (in this case, gratitude).

Victor's subsequent decision not to create another creature is also made in response to a virtual population, and his fear, as McLane stresses, is based on the link between reproduction and population growth. However, Victor now fears the effects of reproduction precisely because he no longer understands a population as a *homogeneous* entity but instead as an aggregate of variations. Though the creature promises to "quit the neighbourhood of man" (158) with his newly created mate, journeying "to the vast wilds of South America" in order to live a life that is "peaceful and human" (157), Victor concludes that even were the creature (and presumably also his mate) to honor their word, "one of the first results of those sympathies for which the daemon thirsted would be children, and a race of devils would be propagated upon the earth, who might make the very existence of the species of man a condition precarious and full of terror" (174). In this scenario, progeny function not, as in Victor's initial approach to population, as additional sources of the same homogeneous emotion (gratitude) but rather as sources of variation and difference: No matter what the creature and his mate might promise, his children are likely to act differently. Though both Malthus and Victor link populations to reproduction and both fear that population growth will lead to violent competition, they nevertheless understand the nature of population quite differently. For Malthus, the reproduction of population brings simply more of the same, while for Victor, reproduction of population is a source of difference.

Since the narrative of *Frankenstein* describes the unhappy consequences of Victor's shift from one model of population to another, it is tempting to conclude that the novel critiques one or both of these models. We might conclude, for example, that had Victor only realized from the start that populations are aggregates of variations that cannot be controlled, he would never have sought to create a new kind of population, and he would thus have spared his family (and their servant Justine) much suffering. From this perspective, *Frankenstein* would indeed function as a technology of socialization, one that valorizes normative beliefs and practices by treating its readers to lessons that reveal the horrifying consequences of improper beliefs and norms.[28]

As tempting as it is to understand *Frankenstein* as providing either a direct or indirect lesson, such lessons become extraordinarily complicated when they bear upon the topic of population. For example, *had* Victor realized from the start that populations are aggregates of variations that cannot be controlled, he would then also presumably have realized that, since he himself was a member of an existing population, he could not protect himself and his family from the uncontrollable effects of populations simply by choosing or not choosing to create a new population. He might even have concluded that his anomalous wish to create a new population was an instance of those infrequent but nevertheless predictable outlier behaviors that one expects in a large population that lives in a society that allows mobility and self-directed education; as a consequence, even if *he* had destroyed his materials before creating the first creature, another Victor-like autodidact interested in creating life would likely emerge somewhere else. And the only solution to *that* kind of problem, it seems, would be rigid, authoritarian, and disciplinary structures that locate and destroy those far-from-normal instances of individuality that Victor represents.

However, before we arrive at the counterintuitive conclusion that *Frankenstein* endorses, via negative example, authoritarian and conservative social norms, it makes more sense to read the novel's task less as valorizing one understanding of population over another and more as a matter of helping its readers in the more primary task of learning to see the world in terms of populations. Looking at the world in terms of populations means looking for collective surfaces capable of holding variations and receiving the action of selective forces; it also means locating points at which dynamic relations between surfaces and selective forces might be slowly altered. This can mean making assumptions about hidden aspects of human biology, such as population growth or disease mortality rates. However, it can also mean looking for variations, surfaces, and selective forces in cultural phenomena, such as choices people make about work, consumption, and pleasure. Thus, rather than providing a didactic lesson about a specific model of population, *Frankenstein* instead provided its readers with tools for identifying aspects of the world that can be understood in terms of populations. It did so by providing two population models (population as homogenous aggregate and population as a heterogeneous aggregate) and a series of dramatic schemata, such as Robert's and Victor's desires for glory, the creature's search for sympathy, and Justine's legal troubles. These dramatic schemata not only focus attention on points in the social field at which thinking in terms of populations can have effects but also propose specific individual variations, such as the desire for glory or sympathy, or willingness to break laws, that make a difference.

Readers committed to an understanding of literature as a technology of socialization may not be convinced by my distinction between "didactic lesson" and "tools." Could not every normative "lesson" be redescribed as a kind of "tool"? And does not providing readers with population models necessarily mean socializing readers into a normative way of seeing the world, namely, as "naturally" divided into populations? Both points are valid, but only in a very limited sense. One is already in vexing territory when a purportedly normative way of seeing the world is, as in the case of the modern concept of population, one that itself emphasizes the relativity of norms. Moreover, *Frankenstein* provides its readers with two competing models of population, which emphasizes that facts about populations are always dependent upon *both* what is out there independently of the model *and* the specific model of population that is employed.[29] Moreover, as I will discuss near the end of this chapter, even if populations have to be understood as "natural," they are by no means bound to a biological register, for such models can also be used to locate surfaces that hold variations and forces of selection in those kinds of cultural phenomena to which Robert's and Victor's desires for glory pointed.

Species of Novels: Reviewing *Frankenstein*

The hypothesis that *Frankenstein* encouraged its readers to see the world in terms of populations receives tentative confirmation from the responses of some of Mary Shelley's first and most important readers, namely, those who published reviews of her book shortly after its initial publication. Though it had been commonplace since at least Samuel Richardson to describe novels as a particular "species of composition," *Frankenstein*'s explicit emphasis on the creation of a new biological species allowed reviewers to reframe this literary cliché by considering both the populations that occurred *within* the general species of the novel and the dynamics of and among novelistic species.[30] Walter Scott, for example, noted in his anonymous review of *Frankenstein* that "this is a novel, or more properly a romantic fiction, of a nature so peculiar, that we ought to describe the species before attempting any account of the individual production."[31] Thinking of the novel not simply as *a* species of composition but rather as a genus or metaspecies—that is, a collective corpus made up of species—helped reviewers (and presumably readers) in several ways.[32] First, it helped reviewers and readers identify and assess the criteria that ought to guide the reading of a particular novel. Many reviewers, for example, understood *Frankenstein* as an example of the "Godwinian" species of novel established by Mary Shelley's father.[33] Second, the assumption that the novel contained many subspecies helped reviewers and readers make sense of novels that seemed to offer *new* kinds of reading experiences. Scott, for example, suggested that *Frankenstein* was a new species of novel, one that "excites new reflections and untried sources of emotion" and thus "enlarge[s] the sphere" of the "fascinating enjoyment" of reading novels.[34]

Understanding the novel as a surface made up of species also allowed reviewers to speculate on the forces that encouraged some kinds of variations and discouraged others. Some reviewers, for example, interpreted the departure of *Frankenstein* from the Godwinian norm via the concept of monstrosity, attributing the peculiarity of this novel to a more general contemporary tendency toward exaggeration. The reviewer for the *Edinburgh Magazine and Literary Miscellany*, for example, claimed that *Frankenstein* represented "one of the productions of the modern school in its highest style of caricature and exaggeration" and sought to identify those elements of the social milieu that encouraged these variations, describing the central premise of the novel as one of "those monstrous conceptions" produced by "the wild and irregular theories of the age."[35]

The critical tone of these latter comments underscores the fact that reviewers did not seek simply to provide objective taxonomic descriptions of literary productions but also sought to locate points that would allow them to intervene in these dynamics. The reviewer for the *Edinburgh Magazine and Literary Miscellany*, for example, sought via the genre of the review essay to discourage interest in the genre of "system" (what we would now call "theory"). The form of the review itself, moreover, was intended to encourage or discourage book sales and, in this way, to affect indirectly an author's ability to continue to publish. This latter goal was also pursued by means of the acid wit of many nineteenth-century reviewers, which exploited that desire for glory—and corresponding fear of shame—that Shelley had emphasized as motivating both Robert's and Victor's endeavors and that certainly motivated many Romantic-era authors.[36]

If *Frankenstein* encouraged some of its readers—namely, Shelley's first reviewers—to see the world in terms of populations, this had certainly become a relevant task by the time Shelley's novel was published in 1818.[37] By this point, an ever-increasing number of population models were available, and as Malthus's attack on the Poor Laws had demonstrated, many of these models had significant implications for daily life. When Shelley published her novel, preeminent among these population models were both the traditional model of national population growth as a virtue and Malthus's inversion of that model (population growth as a threat). The first model of population supported multiple claims for the proper nature of the polity. In the sixteenth, seventeenth, and eighteenth centuries, this model of population was linked to an absolutist model of political sovereignty through claims that the sovereign's power increased to the extent that the national population grew. Yet the same model was then deployed in the eighteenth century in liberal *critiques* of absolutist monarchies. This strategy was exemplified by Montesquieu's suggestion that the population had declined under the absolutist rule of Louis XIV and the linked claim that populations grew most swiftly where personal liberties were greatest. It was also evident in David Hume's suggestion that, though Montesquieu was likely wrong in supposing that the global population of the modern world was smaller than that of the ancient world, it was nevertheless true that national population grew most swiftly when trade was encouraged.[38] Richard Price provided a republican variant of the model of positive population growth by arguing that the English population had decreased in the last century because of a financial policy that favored national debt and luxury over the simple life of small property owners.[39] As I have noted, there were also additional models that focused on *differences* among members of populations, rather than the overall size of the population. These included accounts of populations inoculated (and by 1818, vaccinated) against smallpox and populations of those who wished to purchase life insurance, to name just two important eighteenth-century versions of this latter model.[40] The development of statistical methods in the nineteenth century enabled a veritable explosion of these latter kinds of population models, with researchers tracking the incidence of suicide, diseases, injuries, accidents, crimes, and many other variables within national and regional populations.[41]

When Shelley published her novel, in other words, the term "population" did not have a single referent but rather stitched together multiple (and often incompatible) models and theories of what a population was, how one gathered information about population dynamics, and how that knowledge related to political action. Though by 1818 all of these population models were biopolitical, in the sense that political policies were supposed to be grounded in facts about populations, discerning the differences among these different models and their political implications was not an easy task. In such a milieu, the virtue of *Frankenstein* was that, through both its plot and its references to potential alternative plots (for example, a population of creatures multiplying in South America), it enabled readers to recognize and think through the implications of multiple models of population.

Populations and Literary Study

If *Frankenstein* helped nineteenth-century readers engage a milieu characterized by the multiplication of models of populations, what could it mean for literary critics in our even more thoroughly biopolitical twenty-first century to follow the lead of Shelley's reviewers by understanding literary texts in terms of populations? Franco Moretti has provided one contemporary answer to this question, arguing that literary critics ought to focus on the "literary evolution" of populations of texts such as nineteenth-century short stories and novels.[42] Though Moretti has since explicitly abandoned this approach in favor of computer-mediated, quantitative processes of "distant reading," an account of both the promise and pitfalls of his earlier approach helps us refine what it might mean for literary critics to think literature and its readers in terms of populations.

As Moretti tells his story, from 1987 until roughly 2000, "evolutionary theory was unquestionably the most important single influence" on his work, and during this period, Moretti's reading of evolutionary theory—which, as it turns out, was primarily based on Ernst Mayr's work on speciation—encouraged him to treat individual texts as "variations" that are exposed to forces of selection within "ecosystems" of readers.[43] These ecosystems encouraged the publication of many variants of a given kind of text. However, as a consequence of strong selective forces, only a small number of the "fittest" variants survived, in the sense that they continued to be read and published. The kinds of texts considered and the specific nature of the selective forces differ among Moretti's various articles and books in which he pursued this approach, but the basic schema remained fairly constant. In "On Literary Evolution" (1988), for example, Moretti argued that, following a period of "random variation," in which a profusion of novel-like forms were generated in the eighteenth century, there was then a period of "necessary selection" exercised by cultural forces, which had the effect of culling out all but one of these variations, the *Bildungsroman*, which then "dominate[d] the narrative universe" of the nineteenth century.[44] In "The Slaughterhouse of Literature" (2000), Moretti used a similar schema to explain why Arthur Conan Doyle's Sherlock Holmes detective stories became popular and paradigmatic, whereas the similar stories of competitors did not. He described the stories of both Conan Doyle and his competitors as variants within the ecosystem of the literary marketplace, with readers serving as the forces that "selected" the most fit of the variants, namely, those that contained formally compelling "clues."[45] And in *Graphs, Maps, Trees* (2005), Moretti drew on evolutionary accounts of populations and speciation to explain, among other things, the emergence and then extinction of nineteenth-century novelistic genres now long since forgotten by anyone but period experts, such as the Anti-Jacobin novel, the Evangelical novel, the Newgate novel, and many others (see Figure 2).[46] Noting that neither random distribution nor the chronology of political events can explain well the cyclical progression of genres revealed by his chart, Moretti proposed that the causes of the "six major bursts of [genre] creativity" we see in the "late 1760s, early 1790s, late 1820s, 1850, early 1870s, and mid-late 1880s"

> must thus be *external* to the genres, and common to all: like a sudden, total change of their ecosystem. Which is to say: a change of their audience. Books survive if they are read and disappear if they aren't: and when an entire generic system vanishes at once, the likeliest explanation is that its readers vanished at once. (20)

Just as for Mayr a new biological species can appear when a population of a species shifts to a new environment, so too for Moretti do new novelistic species (that is, genres) appear when a new generation of readers provides the literary analogue of a new environment.[47]

Moretti's Darwin/Mayr-inspired accounts of populations of literary variants are interesting and provocative for thinking about the dynamics of literary change. The application of evolutionary logic in "The Slaughterhouse of Literature," for example, allows Moretti to provide not only a

Figure 2. Map of British novelistic genres, 1749–1900. From Franco Moretti, *Graphs, Maps, Trees: Abstract Models for Literary History* (New York: Verso, 2005). Copyright © 2007 by Franco Moretti.

compelling new account of the relationship of the literary element of the clue to the genre of the detective story but leads as well to intriguing proposals about both the nature of genres and the ways that authors orient themselves toward the literary marketplace. And the method of genre analysis Moretti proposes in *Graphs, Maps, Trees* leads to an intriguing sketch of the relationship among readers' preferences, markets, human generations, and literary genres, and one hopes that other researchers will expand this sketch in more detail at some point.

At the same time, though, Moretti's use of Darwinian evolutionary theory arguably hinders his attempt to understand the dynamics of populations of texts. In a compelling and scathing review of the essays collected in *Graphs, Maps, Trees*, Christopher Prendergast has pointed out many of these problems. As Prendergast notes, Moretti's method of understanding literary-historical dynamics through theoretical terms drawn from evolutionary theory (populations, speciation, variations, selections, and competition) "places a very large bet on bringing the laws of nature and the laws of culture closer than they are normally thought to be" (56). Prendergast implies that this bet is unlikely to pay off even were Moretti to map the concepts of natural selection carefully onto literary dynamics and argues explicitly that it does not pay off if natural selection is identified with "the activities of the market" (60). Prendergast underscores the problems of such an approach for someone who, like Moretti, claims to be working from a Marxist perspective: "Philosophers of the market like to think of it as a cognate of Nature. I cannot recall a single 'Marxist' who does so. The equation of market and nature under the aegis of evolutionary biology is exactly the move of social Darwinism" (61). What such an approach prevents, Prendergast argues, is *interpretation* of literary history in terms of social and cultural dynamics. Instead of explanation, Moretti can only suggest that "if certain texts are [now] lost to us, that is because they are natural born losers," that is, not sufficiently fit for the market (61). Just as modern biologists do not seek to "explain" natural selection—it simply is, and it is not to be confused with, say, nature's pursuit of complexity, perfection, etc.—the market becomes, in Moretti's account, the unexplainable, quasi-natural force that separates winners from losers.[48]

Prendergast's critique of Moretti's identification of evolutionary and market dynamics also helps us isolate the many divergences of Moretti's model from the evolutionary theory that he cites as inspiration. In Mayr's evolutionary model, for example, speciation can occur when differences in geography encourage *an entire population* to drift away genetically from another, still existing population of the original species located somewhere else. Neither the members of the original population nor the members of the second population are in competition with one another in the sense that Moretti uses that term, nor are the two populations in competition with each other. For Moretti, though, members of a single population of literary variants are always in competition with one another, and this Malthusian, internecine conflict inevitably eliminates all but one (or at most a few) of the members of the initial population. In "On Literary Evolution," for example, the *Bildungsroman* conquers all rivals, while in "The Slaughterhouse of Literature," Conan Doyle's detective stories destroy all competitors. Even Moretti's account of the succession of genres in the nineteenth century employs this same schema of a slaughterhouse in which the vast majority of the members of a population are destroyed. Though Moretti's earlier claim that the *Bildungsroman* dominated the nineteenth century gave way to his later image of multiple genres, his chart suggests that five to ten genres always dominate a generation of readers and that these are then extinguished to make way for the next five to ten genres that command the attention of the next generation of readers. In each of these accounts, populations of literary variants do not move into new terrain, as in Mayr's account of speciation, but serve as the pile of textual corpuses upon which the "fittest" can stand.

It is also difficult to coordinate Moretti's image of large populations being pared down to a few victors with his important claims about the expansion and subdivision of the literary market during the nineteenth century. He notes, for example, that around 1820, "the internal composition of the [literary] market changes," with the following consequences:

> So far, the typical reader of novels had been a "generalist"—someone "who reads absolutely anything, at random," as Thibaudet was to write with a touch of contempt in *Le liseur de romans*. Now, however, the growth of the market creates all sorts of niches for "specialist" readers and genres (nautical tales, sporting novels, school stories, *mystères*): the books aimed at urban workers in the second quarter of the nineteenth century, or at boys, and then girls, in the following generation, are simply the most visible instances of this larger process, which culminates at the end of the century in the super-niches of detective fiction and then science fiction.[49]

One would think that this emergence in the nineteenth century of new kinds of readers—urban workers, boys, and girls, to draw on the examples Moretti cites—would encourage the proliferation of new populations of novels, rather than the *elimination* of most members of a population of literary variants. Yet Moretti's focus on elimination is a consequence, as

Prendergast correctly notes, of the fact that he seems to understand competition through an economic lens, rather than the biological approach promoted by Mayr. Moretti's tendency to understand natural selection through the lens of the market also explains the otherwise rather baffling omission from his model of any analogue for sexual reproduction, which would seem to be a sine qua non for population-oriented theories of natural selection.

Yet these criticisms seem to me more a reason to detach Moretti's concept of populations of literary variants from the framework of evolutionary theory than for rejecting the population approach itself. Or, to put this another way, Moretti simply chose the wrong model of population. Moretti clarifies that he chose evolutionary biological models of populations in order to make literary history more scientific, which suggests that evolutionary biology was for Moretti not a model but simply unequivocal scientific truth.[50] However, as I have noted, *every* claim about population is based on a model, and—for reasons that Prendergast discusses and that I have supplemented—the evolution-as-market-logic model of populations simply does not fit Moretti's interests especially well. Choosing (or creating) another population model might allow us to understand better how those populations of literary variants to which Moretti has drawn our attention relate to populations of readers. It might allow us to take into account, for example, both that rise of many new populations of readers to which Moretti points and also the possibility of monstrous crossings of genres noted by some of the first reviewers of *Frankenstein*.[51]

Explicitly treating population theories as models encourages us to relate literary-critical population models to the numerous other kinds of population models employed by eighteenth- and nineteenth-century authors, such as populations of smallpox sufferers, populations of those insured by life insurance contracts, Malthusian populations, and colonial populations. Each of these population models is intended to intervene biopolitically in a specific way, though the population dynamics proposed by each model differs from those of others. Keeping this long history of concepts of population in mind—a history in which, moreover, literature itself played a key role—is vital in order to avoid either reducing or naturalizing the concept of population. Where Moretti arguably failed because of his aspiration to bring to the study of the eighteenth- and nineteenth-century literary market "properly" scientific concepts of population and speciation, I urge us instead to begin with those models of populations developed *within* eighteenth- and nineteenth-century literary texts, including both *Frankenstein* and the responses of reviewers to that novel.

Conclusion: Neoliberal Populations and Markets

Prendergast notes that Moretti's identification of natural selection with the market is eerily reminiscent of "the move of social Darwinism." While Prendergast likely had in mind social Darwinists of the late nineteenth century, Moretti's linkages among natural selection, markets, and populations resonate even more strongly with twentieth- and twenty-first-century neoliberal characterizations of the market. Yet it is worth trying to separate the use of population concepts for literary critics, and even parts of Moretti's method, from these echoes of neoliberalism. I will thus conclude by describing the informatic lens that enables neoliberal theorists to describe markets as the *only* structures able to solve the kinds of information problems faced by modern societies, so that we can better distinguish this neoliberal logic from the possibilities that Shelley's approach to population opens up.[52]

Friedrich Hayek's neoliberal conceptualization of the market in the 1940s and 1950s is an important twentieth-century site for the deployment of population logic. Writing immediately after the Second World War and in the context of proposals in the United States and the United Kingdom to continue wartime centralized economic planning into peacetime, Hayek contended that the economic information one needs in order to plan centrally—information about, say, raw materials, production costs, and consumer preferences—can never be gathered together at one single point but rather exists "solely as . . . dispersed bits of incomplete and frequently contradictory knowledge which all the separate individuals [of an economy] possess."[53] Hayek's suggestion that economic knowledge is "dispersed among all the people" presumes that each individual differs from one another, in the sense that each individual is situated in, and has the most knowledge of, his or her own particular "time and place" and his or her "local conditions" (521–22). As a consequence, "practically every individual has some advantage over all others in that he possesses unique information of which beneficial use might be made, but of which use can be made only if the decisions depending on it are left to him or are made with his active cooperation" (521–22). Hayek argued that this distributed knowledge is especially important in the context of changing economic conditions, such as rising or falling production costs or changes in the availability of raw materials (523). The only possibility of "planning" in this state of distributed knowledge is to enable economic competition; that is, "competition . . . means decentralized planning by many separate persons" (521).[54] Or, to put this another way, the price system of capitalist competition is a mechanism by which distributed individual perspectives are brought together and economic problems are "solved" (525). However, Hayek stressed that "the whole acts as one market, not because any of its members survey the whole field, but because their limited individual fields of vision sufficiently overlap so that through many intermediaries the relevant information is communicated to all" (526).

For neoliberals, market relations are thus a kind of population-based computing that has "evolved" within human relations "without [conscious human] design" (527). Hayek's approach resonates with Mayr's slightly later account of biological populations, for both Mayr and Hayek emphasize large aggregates made up of unique individuals who each relate to their environments in ways that differ slightly from that of their fellows, and both stress that this system of differences allows the aggregate to change as its environment alters. For Mayr, sexual reproduction allows the characteristics of the population to change, while for Hayek, the price system connects unique individual perspectives in such a way that economic production, distribution, and consumption can shift as the milieu of the economy changes.[55]

This understanding of the price system as an institution that has evolved autonomously within human relations has encouraged neoliberals to treat the market as a metasurface that ought to contain all other population-oriented surfaces. Neoliberals tend to treat any institution in which they detect population logic—for example, the peer-review system of scientific research, which employs competition among many researchers in order to fund and publish the results of only a small subsection of these—as an implicit market (a "marketplace of ideas") and suggest that such de facto marketplaces would be more efficient as de jure markets.[56] This claim is seductive because it acknowledges resonances between the neoliberal concept of the market and other population approaches. Yet this neoliberal approach ultimately confuses structural similarity with identity: What links the neoliberal concept of the market with the structure of knowledge production in the sciences, for example, is that both employ population logic, rather than that both are markets. Moretti risks a similar conflation between Mayr's evolutionary biological model and the model of the market.

If, as I noted at the start of this chapter, neoliberalism is bound up with an uncanny, neo-Romantic return to the Malthus-Godwin debate, we should not respond by rejecting the logic of populations, though we should not limit this logic to the register of biology (and especially not the coarse Malthusian axis of "reproduction"), nor should we confuse the logic of population with that of the market. We should instead follow Shelley's lead.

This means, in part, recovering and fleshing out the numerous population models that emerged in the eighteenth and nineteenth centuries (many of which have continued currency, as both recent invocations of Malthus and the contemporary dominance of actuarial logics make clear). Emphasizing the multiplicity of population models has the virtue of relativizing the neoliberal model of markets as simply one population model among many (and, as Prendergast's critique emphasizes, often a rather restricting one for understanding the experience of creative literature and the dynamics of literary production since the nineteenth century).

Following Shelley's lead would also mean understanding creative literature not simply as subject to the population dynamics of literary markets, as in Moretti's analyses, but also as a space within which existing population models can be gathered for the sake of generating new population models. In the case of *Frankenstein*, the milieu of models included the older model of national population growth as a virtue (in its various political parsings: for example, absolutist, liberal, and republican), the Malthusian inversion of the traditional model, and difference-oriented population models. *Frankenstein* did not endorse any one of those models but brought these into relationship with one another and, in this way, made possible *new* difference-oriented models of population. Such models may indeed help us understand better how, in a century of ever-expanding readership, novelistic genres crossed and hybridized, creating ever more genres. But to do so, they will likely also force us beyond the conflict between Godwinian and Malthusian subjects—that is, the contest between an infinitely malleable subject always threatened by institutional inscription and a subject unalterably fixed in its biological nature—and toward population subjects that relate to collectivities *through* individual differences.[57] This would in turn mean understanding both Romantic and post-Romantic literature not solely as institutions that create and enforce norms but also as occasions for the invention of new methods of locating and experiencing non-normative variations.

3. Freed Indirect Discourse

Biopolitics, Population, and the Nineteenth-Century Novel

In *The Human Condition*, Hannah Arendt includes an intriguing aside on the eighteenth-century emergence of the novel, suggesting that this mode of writing was tied to the near-simultaneous emergence of the concept of "society" in the work of authors such as Jean-Jacques Rousseau. Arendt argues that intrinsic to the concept of society is the "deman[d] that its members act as though they were members of one enormous family which has only one opinion and one interest."[1] Arendt suggests that the concept of society is inextricably tied to "behavior," a concept that focuses attention on the ways that people in the aggregate habitually comport themselves, rather than those unusual or rare words or deeds—what Arendt describes as "actions"—that change our collective relations to one another. Arendt contends that modern social sciences, such as economics, depend on the concept of society insofar as they focus on how people behave, rather than act, and she emphasizes that such descriptions of behavior have normative force, for individuals in fact conform to descriptions of typical behavior (41–42). Arendt suggests that the literary form of the novel is, like economics, a consequence of this emergence of the concept of society, for the novel is "the only entirely social art form" (39). More specifically, the concept of society produces, as a reaction to its normalizing tendencies, the opposed concept of "intimacy," which functions as the individual's supposed refuge from society, and the novel is the art form dedicated to the (impossible) task of mediating between the normalizing tendencies of society and the individualizing possibilities of intimacy.

Arendt's brief note about the novel is compelling, no doubt in part because her account resonates with a number of now-classic literary-critical accounts of the emergence of this literary form, such as Ian Watt's *The Rise of the Novel*, John Bender's *Imagining the Penitentiary*, Nancy Armstrong's *Desire and Domestic Fiction*, and D. A. Miller's *The Novel and the Police*.[2]

These histories of the novel have underscored, in various ways, the apparent paradox to which Arendt's account points: Even as the novel may seem to defend the sphere of individualizing intimacy against the normalizing tendencies of society, it also functioned as a vehicle for normalizing behavior and, in this sense, inhibited what Arendt calls action as much as did the science of economics.

While this chapter does not call into question this critical consensus about the relationship among novels, society, and normalization, I approach the interrelationship of these terms differently. My account takes its starting point from another important eighteenth-century concept, population, which is related in complicated ways to concepts and practices of society and normalization. I propose that we see the nineteenth-century novel less as an attempt to mediate between the contrasting demands of society and intimacy and more as seeking to mediate between the concepts of society, on the one hand, and difference-oriented concepts of population, on the other. The concept of society and difference-oriented concepts of population diverge because they focus on fundamentally different registers: Where the concept of society understands individuals as malleable subjects who respond more or less exclusively to *social* influences, difference-oriented concepts of population seek those points at which individuals respond to opaque, *non*social forces, such as those that emerge from our collective biological existence. There is a structural similarity between my argument and Arendt's claim about the novel, for I suggest that the conflict within the novel between the normalizing premises of the concept of society and the individualizing premises of the concept of population is like that tension between social normalization and individualizing intimacy to which Arendt pointed. However, where Arendt reads the novel as an ultimately unsuccessful attempt to guard against or compensate for the normalizing force of "society," I read the novel as a technology that mediated between social and population logics. The novel mediated between these two logics both in the sense that it helped readers link these two logics and in the sense that novels sought to locate, and valorize, forms of individual difference that could not be contained within a social logic. In this sense, I suggest, the novel had, since at least the nineteenth century, an essentially biopolitical vocation.

I pursue this proposition by focusing on the relationship between two literary devices that underwent significant development in this period: first, the massive expansion of character-systems in nineteenth-century novels, and second, the emergence and development of free indirect discourse. I argue that the emergence of free indirect discourse in the nineteenth century was a tool by means of which nineteenth-century novelists developed "surfaces" that allowed them to search, quasi-scientifically, for forces that determine the characteristics of populations.[3] Novelists often presented these forces, which included habitual comportments and what the French novelist Émile Zola described as hereditary "cracks," as nondiscursive causes of both regularities and differences among members of a population and used free indirect discourse to document the effect of these forces on the thoughts and feelings of individuals. Understanding free indirect discourse as a method by means of which novelists searched for nondiscursive forces requires that we keep this device connected to character-systems, since free indirect discourse can only function for novelists as a surface for detecting hidden forces when this literary device is tied to the population illuminated by a specific character-system.

My argument has five parts. I begin by expanding on what I mean by difference-oriented concepts of population. I then turn to Alex Woloch's notion of a novelistic character-system, arguing that his formulation must be altered to take into account the actual systems of characters that populate nineteenth-century novels, which included not just humans but multiple nonhuman agents. In the third section, I take up free indirect discourse, emphasizing the ways that nineteenth-century novelists used this device to register far more than simply thoughts, feelings, and minds, but also—and perhaps more significantly—nondiscursive forces that bear upon collective biological existence. I then consider the relationship between novels, on the one hand, and Michel Foucault's trinity of population, security, and territory, on the other, arguing that, though nineteenth-century novels engaged both populations and territories in a quasi-scientific manner, they did not aim at security. I conclude with a reflection on how my account of the connection between free indirect discourse and character-systems in the nineteenth-century novel helps us think about relationships between science and literature more generally.

Before beginning, I have two caveats and one terminological specification. My first caveat: This chapter focuses less narrowly on the Romantic era than did the preceding two chapters, for several of the novels that I discuss appeared well after the Romantic period proper. Yet the basic elements that I consider here—the creation of population models, expansive novelistic character-systems, and the development of free indirect discourse—first emerged, and were merged, in the Romantic era. My second caveat is that I focus closely on a small number of passages from a small number of novels. However, I see these as exemplifying more general trends within nineteenth-century European and American novels. Though not all nineteenth-century novels employ character-systems and free indirect discourse in the ways I describe, quite a few do. Finally, my terminological specification: Though the novels that I consider here do not belong to one genre, I nevertheless describe them as realist, by which I mean that the fictional worlds that they describe are intended to be understood by readers as referring to, and capturing key aspects of, the real world in which the reader lives. (And thus, as I note below, my account troubles the distinction, solidified by Georg Lukács, between "realism" and "naturalism.")

Concepts of Population and Biopolitics

As I have noted in previous chapters, populations can be understood in two quite different ways. First, population is often understood as a primarily quantitative term, which refers to the *number* of individuals living within a given geographic region, whether a city quarter, a country, or the earth as a whole. It was in this sense, for example, that Francis Bacon referred in the seventeenth century to population as one of several factors by means of which one could determine the "true Greatness of Kingdoms and Estates," and this primarily quantitative sense of the term also underwrote Thomas Malthus's *An Essay on the Principle of Population* (1798).[4] For this sense of the term, the relevant aspects of a population are the borders, usually political, within which the population exists, and the total size of the population. Differences among individual members of a population are of minimal or no importance for this concept of population. For Malthus, for example, the key facts about the individuals in a human population are that each has a belly that needs to be filled, and each can contribute, via sexual reproduction, to the creation of more individuals.

Population can also be understood in a second way, as a large collection of individuals who *differ* from one another in key ways. This latter sense of the term also has its remote origins in the seventeenth century, as early political arithmeticians sought to distinguish between categories of people, such as the "productive" and the "unproductive" individuals in a national population, and proposed policies intended to increase the number of productive individuals and decrease the number of unproductive individuals.[5] This second sense of population became even more important in the late eighteenth and early nineteenth centuries, as the relevant differences among individuals were often figured not as moral but as biological differences that were beyond the direct reach of consciousness or laws. This premise of biological differences among members of a population enabled the emergence of what Michel Foucault describes as biopolitics, which aimed to locate regularities and normal distributions among these differences so that one could develop techniques and technologies for changing those norms. To draw again on one of Foucault's examples, eighteenth-century advocates of smallpox inoculation sought to locate the relevant differences among individuals—for example, the age at which one was inoculated—that both pointed to norms and at the same time enabled those norms to be altered. For example, if individuals under the age of three who had been inoculated died more frequently than those over the age of three, one could delay inoculation a bit to change the norm of survival for those under the age of three.[6]

This second concept of population presumes that relevant differences among members of a population cannot be directly controlled solely by traditional political measures, such as legislative decree.[7] Though my example of inoculation focuses on the specifically corporeal difference of susceptibility to smallpox, the biopolitical logic of this second concept of population held for other kinds of differences as well. As Foucault noted, eighteenth-century physiocratic and political-economic theory employed a difference-oriented theory of population in order to promote an indirect channeling of differences that occurred in the realm of self-consciousness—namely, economic choices about investing and purchasing—but that at the same time could not be directly (or at least effectively) determined by legal pronouncements, in the sense that many individuals often sought successfully to find ways around these laws.

This second concept of population enabled the development of what Foucault called biopolitics when its premise about the importance of individual difference was linked to three technologies. First, one needed a specific population *model* that identified the relevant differences among individuals for a given problem or issue. In the smallpox example, this might be a model that stressed the importance of age differences among individuals. In the case of French physiocratic and British political-economic theory, this meant a model that stressed the different decisions about purchasing and investing made by a population of individuals. Second, biopolitics required the construction of *mechanisms able to capture data* about the individual differences proposed by the model. In the case of smallpox, such data could be captured by linking the observations of individual doctors about their patients, so that one could locate norms across many individuals.[8] For political economy, this data might take the form of stock or market prices that connected individuals interested in buying and selling goods. Finally, the development of biopolitics required the invention and deployment of *small-scale switching mechanisms* that changed, often in minimal ways, the behavior of at least some individuals so that one could slowly alter the population-wide norms that were initially observed. In the smallpox example, these mechanisms included inoculation campaigns and methods for ensuring compliance; for political economy, they included removing legal restrictions on the import and export of goods and money.

These models, data-capture mechanisms, and small-scale switching mechanisms depended in part on juridical-political measures, such as changes in the laws governing who can or must be inoculated or dictating when goods or money can be moved from one location to another. But they depended equally on mechanisms that convinced individuals to alter the expectations, hopes, and fears that patterned daily life. As I noted in the Introduction, Foucault described these technologies of the self as "operations" by means of which individuals employed truth-claims in order to transform "their own bodies, their own souls, their own thoughts, their own conduct, and this in a manner so as to transform themselves, [and] modify themselves, and to attain a certain state of perfection, happiness, purity, supernatural power."[9]

The question that motivates this chapter is this: What if we were to think of the nineteenth-century novel as one of the sites at which these three biopolitical technologies—population models, data gathering, and switching mechanisms—were integrated? That is, what would the nineteenth-century novel look like if we thought of it as analogous in spirit and effects to an inoculation campaign or a liberal political-economic program? In what ways could novels establish themselves as sites of truth-claims about populations, which could then be employed by a reader as a technology of the self?

Novels, Populations, and Character Systems

We can begin to answer this by considering connections between the literary devices of character-systems and free indirect discourse. The concept of character-systems has been employed by the literary critic Alex Woloch to theorize the development, in nineteenth-century novels, of increasingly numerous and complex casts of characters, at least in comparison to the eighteenth-century novel. Or, as Woloch puts it,

> the nineteenth-century novel contains a greater quantity of characters than
> most previous literature—a huge variety of individuals who get crowded

together into a single story. The omniscient totality of the nineteenth-century novel compels us to "connect" these individuals—to comprehend forms of social relation which can encompass the diverse populations that people these novels.[10]

Woloch emphasizes that we should understand nineteenth-century novelistic character-systems in terms of both their "structural aspects" and "referential" dimensions (15–19). By "structural aspects," Woloch means that we should attend to how major and minor characters form an abstract system, in which each character is implicitly defined against all of the other characters in the system. The "referential" dimension of a character-system, by contrast, emphasizes that even minor characters seem to refer outside the novel to a "compelling human singularity" that we otherwise associate only with actual human beings (17).

Yet when it comes to explaining *why* novelistic character-systems in the nineteenth century expanded, Woloch is arguably less helpful, for he sees this expansion as a response to the entirely political imperative of expanding the voting franchise. "I want to argue," Woloch writes,

> that the realist novel is structurally destabilized not by too many details or colors or corners, but by *too many people*. It is the claim of individuals who are incompletely pulled into the narrative that lies behind the larger empirical precision or realist aesthetics. As the logic of social inclusiveness becomes increasingly central to the novel's form . . . this problem becomes more pressing. The novel gets infused with an awareness of its potential to *shift* the narrative focus away from an established center, toward minor characters. (19)

Woloch suggests that the nineteenth-century "novel's sense of the potential to shift narrative attention [from major to minor characters] is intertwined with a specific notion of human right": namely, that each character (and so, by implication, every actual person) "has a 'case,' an originating consciousness that, like the protagonist's own consciousness, could potentially organize an entire fictional universe" (22). These claims of "minor characters on the reader's attention," Woloch argues, "are generated by the democratic impulse that forms a horizon of nineteenth-century politics" (31).

Many of the novelists whom Woloch considers, such as Eliot, Dickens, and Zola, were indeed concerned with questions of political representation and democracy. Yet Woloch's explanation does not seem able to account for the fact that among the ever more numerous characters of nineteenth-century realist novels were some, such as children, animals, and even plant crops, diseases, and machines, that would not have been political subjects even under the most capacious of nineteenth-century philosophies of political franchise.[11] Consider, for example, the character-system of Émile Zola's *Germinal* (1885) (Figure 3). It is difficult to miss the political and democratic orientation of Zola's novel, which focuses on an extended strike by miners who seek at least minimal improvements in their dehumanizing living conditions and which includes characters who discuss the work of political theorists such as Karl Marx.[12] Yet *Germinal* includes among its important characters not only humans but several animals (for example, the workhorses Battle and Trumpet, who live in the mine, and a pet rabbit named Poland) and also food crops, such as wheat and beets. The characters in *Germinal* also arguably include a mob and a hereditary "flaw" or "crack" within the protagonist of the novel, Étienne Lantier, which operates "beyond his power to control it" and causes him to become uncontrollably angry when he drinks alcohol.[13] This hereditary crack implicitly connects the character Étienne to other Lantier characters in additional novels within Zola's Les Rougon-Macquart series, such as *The Human Beast* (*La bête humaine*), in which the crack also plays a key role.[14] While many of these nonhuman entities or forces function as characters in this novel, there is nothing in *Germinal* to suggest that an expanded political franchise is the enabling frame or aspiration that encouraged Zola to include characters such as horses, rabbits, and hereditary cracks.

The ontological diversity of characters in Zola's *Germinal* is not anomalous within nineteenth-century novels. Nor is this ontological diversity simply a byproduct of the expansion of the number of characters in novelistic character-systems in the nineteenth century. Compare, for example, Daniel Defoe's *Robinson Crusoe* (1719), which stands at or at least near the start of the English novel tradition, with Herman Melville's *Moby-Dick*, published in the middle of the nineteenth century (1851). Both novels employ a first-person narrator who retrospectively seeks to understand the events that earlier befell him; in both, the majority of the plot takes place in a setting that limits significantly the number of possible characters (in *Robinson Crusoe*, a deserted island; in *Moby-Dick*, a whaling ship that was supposed to remain at sea for several years); and each novel includes both human and animal characters as well as a quasi-supernatural agency (in *Robinson Crusoe*, an innate "inclination" that Crusoe retrospectively intuits as having determined his behavior; in *Moby-Dick*, a "monomaniac" idea, itself prompted by "malicious agencies," that possesses the narrator).[15] Yet in comparison to *Robinson Crusoe*, the cast of characters in *Moby-Dick* is vast (see Figures 4 and 5). Even more significant, where all the characters in

Robinson Crusoe—including animals—have importance only insofar as they contribute to the narrator's ability to understand his story as one of personal spiritual salvation, *Moby-Dick* is simultaneously a novel about Ishmael's reflections on Captain Ahab's pursuit of a white sperm whale, an attempt to contribute to scientific knowledge about whales in general and sperm whales in particular, and an account of how populations of whales relate to populations of humans. To return to Woloch's account, the use of non-human and even nonanimal characters within the character-systems of nineteenth-century novels such as *Moby-Dick* suggests either that novelists were under the sway of a democratic and inclusionary politics far in advance of anything imagined in the nineteenth or even twenty-first centuries or—and more likely—that they included so many types of characters for a different, or at least additional, reason than the exclusively political grounds Woloch suggests.

I propose that we understand the expansion of nineteenth-century novelistic character-systems as part of an effort to create *population models*, rather than simply models of an expanded political franchise.[16] Understanding the inclusion of more and more characters in nineteenth-century novels as a response to an imperative to model populations requires that we reconsider two parts of Woloch's account: first, the nature of the pressure that produced the expansion of character-systems, and second, the goals served by that expansion of character-systems. Though Woloch correctly stresses that nineteenth-century novelists suggested that even minor characters were unique individuals, we need to understand this uniqueness in ontological rather than solely in political terms. That is, rather than functioning solely as an attempt to include more voices, perspectives, or personalities within a democratic franchise, expansion of character-systems in nineteenth-century novels was an effort to model the agents that connect the human, animal, vegetable, and even mineral worlds and within which population-level changes are articulated.

Thomas Malthus had already provided, at the end of the eighteenth century, a rather coarse image of these kinds of affiliations, stressing a population dynamic that linked the growth rate of humans to that of food crops. Yet in *An Essay on the Principle of Population*, Malthus sought to limit population thinking to the model of a tragic antagonism between a major character (the swiftly expanding human population) and a minor character (the more slowly expanding population of wheat). In the same year, the physician Edward Jenner developed a more capacious scientific character-system. In Jenner's account of inoculation, a three-character system of humans, cows, and smallpox forms a complex of interspecies

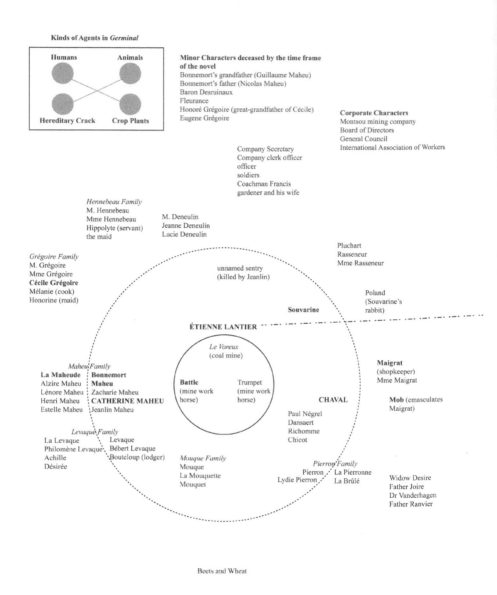

Kinds of Agents in *Germinal*

Humans	Animals
Hereditary Crack	Crop Plants

Minor Characters deceased by the time frame of the novel
Bonnemort's grandfather (Guillaume Maheu)
Bonnemort's father (Nicolas Maheu)
Baron Desruinaux
Fleurance
Honoré Grégoire (great-grandfather of Cécile)
Eugene Grégoire

Corporate Characters
Montsou mining company
Board of Directors
General Council
International Association of Workers

Company Secretary
Company clerk officer
officer
soldiers
Coachman Francis
gardener and his wife

Hennebeau Family
M. Hennebeau
Mme Hennebeau
Hippolyte (servant)
the maid

M. Deneulin
Jeanne Deneulin
Lucie Deneulin

Pluchart
Rasseneur
Mme Rasseneur

Grégoire Family
M. Grégoire
Mme Grégoire
Cécile Grégoire
Mélanie (cook)
Honorine (maid)

unnamed sentry
(killed by Jeanlin)

Poland
(Souvarine's
rabbit)

Souvarine

ÉTIENNE LANTIER

Le Voreux
(coal mine)

Maigrat
(shopkeeper)
Mme Maigrat

Maheu Family
La Maheude : **Bonnemort**
Alzire Maheu : **Maheu**
Lénore Maheu : Zacharie Maheu
Henri Maheu : **CATHERINE MAHEU**
Estelle Maheu : Jeanlin Maheu

Battle
(mine work
horse)

Trumpet
(mine work
horse)

CHAVAL

Paul Négrel
Dansaert
Richomme
Chicot

Mob (emasculates
Maigrat)

Levaque Family
La Levaque   Levaque
Philomène Levaque   Bébert Levaque
Achille   Bouteloup (lodger)
Désirée

Mouque Family
Mouque
La Mouquette
Mouquet

Pierron Family
Pierron : La Pierronne
Lydie Pierron : La Brûlé

Widow Desire
Father Joire
Dr Vanderhagen
Father Ranvier

Beets and Wheat

Figure 3. Character-systems of Émile Zola's *Germinal* (1885) and *The Human Beast* (1890). This map represents a small portion of the massive population of characters of Zola's *Les Rougon-Macquart* novels. The left half represents the character-system of *Germinal*. The three major characters are bolded in capital letters, midlevel minor characters are bolded in title case, and minor characters are unbolded. The dotted-line circle surrounds those characters (and families of characters) who habitually go into the mine. Characters in parentheses and

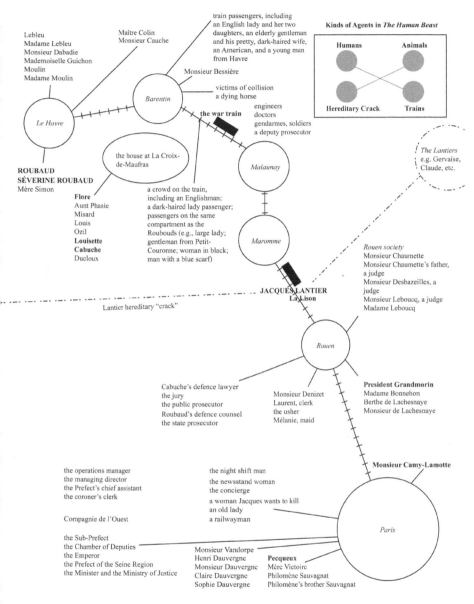

italics have died by the start of the novel's action. The right half represents the character-system of *The Human Beast*. The three major characters are bolded in capital letters, midlevel minor characters are bolded in title case, and minor characters are unbolded. The map structure is provided by the train track from Le Havre to Paris, and characters are situated near the city or site in which they spend most of their time. (My thanks to Catherine Lee for her help with both of these figures.)

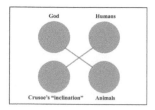

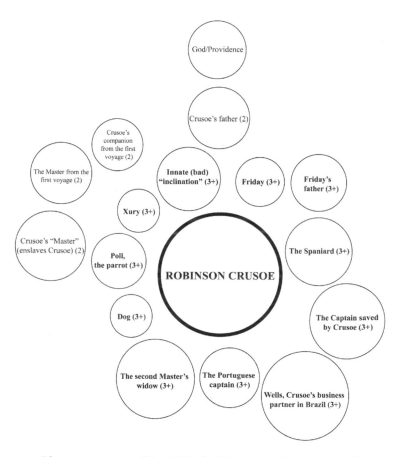

Figure 4. Character-system of Daniel Defoe's *Robinson Crusoe* (1721). Though many individuals are mentioned in *Robinson Crusoe*, the novel nevertheless feels sparsely populated because Robinson Crusoe has one or more *direct* interactions with only a limited number of individuals. In keeping with the novel's stress on spiritual isolation, characters with whom Crusoe has more than one interaction are placed around him, and those with whom he has the most interactions are closest to his own sphere. (The one exception is God, with whom Crusoe does not have any direct interaction but who is for Crusoe always present). The most

Pre-Island Minor References

- Crusoe's mother
- Crusoe's brother, a lieutenant collonel
- Crusoe's brother, no information given
- the Mate from the first voyage
- the Boat-Swain from the first voyage
- the Master from the second voyage, who dies
- the Master from the third voyage
- the men from the Rover
- the English carpenter who has been made a slave
- a Scots sailor in the Portuguese Captain's ship
- the Castle
- Moley
- a monster lion
- a man with lance
- the other creature
- a leopard
- a man Crusoe lives with
- Crusoe's 2 European servants
- Crusoe's black slave
- the master of the ship which shipwrecks, the master's boy, and 14 men, including the man who dies of the calenture, the man who is wash'd overboard, the man who cries out land, and the mate of the vessel

Island Minor References

- cats
- goats
- a kid
- goats and hares that eat Crusoe's crop
- fowls
- an old he-goat and three kids
- Crusoe's cats
- Crusoe's "kids"
- other parrots
- tamed sea-fowls
- the old injured goat
- 9 naked savages
- a drowned boy
- a dog from the ship
- 2 drowned men
- Benamuckee and Oowocakee
- the Mate of the Captain saved by Crusoe
- the Passenger
- the Boatswain, leader of the crew that mutinied
- other men who mutinied, including innocent men and those who stay
- the Spaniard's comrades

Post-Island Minor References

- the Portuguese Captain's son
- the Trustees
- the Procurator Fiscal
- the King
- the Monastery of St. Augustine that works for the poor and on conversion of Indians
- the Steward of the King's Revenue
- the Steward of the Monastery
- the Survivors of the Trustees
- a Merchant who knows the Portuguese Captain
- the Notary
- a Merchant in Lisbon
- the Prior of St. Augustine
- Crusoe's Partner's Wife and two Daughters
- an English Gentleman, the son of a Merchant in Lisbon
- 2 English Merchants
- 2 young Portuguese Gentlemen
- 5 servants, one of them an English sailor

- 4 French Gentlemen
- a Guide
- 12 other gentlemen
- 3 wolves
- the Bear
- other wolves, men who get eaten by wolves
- a new guide
- a Correspondent in Lisbon
- 11 men and 5 women prisoners
- 20 young children
- a Carpenter
- a Smith
- 7 women
- Crusoe's sisters
- 2 children of one of Crusoe's brothers
- Crusoe's wife
- Crusoe's children (2 sons, 1 daughter)

significant characters (3+ interactions) are bolded; italics indicate the least significant (1 interaction). The remaining individuals mentioned in the novel are collected into the three categories of "Pre-Island Minor References," "Island Minor References," and "Post-Island Minor References." I have not included collective entities such as the Spaniards, unnamed people on the shore, unspecified magistrates, creatures, fellow planters, etc. (My thanks to Catherine Lee for her help with this figure.)

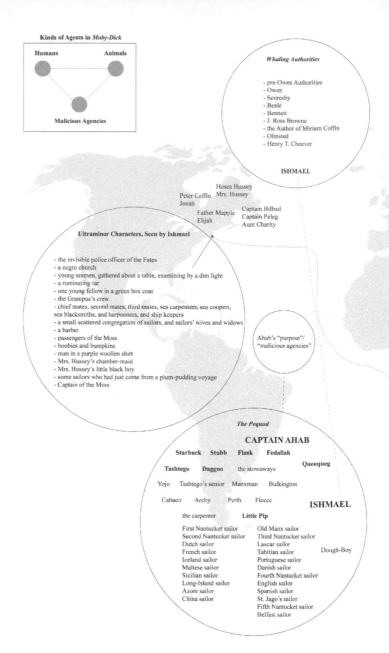

Figure 5. Character-system of Herman Melville's *Moby-Dick; or, the Whale* (1851). In comparison to *Robinson Crusoe*, the cast of characters in *Moby-Dick* is vast, in part because animals play more prominent roles in *Moby-Dick* and in part because *Moby-Dick* is simultaneously a novel about Ishmael's reflections on Captain Ahab's pursuit of a white sperm whale and an attempt to contribute to scientific knowledge about whales in general and sperm whales in particular. There are thus two registers of characters in this novel: characters with whom Ishmael came into contact during the time immediately preceding, and during, his time aboard the *Pequod*; and real (i.e., extratextual) authorities on whales and whaling with whom Ishmael is often in argument. I have placed the authorities in their own sphere and connected this sphere via a dotted line to the time and space traversed by the *Pequod*.

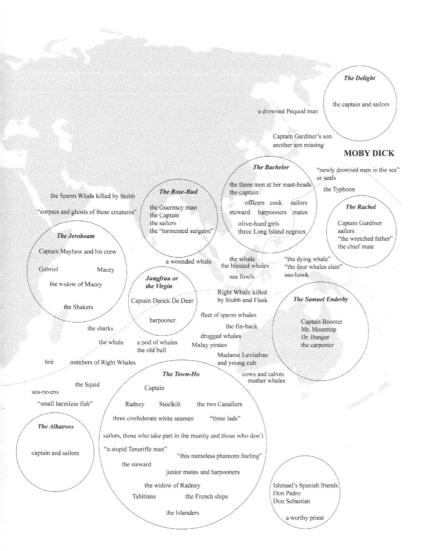

The Delight
the captain and sailors

a drowned Pequod man

Captain Gardiner's son
another son missing

MOBY DICK

The Bachelor
"newly drowned men in the sea"
or seals
the three men at her mast-heads
the captain
the Typhoon

the Sperm Whale killed by Stubb

The Rose-Bud
the Guernsey man
the Captain
the sailors
the "tormented surgeon"

officers   cook   sailors
steward   harpooners   mates

olive-hued girls
three Long Island negroes

The Rachel
Captain Gardiner
sailors
"the wretched father"
the chief mate

"corpses and ghosts of these creatures"

The Jeroboam
Captain Mayhew and his crew

Gabriel   Macey

the widow of Macey

the Shakers

a wounded whale

the whale
the blasted whales

sea fowls

"the dying whale"
"the four whales slain"
sea-hawk

Jungfrau or
the Virgin
Captain Derick De Deer

harpooner

Right Whale killed
by Stubb and Flask

The Samuel Enderby
Captain Boomer
Mr. Mounttop
Dr. Bunger
the carpenter

the sharks

the whale   a pod of whales
the old bull

fleet of sperm whales
the fin-back
drugged whales
Malay pirates

brit   numbers of Right Whales

Madame Leviathan
and young cub

The Town-Ho
Captain

cows and calves
mother whales

the Squid

sea-ravens

"small harmless fish"

Radney   Steelkilt   the two Canallers

three confederate white seamen   "three lads"

The Albatross

captain and sailors

sailors, those who take part in the munity and those who don't

"a stupid Teneriffe man"

the steward

"this nameless phantom feeling"

junior mates and harpooners

the widow of Radney

Tahitians   the French ships

the Islanders

Ishmael's Spanish friends
Don Pedro
Don Sebastian

a worthy priest

(Ishmael thus appears twice as a character on this map.) I have not included acquaintances Ishmael mentions but who do not figure in the voyage of the *Pequod* (e.g., Peter Coffin), individuals whom Ishmael did not personally meet (e.g., Nathan Swain), or Ishmael's memories of earlier encounters with animals (e.g., the first albatross he saw). I have also not included numerous real individuals, including explorers (e.g., Mungo Park), political figures (e.g., Napoleon), and literary authors (e.g., Coleridge). Nor have I included biblical personages (e.g., Jonah), mythological characters (e.g., Perseus), or humans who appear only in dreams (e.g., Ishmael's stepmother in the dream described in chapter 4). I have included only those whales encountered by the *Pequod* and have not included corporate and collective entities and groups mentioned by Ishmael (e.g., the Greeks, the Feegees, temperance societies, etc.).

Figure 6. James Gillray, *The Cow-Pock—or—the Wonderful Effects of the New Inoculation!* (1802). Source: Library of Congress Prints and Photographs Division.

allegiances and alliances that cannot be reduced to that much simpler tragic schema by which Malthus connected humans and their food. As is evident in James Gillray's 1802 satirical cartoon (Figure 6), the shape of the plot generated by this character-system was not immediately clear to contemporary observers: Would a character-system that linked cows, humans, and smallpox lead to "comedic" progress, tragic destruction, or a future that could only be described as farce?

Perhaps not surprisingly, nineteenth-century novelists proved themselves to be far more supple thinkers about the logic of population than Malthus, Jenner, or their critics. Even in the case of a Malthusian character such as food, a more complex sense of wheat as a character emerges in novels such as Frank Norris's *The Octopus* (1901), the first volume in Norris's unfinished trilogy *The Epic of Wheat*. The fate of the novel's main human characters is bound to a nexus of railroads, business monopolies, and tens of thousands of acres of wheat planted in California, which latter "welte[r] under the sun in all the unconscious nakedness of a sprawling, primordial Titan."[17] Wheat is also arguably a minor character in Thomas Mann's *Buddenbrooks* (1901), which maps, with quasi-scientific clinical detachment, those intersections of intergenerational familial dynamics, changing business practices, and vagaries of wheat production that make it possible for a formerly successful commercial family to be destroyed completely through the coincidence of a deadly case of typhus, a failed crop, and a stroke. Nineteenth-century novelists also included as characters many kinds of animals, diseases, and nonliving agents that bear upon population dynamics. In addition to the horse Battle, the rabbit Poland, and a mob in Zola's *Germinal*, these included the white whale of Melville's *Moby-Dick* (1851), which connects the crew of the *Pequod* to the decimation of whale populations so that blubber can be extracted and turned into fuel for increasingly numerous urban human populations; the lobster and the squid held in a fish-market aquarium in Theodore Dreiser's *The Financier* (1912), which enable the young protagonist Frank Cowperwood to witness "a tragedy [the lobster killing the squid over the course of several days] which stayed with him all his life"; the disease typhus in Mann's *Buddenbrooks*, in which an entire chapter is given over to a characterological sketch of this illness; and the train La Lison in Zola's *The Human Beast*, which, as character, bears a synecdochal relationship to the vast train system that connects various local populations in France with other countries.[18]

To describe plants, animals, diseases, mobs, and trains as characters, minor or otherwise, raises several questions. First, how does one define and recognize novelistic characters, such that characters can be distinguished from noncharacter novelistic entities? (For example, what is the "other" of characters: objects? elements? forces? geographies? milieux?) Second, in treating nonhumans as characters, am I simply returning to an earlier structuralist emphasis on narrative "functions" that explicitly eschews any important referential dimension—and if so, does this approach risk, as Frances Ferguson has noted, losing sight of the centrality of human actors in nineteenth-century realist novels?[19]

It is tempting to address the first question by nominating both a proper name and some indication of subjective interiority as the minimum criteria for status as a character within nineteenth-century realist novels, especially given that some of the nonhuman entities I have noted possess both of these characteristics. In Zola's *Germinal*, for example, the workhorse is not only given a name, "Battle" (*Bataille*), but the narrator also uses free indirect discourse to provide readers access to what are apparently Battle's thoughts and feelings, as in the italicized passages in the following two examples:

> Old age was now approaching, and [Battle's] cat-like eyes sometimes clouded over with a look of sadness. Perhaps he could dimly remember the mill

where he had been born. . . . *There had been something else, too, something burning away up in the air, some huge lamp or other,* but his animal memory could not quite recall its exact nature. And he would stand there unsteadily on his old legs, head bowed, vainly trying to remember the sun.

It was Battle. After leaving the loading area he had been galloping along the dark roadways in a state of panic. He seemed to know his way round this underground city which had been his home for the past eleven years. . . . Turning after turning came and went, paths would fork, but he never hesitated. Where was he heading? Towards some yonder horizon perhaps, towards his vision of younger days. . . . And his old docility was swept away by a new spirit of rebellion against *a pit that had first taken his sight and now sought to kill him.*[20]

Yet this apparently easy solution to the problem of identifying characters—simply look for names and narratorial ascription of thoughts and feelings—is troubled by many of my other examples, in which nonhuman entities are neither given names nor ascribed any kind of subjective interiority. This then returns us to the question: Why describe these latter entities as *characters*?[21]

I again follow Woloch partway by proposing that within nineteenth-century novels, characters are both defined immanently within a specific novel *and* constituted through a referential relationship to real entities within the reader's world. Woloch argues that major characters emerge as such only insofar as the plot and other literary devices define and distinguish them against, and from, minor characters. He also argues that characters refer outward beyond the novel, providing readers with a sense of a character as a real person (that is, an "implied human personality").[22] Woloch suggests that both the immanent and referential understandings of character are right, because

a literary character is itself divided, always emerging at the juncture between structure and reference. . . . By interpreting the character-system as a distributed field of attention, we make the tension between structure and reference generative of, and integral to, narrative signification. The opposition between the character as an individual and the character as part of a structure dissolves in this framework, as distribution relies *on* reference and takes place *through* structure. (17)

Drawing on Woloch's point that characters are both defined immanently within a novel yet also refer to realities of the reader's world, I describe as a "character" any entity (a) that manifests a form of agency that is recognized as such within the novel, either by other human characters in the novel or by the narrator, and that at the same time (b) refers beyond the novel to an extranovelistic entity that would be recognized by a contemporary reader.

By "agency," I mean the capacity to pursue a goal that can affect some significant portion of the other characters (that is, other agents) in the novel. As Woloch's account underscores, human beings with personal names and subjective interiority are the paradigm through which readers tend to understand agency. And as the example of Zola's horse-character Battle reveals, a novelist can create a nonhuman character simply by mapping this paradigm (name and subjective interiority) onto an animal or other nonhuman entity. Yet personal name and subjective interiority are simply one means, rather than necessary conditions, by which a novelist can make clear that something is the kind of entity that can formulate and pursue goals.[23] For example, La Lison, the train at the center of Zola's *The Human Beast*, has a personal name but is not ascribed any thoughts. Rather, its status as the kind of entity that can have intentions is expressed indirectly, in her use of too much grease:

> And [Jacques] had only one thing with which to reproach her, she needed too much greasing: the cylinders especially devoured unreasonable quantities of grease, a constant hunger it was, absolute gluttony. Vainly he had tried to get her to moderate her appetite. But she became breathless at once, that was just the way she was.[24]

Because La Lison is the kind of entity that can "hunger," it can also be loved (254 [1228]), become sick (254 [1228]), and, rather than simply being destroyed, is instead "disemboweled" (*éventrée*) and dies (*mourir*) (299 [1267]). To take an even less human-like example, the typhus of Mann's *Buddenbrooks* has neither a personal name nor a subjective interiority, yet it pursues a goal, in the sense that it develops itself in patients through several phases and a period of crisis and can be "fought" (*bekämpft*) through various means. Its status as agent is signaled in part by the fact that it alters a character's own subjectivity and goals: The onset of the disease is subtly announced through a "spiritual discord" (*seelische Missstimmung*) in a character, and whether or not the disease kills the patient depends on whether the patient can reoccupy his or her own position as agent by following the voices that lead him back to his duties and pleasures or instead shies away from this sense of life by pursuing the "path" (*Pfad, Weg*) laid out by the disease.[25] Readers clearly take for granted that humans are characters, and as Woloch notes, some novelistic characters come to be full, round agents only because they are contrasted with "minor" characters who are made to remain flat. Yet novelists such as Zola and Mann could also employ the immanence of a character system—that is, the fact that characters are defined against one another within a novel—to expand the roster of characters beyond human beings. They did so by resolving characters into their more primary identity—agents who can pursue goals—and then finding formal means of depicting agency that operates in the absence of either personal name or subjective interiority (or both).[26]

Even as human and nonhuman characters emerge immanently within a given nineteenth-century novel, characters also refer outward to the reader's own world, as Woloch notes. Since, for Woloch, only human beings can be characters, this reference has for him only a democratic-political dimension (characters are, in essence, human beings who should be able to vote). However, if, as I suggest, nineteenth-century novelists included among their characters many nonhuman beings, such as animals, plants, and trains, then the referential dimension of novels must go beyond this political reference. Malthus's and Jenner's early scientific character-systems are again helpful here, for their efforts to locate new real characters were guided by their attempts to locate the agents that determined and altered population dynamics. This same basic approach is especially clear in the work of Zola, who explicitly presented his novels as "scientific" in character.[27] For Zola, the train La Lison is a character in *The Human Beast* because it is a part of an international transportation network that moves vast populations of people from place to place, creating numerous new forms of contact among individuals, while in *Germinal*, the workhorse Battle is a character because he bears a key relationship to the population dynamics of the novel. (For example, Battle serves as a form of free labor that makes it possible to exploit more fully the paid labor of the miners.)[28] These characters of train and workhorse refer outward to real elements of the social world in which the reader lives and identify these elements as important factors in the reader's relations to the populations of which he or she is a part.

My use of the term "agency" to expand what can count as novelistic characters has significant resonance with Bruno Latour's and Michael Callon's actor-network theory. At the same time, I am hesitant to draw too close a parallel here, lest the specificity of my point about novels be lost. In *The Pasteurization of France*, Latour defines an "actant" as anything that "resists trials" and, as a consequence, can enlist allies among other actants.[29] As is clear even from this very minimal description of Latour's theory, the basic point of his approach is that anything (and everything) can be an actant: atoms, elements, viruses, plants, animals, people, stars, etc. Applied to literary criticism, this suggests—in one sense correctly—that anything in a novel *could* be a character. However, my point is that each novel posits a specific immanent character-system, within which some things are, and some things are not, characters. We should understand the character-system of each novel that I have cited as a *model* of the specific actants that are relevant for whatever problem that novel poses; that is, each novel is a proposition about precisely which actants are relevant. My approach thus resonates less with the second half of *The Pasteurization of France*, in which Latour seeks to convince us of the general point that anything and everything can be an actant, and more with the first half of his book, in which Latour seeks to document a very specific character-system that included Louis Pasteur, French hygienicists, French and German physicians, French farmers, French lab technicians, and—eventually—the anthrax bacillus itself as agent/character.[30]

To think of character-systems as population models rather than as models of the democratic franchise is a difference that makes a difference. It means, for example, that what Woloch describes as the struggle endemic to the character-systems of the nineteenth-century realist novel was not (or was at least not primarily) the political-theoretical struggle of how to include all humans within a democratic franchise—that is, how to imagine all characters as potentially "major" characters—but rather the question of how to identify and model those variables that determine population norms and that enable the latter to be altered. Political forms and relations will certainly be extraordinarily important to such a project. However, political forms and relations are not the end of the story but are instead elements within the large dynamics of population. To return to the example of Zola's *Germinal*, we can read this novel's focus on questions of both basic subsistence needs and the sexual relations of miners (and mine owners) as an attempt to determine precisely how political-social relations intersect with Malthusian population dynamics. The novel proposes, though, that it is only by treating nonhuman entities, such as the horse Battle and Etienne's hereditary crack, *as* characters that we can determine how the dynamics of society and population actually relate to each other.

As Georg Lukács's early-twentieth-century attacks on Zola underscore, theorists of the novel have tended to shy away from a population-oriented approach.[31] Lukács valorized the "realism" of authors such as Honoré de Balzac over Zola's "naturalism," and he grounded that distinction in an author's willingness to focus solely on *social* relations: Authors who restricted their focus to social relations were able to "narrate" human relations properly, while those who appealed to nonhuman forces limited themselves to mere "descriptions." For Lukács, Zola fell into the latter camp because he divided the determining forces of human action into two realms: on the one hand, the "social environment," and on the other, "diverse and heterogeneous forces, like heredity, which affect men's thinking and emotions with a fatalistic inevitability."[32] For Lukács, there was no "organic" way of connecting these two realms of causality, with the result that, in a novel such as *Germinal*, Etienne's hereditary crack "causes explosions and calamities with no organic connection to Etienne's character" (123). While theorists of the novel have tended to follow Lukács's lead, my proposal is that if we understand these two realms in terms of social dynamics and population dynamics, respectively, then it is precisely the nature of their interrelation that is at stake in Zola's work as a whole.

My account of character-systems as population models fits best the works of nineteenth-century novelists such as Balzac, Zola, and Mann, who were themselves explicitly interested in a "scientific" understanding of society and in questions of milieu, environment, and heredity. Or, to put this another way, my claim is not that *every* novel in the nineteenth century took an explicit interest in questions of, say, sustenance or disease or included animals, plants, and diseases among its system of characters. Rather, my point is that understanding nineteenth-century novelistic character-systems (and especially the implied reference of such character-systems to unique individuals) as population models, rather than solely as models of political relations, helps us account more fully both for the kinds of characters that appear in nineteenth-century novels and the not-exclusively-social logic that motivates the plots of many nineteenth-century novels.

Free(d) Indirect Discourse

Understanding character-systems as population models also helps us understand better another important formal aspect of the nineteenth-century novel, namely, the emergence and development of free indirect discourse in the work of novelists such as Jane Austen, Stendhal, George Eliot, Balzac, and Zola. If expanding character-systems represented the nineteenth-century novelist's efforts to develop population models, free indirect discourse represented the attempt to establish a "surface" by means of which novelists could locate and register those forces that lay outside consciousness and laws and that were expressed differently across a population. Where the character-system of a specific novel mapped out the territory within which these generally nonconscious and extrapolitical forces operated, free indirect discourse revealed the effect of these forces on individuals and on collective dynamics.

My claim that free indirect discourse served many nineteenth-century novelists as a technique for registering forces that lay *outside* both consciousness and laws (whether laws are understood as legislative decrees or communal norms) may seem counterintuitive. Much of the best work on free indirect discourse, whether from narratological or historicist perspectives, has stressed that this literary device reveals for readers the thoughts or feelings *of* characters, rather than revealing forces that occur below the level of consciousness or sentiment. That is, free indirect discourse has generally been understood as a novelistic method of making the mind of a character "transparent," to paraphrase the title of Dorrit Cohn's well-known book on this literary device, rather than as a device that reveals the effects of something extramental on the mind.[33]

Yet even as free indirect discourse *was* often used by nineteenth-century novelists to represent the thoughts and feelings of characters, it was also often used to represent something that could not be explicitly thought or felt by a character. In some of its earliest instantiations in Jane Austen's novels, for example, free indirect discourse is employed to reveal a character's unconscious moral comportment, with the goal of establishing for the reader the moral telos toward which the character must aspire. In *Emma* (1815), Austen's narrator uses free indirect discourse to reveal to readers, at the start of the novel, the contours of Emma's unconsciously self-centered behavior. Consider the following, from early in the novel:

> [Harriet's] early attachment to [Emma] was very amiable; and her [Harriet's] inclination for good company, and power of appreciating what was elegant and clever, shewed that there was no want of taste, though strength of understanding must not be expected.[34]

Though this sentence is technically (grammatically) the narrator's objective description of Harriet, rather than something that Emma says or something Emma thinks, readers do not encounter it as an objective statement about Harriet (we do not believe that Harriet really possesses taste). Rather, the sentence reads, especially through its syntax, as the narrator's rendering of the logic of Emma's unconscious comportment toward the world. The carefully placed semicolon suggests that all of the seemingly positive objective facts about Harriet ("inclination for good company," "power of appreciating what was elegant and clever," "taste") are in fact conclusions, masquerading as observations, based solely on the facts that Harriet is attached to Emma and that Emma believes herself to be an astute judge of character. This is, of course, a rather self-centered and self-flattering form of "proof," one not likely to be convincing to others (including the reader). Equally important, Austen's sentence describes Emma's self-centered comportment in such a way that, were Emma herself to read this description of her behavior, she would likely become uncomfortable, for she is sufficiently intelligent that the discursive representation of her self-centered comportment would become visible to her as well. Hence, passages like this at the start of *Emma* establish that the novel can come to an end when Emma herself has matured to the extent that, were she to read this novelistic account of her life, she could recognize the irony of the passages in free indirect discourse at the start of the novel.[35] In this case, then, free indirect discourse does *not* represent Emma's explicit thoughts and feelings but instead the unconscious logic that, at the start of the novel, frames and enables her explicit thoughts and feelings.

Focusing on free indirect discourse as a means by which nineteenth-century novelists represented unarticulated logics, comportments, and forces opens up a new way of approaching this literary device. In the case of a novelist such as Austen, of course, the inarticulate and unfelt forces *are* ultimately convertible to consciousness and active judgment; *Emma*, for example, tracks precisely how Emma's initial unconscious comportment becomes the subject of her attention and is thereby transformed. However, free indirect discourse could be used by novelists to search out different kinds of forces that determined individual feelings, thoughts, and actions but that could never be transformed by a character into conscious thoughts and feelings.[36] In *The Human Beast*, for example, Zola often employs free indirect discourse to locate character comportments that emerge from a biological dimension of human existence. In the following, free indirect discourse expresses the insight and violent thoughts that emerge in Jacques Lantier but that themselves result from Lantier's hereditary "crack":

> And the inner agitation that had quickened [Jacques's] steps, the horrible fascination that kept him standing there, culminated in one piercing insight [*pensée aiguë*] that burst from the depths of his being: that man, the one he'd seen with the knife in his fist, he had dared! that man had travelled the distance of his desire, that man had killed! Oh! to stop being a coward, to have satisfaction at last, to plunge the knife in! And what about him, who'd spent the last ten years desperately wanting to do just that! There was, in the midst of his fevered interest, a measure of self-contempt, of admiration for the other man.[37]

Though there is a conscious "insight" here, as well as thoughts that both flow from and express that insight, Zola uses free indirect discourse here as a surface by means of which he can register and isolate the interaction of biological instincts and social relations. In other words, what is represented here are not precisely Jacques Lantier's thoughts but instead the Lantier hereditary crack as parsed through Jacques Lantier.

As Gilles Deleuze pointed out in his discussion of Zola's novel, the relationship of biology to consciousness in such passages should not be understood as deterministic. For Zola, the hereditary crack was not a channel through which a specific atavism was transmitted but that which forced a character to discover and create his or her own particular relation to his or her milieu. Thus, as Deleuze noted, "it is important that Jacques Lantier . . . be sound, vigorous, and in good health, for the [hereditary] crack does not designate a route along which morbid ancestral elements will pass. . . . Heredity is not what passes through the crack, it is the crack itself." As a consequence, the hereditary crack "is not tied to a certain instinct, to an internal, organic determination, or to an external event that could fix an object. . . . Transmitting only itself, it does not reproduce that which it transmits."[38] Because the hereditary crack does not transmit a particular content but rather an impulsion for a character to address this crack in some way, the particular way in which the crack is expressed in each character— Etienne Lantier's alcoholism in *Germinal*, Jacques Lantier's violent relationship to women in *The Human Beast*, Claude Lantier's self-destructive obsession with completing his painting in *The Masterpiece* (1886)—only develops in the context of dynamic and shifting interactions among the individuals who characterize a population.[39]

This understanding of heredity leads Zola to employ free indirect discourse neither to reveal the transformation of an unconscious moral comportment into consciously chosen acts, à la Austen's *Emma*, nor to highlight some sort of nonhuman force that "determines" human thoughts and feelings, but rather to identify points of intersection between biological forces and social relations at which pressure can be exerted in order to shift relationships between seeming biological givens and more properly social relations. In the case of *The Human Beast*, for example, both the character-system of the novel and Zola's use of free indirect discourse suggest that Jacques's specific pathological relationship to women cannot be understood apart from, and in fact emerges as a consequence of, the vast transportation network and administrative bureaucracy created by the train system, which is itself tied directly to the corrupt judicial and political institutions of the French Second Empire.[40]

Though Zola's use of free indirect discourse to capture the flexible inter-section of biology and sociopolitical relations is an unusual example of nineteenth-century uses of this literary device, my larger point is that free indirect discourse served many novelists as a surface for capturing relation-ships between nonconscious forces and conscious thoughts and feelings. In Austen, free indirect discourse often captures a relationship between uncon-scious comportments and conscious thoughts and feelings, and the plot of a novel such as *Emma* reveals the complete transformation of the former into the latter, at least for the protagonist.[41] In Zola, by contrast, free indirect discourse captures the intersection of the biological dimension with social sources, an intersection that can never be fully accessible to a character's consciousness, since that intersection determines the objects toward which conscious thoughts and feelings are directed. For an author such as Balzac, free indirect discourse captured the ability of a character to mirror her milieu and then exploit that reflection financially, while for an author such as George Eliot, free indirect discourse revealed the kinds of half-conscious, and often erroneous, "inferences" necessary for the stability of "civilization."[42]

Yet if nineteenth-century novelists were interested in revealing the relationship of nondiscursive forces to social relations, why would they have chosen a literary device such as free indirect discourse, which trans-lates everything into a kind of discourse? Nineteenth-century novelists certainly could have followed Malthus's lead by treating these nondiscur-sive forces as threatening, alien forces characterized only by mute pressure. Free indirect discourse, by contrast, brings these forces into the novel in the form of discourse—that is, in a form that readers necessarily engage as like the forms of human discursive thought or articulate speech—rather than granting these nondiscursive forces their own, nonlinguistic forms of agency. Why would this have been the case?

Nineteenth-century novelists used free indirect discourse to register the effects of nondiscursive forces of populations because a novel can establish a coherent "territory" over which it ranges only when it treats all subdivi-sions and elements of that territory as if they were contained within a unified medium. Given that novels are themselves discursive constructions and that nineteenth-century novels were especially committed to explor-ing the possibilities of a unified narrator, a given novel can only bring something—a person, an object, a force—into its purview for serious treat-ment if that thing can be made to produce discourse in some way. This does not mean that nineteenth-century novelists sought to present nonhu-man forces (for example, Zola's "heredity") as really speaking but rather that they concluded that a force or entity must be connected to speech in some way in order to emerge as a significant and determining part of the territory that the novelist engages. This then is the problem for nineteenth-century novelists interested in population dynamics: Something that in reality cannot speak must nevertheless speak within the world of the novel. Free indirect discourse is a solution to this problem, for it can be used to embed within a character quasi-thoughts and quasi-feelings that are not precisely possessed by the character but rather establish both the limits and unchosen trajectories of characters.

This understanding of free indirect discourse as a method of rendering discursively forces that are not actually discursive returns us, in an interesting way, to Arendt's account of the novel. As I noted at the start of this chapter, Arendt describes the novel as "the only entirely social art form," by which she meant that the novel seeks to guard a space of individual, and individualizing, intimacy against the normative, and normalizing, descriptions of society. Because intimacy and society are simply two sides of the same coin, Arendt does not have any faith in the recuperative potential of the novel, and she opposes the concept of "action" to both the concepts of society and intimacy (and, implicitly, to the novel). Action is, for Arendt, the capacity to bring something new and improbable into existence: "The new always happens against the overwhelming odds of statistical laws and their probability, which for all practical, everyday purposes amounts to certainty."[43] However, action is equally the capacity to disclose that newness as the consequence of an agent, and so it requires speech:

> Speechless action would no longer be action because there would no longer be an actor, and the actor, the doer of deeds, is possible only if he is at the same time the speaker of words. The action he begins is humanly disclosed by the word, and though his deed can be perceived in its brute physical appearance without verbal accompaniment, it becomes relevant only through the spoken word in which he identifies himself as the actor, announcing what he does, has done, and intends to do. (177–78)

For Arendt, speech is necessary to action because action is bound to the uniqueness of an agent, and it is through speech that the unique "who" of that agent is revealed.

Arendt's claims about action and speech bear upon my claims about free indirect discourse in two ways. First, they suggest that a novelist such as Zola uses free indirect discourse for nonhuman entities, such as the horse Battle in *Germinal* or the hereditary crack that connects *Germinal* and *The*

Human Beast, precisely because he wants to illuminate these entities as unique agents who create unlikely and unanticipated actions, and linking them with speech is a phenomenologically viable way to do so within the literary form of the novel (that is, it accords with a reader's sense of what is necessary for something to be an agent and to act). Second, and more important, Arendt's concept of action allows us to think of each specific novel itself as an action, in the sense that it is a speech act, made by its unique author, concerning what the author believes are the key elements of the real world, and it is made with the hope of changing that world. From this perspective, the nineteenth-century novels I have considered would not, *pace* Arendt's claim, seek to protect a sphere of intimacy against the normalizing tendencies of society but would instead counter those normalizing tendencies by developing population models of forces that work against normalization. This perspective returns us to an older sense that authors such as Austen, Balzac, Eliot, Zola, and Mann sought, each in their own unique ways, to make a difference in the world through their novels—a perspective that has, in recent decades, taken a back seat to interest in the ways that nineteenth-century novelists unwittingly facilitated disciplinary or ideological work.

Understanding free indirect discourse as an important tool with which nineteenth-century novelists developed and proposed specific population models qualifies Latour's suggestion that the sciences are the key modern means by which nonhuman agents are brought into speech. Latour argues that the sciences serve as the "spokespersons of the nonhumans," in the sense that each science "can define itself as a complex mechanism for *giving worlds the capacity to write or speak*, as a general way of making mute entities literate."[44] He claims that scientists invent, by means of their devices, experimental protocols, and written communications, "*speech prostheses that allow nonhumans to participate in the discussions of humans, when humans become perplexed about the participation of new entities in collective life*" (67). The scientifically enabled speech of nonhuman entities has no more power than any other speech contribution to matters of common concern: "As is the case with all spokespersons, *we have to entertain serious but not definitive doubts about their capacity to speak in the name of those they represent*" (64–65). Or, to put this in the terms of what I have described in this chapter, when "humans become perplexed about the participation of new entities in collective life," scientists propose *models* of relationships between human and nonhuman agents. Yet this turns out to be an apt description of the project of nineteenth-century novelists such as Balzac, Zola, and Eliot (as well as Romantic-era scientists such as Malthus and Jenner), in the sense that each novel proposed a model and sought to give voice to the key agents that mediated between populations and social relations. From this perspective, Latour's recent efforts to establish a "political ecology" is continuous with, and seeks to revive, an approach to the relations among the sciences, speech, and agents first articulated in the work of nineteenth-century novelists who were oriented toward the sciences.

Population, Territory, and Security

Not every novelist who employed free indirect discourse was explicitly interested in questions of population. Nor did fictional free indirect discourse aim to gather real population data in the way that, for example, smallpox patient records could. Rather, I propose that character-systems and free indirect discourse facilitated an experimental attitude among nineteenth-century novelists, encouraging many to search for forces that operated outside consciousness and laws and that were expressed differently across a population. This biopolitical perspective not only helps us make sense of the actual character-systems and uses of free indirect discourse in the nineteenth century; it also moves us beyond somewhat dubious claims that novelistic character-systems expanded only under the pressure of a democratic imperative or that the nineteenth-century history of free indirect discourse can be explained as a dialectical narrative that begins with Austen's "reflective" use of the device and ends in either Flaubert's "objective" style or in the multivocality of modernist novels.[45] A biopolitical perspective on character-systems and free indirect discourse also allows us to relate these literary techniques to the interest of nineteenth-century authors in both the methods and results of the various sciences, such as Balzac's and Zola's focus on scientific milieu theory, Flaubert's interest in "great Art" as something "scientific and impersonal," and George Eliot's interest in evolutionary science.[46]

As I hinted at the end of the last section, interpreting nineteenth-century novelistic character-systems and uses of free indirect discourse through a biopolitical perspective also allows us to understand the effects of novels on readers beyond the somewhat limiting frames of ideological capture and disciplinarity. Since the 1980s, literary critics inspired by Foucault's contrast between punishment and discipline have tended to interpret free indirect discourse as a technology by means of which nineteenth-century novels disciplined readers into, say, a belief in the naturalness of social norms and various institutional mechanisms for policing these norms. For D. A. Miller, for example, "the great prominence [that] the nineteenth-century novel gives to *style indirect libre*" is a function of the more general panoptic structure of nineteenth-century novelistic "omniscient" narration, by which, "respeaking a character's thoughts or speeches, the narration simultaneously subverts their authority and secures its own," while Casey Finch and Peter Bowen argue that, in Austen's *Emma*, free indirect discourse "functions specifically to disguise the ideological imperatives of the novel as the autonomous ideation of one of its characters."[47]

My analysis suggests that free indirect discourse also enabled the novelistic creation of surfaces intended to identify, for both the novelist and the reader, points at which individual feelings and thoughts come into contact with something *other* than individual thoughts and feelings. Critics such as Miller, Finch, and Bowen have adopted their own model for that "other" force, namely, disciplinary policing or ideological imperatives. But rather than reducing all population models of nineteenth-century novels to Foucault's disciplinary model or Marx's model of ideology, I find it more useful to place these novelistic models on the same level as Foucauldian or Marxist literary-critical models. This allows us to understand each of the novels I have described as its own specific model of relationships among three elements: individual feelings and thoughts, something "below" individual feelings and thoughts, and the dynamics of populations. It seems to me difficult to reduce these multiple models to the same schemas of the disciplinary inculcation of norms or ideological imperatives.[48]

To put this another way, understanding novels as population models—or, more precisely, models of the relationship between population and society—allows us to avoid an unwarranted reduction of novels to policing and security mechanisms. With that said, Foucault's key reflections on population appear within a lecture series entitled *Security, Territory, Population*, and given my account of novelistic character-systems as each implicitly proposing a population-oriented "territory," the surface of which could be explored by literary devices such as free indirect discourse, it is reasonable to ask whether this means that novels, like political economy and inoculation campaigns, also aimed at the third term of Foucault's title, namely, "security." To aim at security would mean that novels, like inoculation campaigns, political economy, and Malthusian population control, sought to create knowledge about the internal dynamics of populations with the goal of then making those dynamics more regular (for example, less likely to disrupt commerce). To conclude that novels aimed at security would not quite return us to the disciplinary paradigm adopted by Miller and other literary critics, since security does not share the "panoptical" aspirations of discipline, but it would at least get us within shouting distance of that model.

It is not clear to me, though, that novels aimed at security in the sense that other biopolitical discourses did. Political economy, inoculation campaigns, and Malthusian population control aimed at security in the sense that their advocates linked their population models to battles around specific institutions, laws, initiatives, and regulations. Nineteenth-century novelists, by contrast, had nothing like this shared sense of battle or object, and each novel proposed such distinctive models of the relationships between society and population that it is hard to find much shared space in them. (There is a vast distance, as Lukács stressed, between Balzac's and Zola's models, despite the fact that both authors adopted quasi-scientific concepts of milieux.) The point of each novel that I have discussed was not simply to illuminate a territory and then convince readers that certain relationships were inevitable (or likely, or "realistic") within that territory. That *was* their point, in part, of course. However, the explicitly fictional nature of the medium in which these authors wrote meant that their population models were encountered by readers *as* models, that is, as propositions for understanding the relationships of populations to territories and to society, rather than as scientific truths to which readers *must* submit.

Precisely because nineteenth-century novels emphasized, through their fictionality, the modeling activity inherent in population models, they both illuminated and suspended what we might describe as the "active turn" to passivity that characterized other modes of biopolitics, such as Malthusian population regulation.[49] Other modes of biopolitics actively employed passivity in the sense that less valued members of a population were not killed, but also were not protected, and so allowed to die.[50] Malthus, for example, did not advocate killing anybody when a human population exceeded the carrying capacity of the food supply but instead sought to dismantle institutions that would assist those who were starving, so that "nature" could then end those lives. Insofar as a nineteenth-century novelist such as Zola illuminated in *Germinal* this biopolitical mode of actively employing passivity, he also tended to undercut, or at least suspend, its legitimacy. The "naturalism" of Zola's novels is intended to ensure that readers understand his novels as about the dynamics of the real world, but the fictional status of his novels—the fact that we know that it is Zola who lets some characters live and others die—encourages us to see these real-world dynamics not (solely) as nature's dictates but equally as the result of biopolitical projects and decisions. In this sense, nineteenth-century novels produced a biopolitical form of passive resistance to the active turn to passivity in other modes of biopolitics (and in this way, constituted actions in Arendt's sense of that term).

Conclusion: Literature and the Sciences of Population

My analysis here of the relationships of population concepts to the literary devices of character-systems and free indirect discourse in nineteenth-century novels bears in at least two ways upon the more general question of how literary critics can productively understand relationships between the sciences and literature in this period. It bears, first, on the question of *which* nineteenth-century sciences we wish to bring into relationship to literature of this period and what exactly it means to link literature with these sciences. I have stressed the importance of what I would describe as the biopolitical sciences or, perhaps more accurately, biopolitical concepts and techniques, such as the concept of population, that draw on the authority of science. This was a form of science, or at least the use of the authority of science, that was not necessarily articulated in authoritative books by scientists but instead developed in more subtle ways through the dissemination of terms such as "population" through political economy, through discussions of inoculation and vaccination, and by means of debates about how to collect and interpret statistical knowledge. As a consequence of this approach to the sciences, my account here does not draw heavily on texts by scientists, as, say, Gillian Beer's account of the relationship of Charles Darwin and Victorian novelists does or Nicholas Dames's account of Victorian theories of the physiology of reading does. (Arguably, the closest that I come to discussing "scientists" in this chapter are my very brief mentions of Malthus and Jenner.) While I do not deliberately eschew all mention of scientists, I approach the work of the sciences here less in terms of explicit theories about, say, the organism, reproduction, and evolution and more in terms of how a concept such as population enables multiple kinds of local, tactical, and practical developments in both the sciences and in literature.

My second point is related to this first, for one of my goals in focusing fairly exclusively on novelists is to underscore that the uptake of concepts of population in literature was less a matter of specific scientists influencing novelists and more a matter of novelists pursuing the question of how to understand populations at the same time, but in different ways, than scientists. I am interested in a supervening "population imaginary" that was as important for literature as for the sciences, rather than being something that travels from the sciences to literature. This population imaginary—the premise that one can only fully understand the dynamics of the social field by relating these to those dynamics of population that happen under, or beyond, social relations—was especially compelling to authors, such as Balzac and Zola, who were explicitly interested in the sciences of milieux and in statistics. However, as the recent work of literary critics such as Emily Steinlight, Jesse Rosenthal, and Caroline Levine suggests, something like this population imaginary also underwrote the work of authors such as Dickens.[51] I have focused here on the relationship between this population imaginary and two specific formal features of nineteenth-century novels—namely, character-systems and free indirect discourse—but my hope is that this approach will encourage further work on the ways that this population imaginary encouraged the development of other formal and thematic features in novels of this period.

Part II: Romanticism and the Operations of Biopolitics

Building Beaches

Global Flows, Romantic-Era
Terraforming, and the Anthropocene

Romanticism is best known for its emphasis on intensely local nature. Whether we consider the referents of William Wordsworth's poems on local places and organisms, such as the wild green landscape "a few miles above" Tintern Abbey or the yew trees of Lorton Vale and Borrowdale; or his efforts to preserve areas of the Lake District in the form of a national park; or Gilbert White's supremely local natural history of Shelburne; or John Clare's numerous poems on local flora and fauna, Romanticism often seems equivalent to the commitment to value, protect, and preserve the minute particulars of local spaces. Yet there is another side to Romanticism, one oriented toward transformations of our planetary sphere so extensive that they are shocking to read. The two examples that I consider in this chapter appear in Romantic-era "philosophical poems," and both propose to alter the global climate through the application of science and technology, a process that we would now call "terraforming." In his wildly popular *The Botanic Garden* (1791), Erasmus Darwin urged European nations to form an international fleet that would alter global weather patterns by towing icebergs from the North Pole to the southern oceans, which he believed would redirect global air currents and enable "the vegetation of [Britain] [to] be doubled, as in the moist vallies [*sic*] of Africa."[1] In *Queen Mab* (1813), P. B. Shelley proposed an even more extreme scenario, suggesting not only that polar ice ought to be "loosed" in order to reform Earth's weather and deserts but also that Earth itself could be shifted on its axis in order to produce more calm and pacific global weather.[2] In contrast to a Wordsworthian vision of nature as local and fragile, Darwin and Shelley approach nature as global, malleable flows that can be redirected in accordance with human desire.

From the perspective of our concerns with global warming, it is difficult to know how best to approach this global side of Romanticism. It is tempting to view Darwin's and Shelley's proposals both as examples of ecological naiveté and ideological mechanisms by means of which Europeans naturalized their violent conquest of non-European lands and peoples. I take a different approach here, though, for what interests me is *how* the Romantics came to "think globally"; that is, by means of what operations, protocols, and principles did the Romantics understand phenomena such as weather and population as intrinsically global phenomena? Such an analysis seems to me necessary insofar as many modes of contemporary critique—including forms of ecological and postcolonial critique through which we might be tempted to rap the knuckles of authors such as Darwin and Shelley—themselves depend upon this same ability to think globally. An analysis of Romantic globalization is thus an effort to consider more generally what it means, then and now, to approach questions of politics, justice, and value from the perspective of the globe.

Romantic globalization is also a way of thinking about the relationship of two kinds of global aspirations. On the one hand is a long-running liberal aspiration to transform the globe into a world-system populated by commerce-oriented, freedom-prizing individuals. On the other hand, the destructive effect of liberal growth-oriented global commerce on the natural environments upon which humans depend has encouraged an increasing number of contemporary authors in both the sciences and the humanities to imagine either significant alterations to the liberal global order or completely alternative forms of global human relations. This latter way of thinking has been captured by the concept of the Anthropocene, which, in one of its earliest and still most famous formulations, denotes a "geological epoch"—that is, an epoch analogous to the Paleocene or Pleistocene geological epochs—that is in part determined by human activity.[3] Some critics have charged the concept of the Anthropocene with encouraging "technocratic" solutions to environmental problems that then lead back to a liberal view of the world, but I suggest that comparing Darwin's and Shelley's different Romantic approaches to globalization helps us distinguish between liberal and nonliberal approaches to the Anthropocene.

In the first section of this chapter, I argue that the Romantics learned to think globally by considering phenomena as diverse as weather, population, and magnetism as modes of *flow*. They understood flow by means of two operations and two principles: the operation of *untethering*, by means of which phenomena such as weather were detached from local places; the principle of *circular reinforcement*, which stressed the effects of an untethered flow when it encircled the globe and returned upon itself; the principle of *orthogonal drag*, which allowed Romantics to understand why potentially globe-spanning flows in fact were often limited; and the operation of *global network construction*, which allowed the Romantics to map—and, many hoped, ultimately to control—global flows by establishing the geographies of orthogonal drag. In the second and third sections, I argue that Darwin's and Shelley's philosophical poems participated in this Romantic global revisioning, both because each author understood the world in terms of flow but also because each poem sought to intervene in global flows by inventing a new technology of contiguity. In Darwin's case, this meant hybridizing two existing technologies of contiguity—the botanic garden and the book—while in Shelley's case, it meant creating a "gravitational" technology of contiguity that allowed a reader to be drawn back to the terrestrial globe after imagining a plurality of globes. In the fourth section, I argue that Darwin's and Shelley's approaches clarify the stakes of a recent debate, focused around the historian Dipesh Chakrabarty's work, about whether globalization ought to be understood through the lens of the Enlightenment project of freedom or, instead, in terms of limits on human freedom seemingly implicit in the concept of the Anthropocene. In the fifth section, I note that this debate unfortunately ignores recent neoliberal demands that we think beyond global limits, and I address this aspect of neoliberalism by using Shelley's reflections on an interminable wilderness of planets as a lens through which to read Kim Stanley Robinson's recent science-fiction novel *Aurora* (2015), a thought experiment about escaping from earthly limits by finding another globe.

Franklin's Fennel-Earth: Spheres and Flow

Romantic globalization is characterized above all else by the effort to think the dynamism of flows in terms of the geometry of a sphere. "Globalization" is an intrinsically temporal term; that is, it denotes a process, for which one can determine axes along which change is measured and marked, thresholds that determine the velocity and nature of change, and an overall directionality. The Romantics tended to think globalization through the figure of flows: flows of weather, of water, of magnetism, or of living populations, for example. Understanding and potentially controlling globalization was thus a matter of understanding how flows behaved when they occurred within that relatively thin strip of atmosphere, land, and water that encircled the globe. The Romantics sought to understand flows by means of two operations and two principles: an operation of *untethering*, a principle of *circular reinforcement*, a principle of *orthogonal drag*, and an operation of (global) *network construction*.

Seeing the world in terms of flows meant, first and foremost, untethering forces from local places. This operation was evident in both Romantic-era science and political theory. Late-eighteenth- and early-nineteenth-century weather theory, for example, developed in part by separating itself from late-seventeenth-century "exhalation theory." Where exhalation theory held that many kinds of weather were attributable to "vapors exhaled from a source within the earth which carried mineral deposits into the atmosphere," Romantic-era observers, by contrast, tended to see "weather as a global system of exchange, something that passed from one region to another over local and national borders."[4] Local weather, in other words, must be understood from the perspective of the global atmosphere, which the Romantic-era natural philosopher and protometeorologist John Dalton described in 1793 as the "invisible, elastic fluid which every where surrounds the earth."[5] This elastic fluid was characterized by flows in multiple directions. In *The Botanic Garden* (1791), Erasmus Darwin sought to help his readers understand why "on the eastern coast of North America the north-west winds bring frost"—rather than, as in Britain, the north-east winds bringing frost—by providing instruction in the logic of flows:

> When a sheet of air flowing along from the north-east rises from the [ocean] shore in a straight line to the summit of the Apalachian [*sic*] mountains, a part of the stream of north-east air will flow over the mountains, another part will revert and circulate spirally between the summit of the country and the eastern shore, continuing to move toward the south; and thus be changed from a north-east to a north-west wind.[6]

In *The Climate of Great Britain; or Remarks on the Change It Has Undergone, Particularly within the Last Fifty Years* (1806), John Williams provided similar instruction, contending that the "North easterly winds, so frequently experienced in England in May and June, which and generally accompanied with haze in the Night, and a close warmth during the Day, are probably occasioned by a continuation of the stream of air which flows in a North-easterly direction from the Northern tropic at this Season."[7]

Within political theory, Thomas Malthus performed a similar untethering operation by globalizing the concept of population via the concept of flow. Malthus's *An Essay on the Principle of Population* (1798) was written in large part to contest William Godwin's claim that society could be perpetually improved through the application of individual reason. Malthus contested especially Godwin's claims about reproduction in the more rational future: "It would be of little consequence, according to Mr. Godwin, how many children a woman had, or to whom they belonged. Provisions and assistance would spontaneously flow from the quarter in which they abounded to the quarter in which they were deficient."[8] Malthus's *Essay* was intended to replace what he saw as Godwin's undertheorized account of population flows—for Godwin, spontaneous flows of provisions and assistance magically appear precisely where they are needed—with what Malthus presented as a more scientific theory of flows of populations and provisions. Though Malthus was primarily interested in influencing a specific and local polity, Great Britain, in order to produce local political effects, he cast his basic claim about population—that human population increases at a swifter rate than any possible rate of increase of the supply of food—as a principle that concerned necessarily *global* flows. He suggested that to understand the dynamics of the British population, one had to consider population from the perspective of the globe. In the first pages of his *Essay*, Malthus cited approvingly Benjamin Franklin's reflections on the globalizing ambitions of all living beings:

> It is observed by Dr. Franklin that there is no bound to the prolific nature
> of plants or animals but what is made by their crowding and interfering with
> each other's means of subsistence. Were the face of the earth, he says, vacant
> of other plants, it might be gradually sowed and overspread with one kind
> only; as, for instance, with fennel: and were it empty of other inhabitants,
> it might in a few ages be replenished from one nation only; as, for instance,
> with Englishmen. (14)

Population, like weather, could not be understood if one focused only on the small number of geographical locales in which a given species lived; instead, one had to begin with the implicit global aspirations of every species. That is, one had to understand population as a flow that related to "the whole earth," from which "emigration would of course be excluded" (19).[9]

The Romantic operation of untethering forces from local places led to a first key principle: that in the absence of other forces, a flow on one part of a sphere would seek to assert itself everywhere. This tendency is evident in Franklin's thought experiment of the transformation of a vacant planet into a fennel-earth (or an Englishman-earth). It could also be imagined in the case of weather, for, as Dalton observed in his *Meteorological Observations and Essays*, "were the whole globe covered in water, or the variations of the earth's surface in heat regular and constant, so that the heat was the same everywhere over the same parallel of latitude, the winds would be regular also."[10] A flow that could move around a sphere unimpeded would always return upon itself and, in this way, reinforce its dominance or regularity.

Yet we do not live on a fennel-earth or upon a world on which the winds are constant, which suggested a second principle: that flows were always engendered and partially controlled by forces that acted at orthogonal or oblique angles to the flow. Actual winds, for example, were not autochthonous, self-perpetuating atmospheric phenomena but instead depended upon both solar rays that struck the atmosphere from above and the irregular surface of the earth upon which the atmosphere rested.[11] This was true even for a relatively constant, regular wind such as the trade winds. Dalton contended that trade winds were produced partly by "rarefaction" of the air caused by the sun's heat in the torrid zones and partly by the earth's rotation.[12] The sun's heat caused "two general masses of air" to move in "both hemispheres," yet each air mass was "deflected" as a consequence of the earth's rotation, and "these two masses meeting about the equator, or in the torrid zone, their velocities north and south destroy each other, and they proceed afterwards with their common velocity from east to west round the torrid zone" (89–90). The trade winds were relatively regular because they flowed primarily over water. Yet the variable surface that the earth presented to the atmosphere—sometimes seas, sometimes flat earth, sometimes mountains—altered what would otherwise have been perfectly regular flows:

> We find the irregularities of heat, arising from the interspersion of sea and land, are such, that though all the parts of the atmosphere in some sort conspire to produce regular winds round the torrid zone, yet the effect of the situation of land is such, that striking irregularities are produced: witness, the monsoons, sea and land breezes, &c. . . . (90–91)

Though flows might tend toward global uniformity, so to speak, they were engendered or affected by forces that promoted variability, irregularity, and difference.

From the perspective of this second principle, Malthus's account of population is especially revealing, for he sought to illuminate the *globalization* of a flow, rather than simply accounting for an existing flow, such as weather, that was already global. Thus, in Malthus's text, the tendency of a plant or animal species to expand across the globe is not the true movement that explains population; rather, the true movement of population is its tendency to "repress" itself. Population represses itself because it depends upon a force that operates orthogonally to its direction of flow. To return to Franklin's example of the vacant earth just sown with fennel, the flow of the fennel population as a whole is horizontal, for it moves around and eventually encompasses the globe. Yet each fennel plant relies on roots that grow vertically into the earth, and this vertical rootedness eventually represses horizontal flow. Thus, though a plant species is "impelled by a powerful instinct to . . . increase," the "superabundant effects [of this increase] are repressed afterwards by want of room and nourishment."[13] This same dynamic occurs in the case of animals, some of which are rooted to the living carpet of plants (which are rooted to the earth), while others are rooted to preying on other animals (which are either directly or indirectly rooted to plants, which are rooted to the earth). The spherical nature of the earth establishes a limit to the expansive movement of even just one species: Animated life can expand until it forms a second sphere that surrounds the earth like a skin, but it can go no further. At that point, the pressure that impelled the spread of a species across the globe presses upon animated life itself, and this repression produces the true movement of population, namely, oscillations of births and deaths within this living flow (26). This true movement is impossible to understand if one imagines that individuals can always be sent "elsewhere" (for example, to colonies), for it is only from the imaginary perspective of a global population—a population that finally encompasses the globe and thus meets itself—that one can discern its true dynamic.

This kind of global thinking emerged in Romantic texts that were not oriented toward questions of weather or population. In "Perpetual Peace: A Philosophical Sketch" (1795), for example, Immanuel Kant stressed the importance of the earth's spherical nature for our understanding of long-term political dynamics and, specifically, the possibility of "perpetual peace." Kant noted that though the earth is "divided by uninhabitable parts . . . such as oceans and deserts," humans can traverse uninhabited areas to reach habitable regions. However, "since the earth is a globe, [humans] cannot disperse over an infinite area, but must necessarily tolerate one another's company."[14] Moreover, "in seeing to it that men *could* live everywhere on earth, nature has at the same time despotically willed that they *should* everywhere, even against their own inclinations . . . and nature has chosen war as a means of attaining this end" (111). Like Malthus, Kant believed that the dynamics of human politics became visible only from the perspective of a spherical globe over which humans had been impelled to spread.

The second principle of flow that I have noted—namely, that flow is enabled and controlled by forces that act orthogonally to the direction of flow—implied another principle: Insofar as humans could act on these orthogonal forces, they could affect, and potentially control, global flows. Despite its pessimistic tone and conclusions, Malthus's *Essay* acknowledges this point, for he grants that advances in European agricultural techniques have indeed increased the yield of food per acre, thus facilitating the globalizing spread of population; his argument is simply that there are limits to this increase in yield. Dalton's description of the relationship between weather and the natural architecture of the earth—his claim that the "striking irregularities" of weather are a function of variations in the surface of the globe—suggested that humans might alter weather by altering land structures. As Alan Bewell notes, this implication was supported by the fact, known to late-eighteenth-century and Romantic-era authors, that weather in parts of the Americas had changed since its settlement by Europeans. Eighteenth-century scientists attributed these changes to the actions of settlers, such as the clearing of woods and marsh drainage, which presented atmospheric global weather flows with new geographic variations and irregularities.[15] These human actions had not aimed to produce climatic change, but they suggested the possibility that humans could ameliorate their condition not simply by using small-scale architectural structures (for example, houses) to protect themselves from global weather flows but also by redirecting the flows themselves.

However, channeling and bending flows in a controlled way required that one understood more precisely the nature of the orthogonal forces upon which a specific flow depended, and this in turn required the creation of global observational networks. Observational networks were made up of three elements: (1) individual nodes at which observations could be made and samples collected; (2) links between nodes that allowed observations and samples to be transferred from one node to another; and (3) a node that Bruno Latour has called a "center of calculation," to which observations were brought together in such a way as to do work.[16] Eighteenth-century and Romantic mapping of the magnetic flows of the earth, for example, depended upon a global network of land- and ship-based observers who tracked changes in magnetic declination, inclination, and intensity, while the Romantic science of weather depended upon an equally global network of observers who kept precise measurements about temperatures, barometer readings, and wind speeds.[17] Ideally, a global observational network would map in skeletal form the flow itself, which meant that network nodes were more effective the more closely they were situated to points of inflection (points at which flows changed direction, speed, etc.). Yet because flows were fluid phenomena, figuring out where these points of inflection might lie was by no means a straightforward task. Knowing where to place a node depended upon previous knowledge of how orthogonal forces affected flows, yet this knowledge was precisely what the network itself was intended to determine. Romantic-era flow hunting was thus often a matter of trying to establish sufficient measurements over the entire globe so that one could begin to intuit, through maps and tables, where points of inflection might lie. Several of Cook's sea voyages, for example, were intended to fill in those parts of terrestrial magnetism charts, such as the Pacific Ocean, that earlier observers such as Edmund Halley had left blank.[18]

If each node of the network was a point at which observations could be made and samples collected, and if nodes were connected to one another by links such as shipping routes and postal systems, the center of calculation was the point at which observations or samples were placed alongside one another in such a way as to enable what Latour calls "action at a distance." In the case of some flows, such as those of magnetism or weather, placing observations alongside one another meant assembling tables or maps, such as the comparative tables of barometer readings and rainfall volumes that appear in Dalton's *Meteorological Observations and Essays*.[19] For other flows, such as those of plant or animal life, contiguity was established more literally. Botanical gardens, for example, assembled plant samples that had been potted in distant locales and transported via transoceanic ship routes; these latter served as the vectors by means of which plants from multiple areas of the globe made the leap across otherwise inhospitable oceans in order to be placed alongside living plants from other parts of the globe.

Romantic global networks differed from early- and mid-eighteenth-century naturalist networks insofar as Romantic networks reconfigured the local from the perspective of the global. As Vladimir Janković notes for the case of meteorology and weather observations, for example, where "the eighteenth-century meteoric tradition" had established networks that bound together "provincial naturalists who derived their scientific authority from parochial affiliations and access to local facts"—especially local facts that were anomalous or remarkable—Romantic-era networks depended upon "civil servants and professionals who, by the nature of their occupations, moved between places or had a 'stationary residence' in an alien environment [e.g., a colony] in which their judgment of what counted as 'remarkable' mattered far less than back home."[20] Instead of describing anomalous phenomena narratively, observers were instead to employ standardized equipment to record averages, which were sent back to the center of calculation. From this latter perspective, knowledge of local weather was not an end in itself but rather a "prerequisite for a knowledge of *globally* evolving systems. . . . A rain-gauge in Cornwall was not intended to describe Cornwall, but to aid in a construction of a map of European isolines" (167).

The recursive capacity of the network—the fact that information and samples flowed not only in toward the center of calculation but also outward back to the other nodes—made it possible to exert some control over the material flows themselves. Fulford, Kitson, and Lee emphasize, for example, the ways that the botanical gardens encouraged by the Romantic-era naturalist Joseph Banks served as a center of calculation that enabled the operations of empire:

> A vast trawl of plants came from the newly explored lands to London and to Kew [Gardens], where Banks turned the royal gardens into a centre for classification and cultivation. These specimens gave botanists their first encounter with thousands of plants that had previously been unknown to them. . . . His network of botanic gardens that spread across the empire allowed plants to be taken from one colony and then cultivated under scientific supervision so that they could be transported to another colony. This scientific practice was to intervene in global agriculture on a systematic pattern never before seen by white people. It harnessed exploration and pure science to imperialist priorities.[21]

The same data-gathering network, in other words, that allowed scientists to determine precisely where natural variations of land and water checked the otherwise global spread of a specific plant or animal species also enabled the creation of new, "artificial" flows of plant and animal species, which brought these latter into new geographies. In similar fashion, tables and maps of wind and magnetic flows not only enabled more efficient and predictable shipping (and hence more finely calibrated mappings of magnetism and weather) but also potentially revealed more precisely how changes to landscape caused changes in weather. In this sense, maps and tables—or, more generally, books—could perform the same function as botanical gardens; rather than simply "representing" an external reality, books also functioned as technologies of contiguity that allowed individuals to expand and deepen their ability to act at a distance.[22]

Darwin's Icebergs: Hybrid Technologies of Contiguity

In a globalized world, then, every movement along the sphere could be understood as the result of an interplay of flows: in the case of weather, for example, flows of and within water, flows along land, and flows within the pressurized atmosphere. While some flows were more difficult to bend and channel than others, this was for some Romantics simply a technical problem; that is, simply a problem of establishing the right network with the right center of calculation. Might it not then be possible to intervene in, for example, weather at a global scale?

A number of Romantic-era authors certainly thought so. The implications for approaching local control through global flows were hinted at by Williams in *The Climate of Great Britain*. Williams outlined a system that he argued would allow Britons to control the weather of their entire island and thus "render the Seasons more propitious to the health of our growing crops"—a particularly timely consideration, given a series of bad harvests and attending political unrest in Britain in the 1790s.[23] Williams argued that weather was at least partially dependent upon the electrical state of the atmosphere and that the latter was itself partially dependent upon the vegetation that covered the land underneath the atmosphere, since, he claimed, vegetation served as a conductor. Thus, for example, the

> great cause of our clouded Atmosphere, and frequent storms of thunder, in Summer, arises from the exhaled vapour being partially deprived of its Electricity by the great number of conductors which exist in the form of points, on *marginal Extremities of leaves, the bearded ears of corn,* and *various other appendages which serve to constitute the organization and attire of the vegetable world.*[24]

Williams suggested that one could control the weather by reforming the nature of the vegetable surface against which the atmosphere pressed. However, since humans depended upon vegetable crops and thus were limited in the ways that they could control this surface, another method was to present the atmosphere with a different, more powerful, conducting surface. Williams proposed constructing two buildings in every county, each outfitted with an enormous electrical device, which—when used in concert with one another—would "electrize the whole Atmosphere of Great Britain one mile in height" and in this way control the weather (349). Williams noted that such control would itself require development of a new network: "A [meteorological] Board . . . would be united with other Agricultural Establishments for conducting the process; and the machinery should be made to act simultaneously, and under telegraphic signals; otherwise one county would be counteracting another" (351).

Despite the breathtaking scope of Williams's proposal, his belief that one could alter British weather simply by altering the *British* atmosphere still shows a vestige of the earlier, more localist, understanding of weather. A better example of the Romantic approach to flow, which seeks to dispense with this localist prejudice, is Erasmus Darwin's proposal in *The Botanic Garden* to change British weather by altering global weather flows.[25] *The Botanic Garden* is divided into four cantos, each of which provides scientific explanations of the natural processes that underlie the traditional four elements: Canto I focuses on natural processes involved in the "fiery" parts of nature (for example, volcanoes and electricity); Canto II focuses on earthy processes; Canto III, watery processes; and Canto IV, air processes. In Canto I, Darwin outlined a proposal that he believed would simultaneously decrease the heat of the torrid zones and increase the heat of overly cool zones. In the text of the poem, Darwin commanded the *"NYMPHS!"* to "alight" in the polar regions and

> array your dazzling powers,
> With sudden march alarm the torpid Hours;
> On ice-built isles expand a thousand sails,
> Hinge the strong helms, and catch the frozen gales.
> The winged rocks to feverish climates guide . . . [26]

These otherwise cryptic lines are explained in a footnote, in which Darwin provided a crash course in fluid dynamics and weather formation. Darwin argued that scientific research, as well as accounts of European explorers, suggested that the total volume of ice on the earth was increasing. Darwin also reminded his readers of Robert Boyle's "famous experiment" in the preceding century that proved that "ice evaporates very fast in severe frosty weather when the wind blows upon it" and also reminded them that "ice, in a thawing state, is known to contain six times more cold than water at the same degree of sensible coldness" (I: 59, note to Canto I, l. 529). Darwin concluded from this that one "cannot doubt but that the northern ice is the principal source of the coldness of our winters, and that it is brought hither by the regions of air blowing from the north" (I: 60, note to Canto I, l. 529). That fact, combined with the increase of the total ice volume of the earth, indicated to him that the climate of Britain would become cooler and cooler.

Darwin suggested, though, that knowledge of these processes of weather formation enabled a project of weather reformation. While it would clearly require enormous volumes of ice to change global weather patterns in ways that would affect Britain, Darwin claimed that such a project was not impossible:

> If the nations who inhabit this hemisphere of the globe, instead of destroying their seamen, and exhausting their wealth in unnecessary wars, could be induced to unite their labours to navigate these immense masses of ice into the more southern oceans, two great advantages would result to mankind; the tropic countries would be much cooled by their solution, and our

winters, in this latitude, would be rendered much milder, for perhaps a century or two, till the masses of ice become again enormous. (I: 60, note to
Canto I, l. 529)

Darwin suggested that a cosmopolitan project of ice relocation would
simply mimic natural processes, for natural ice "islands" often float from
the north or south pole on their own, and he recalled for his readers a
recent floe encountered by a ship near Botany Bay in 1789 (I: 61, note to
Canto I, l. 529).

Darwin suggested that an international project of weather reformation
would have two primary effects. First, it would increase agricultural pro-
ductivity in Britain. In the "Additional notes" that follow the fourth Canto,
for example, Darwin took up again the topic of weather reformation and
there recommended further research into the chemical basis of the "wonderful contrivance" (I: 411, note XXXIII) that connects all of the earth's
weather with the goal of creating a perpetual good wind for Britain (I:
208–9, note to Canto IV, l. 320). The result of creating such a wind, Darwin claimed, would be that "the vegetation of this country [Britain] would
be doubled, as in the moist vallies [sic] of Africa, which know no frost" (I:
208, note to Canto IV, l. 320). Second, Darwin implied that the benefits of
weather reformation would not be limited to Britain, for changing the
global system of weather flows might ameliorate landscapes afflicted with
"contagious vapours" (I: 207, Canto IV, l. 306) or with the "pestilential
winds of the east" (I: 207, note to Canto IV, l. 306). Darwin's interest in
unhealthy landscapes was part of what Bewell describes as "medical geography," which emerged in the eighteenth century as a scientific project of
locating, describing—and, if possible, altering—those kinds of landscapes
that produced illness. The practice of medical geography was tied to a
utopian vision, for, as Bewell notes, "once these 'pathogenic places' had
been identified, they could be modified and human beings might rid the
earth of disease."[27] In general, pathogenic places were understood as those
in which Europeans got sick, and the tropical, or "torrid," areas were of
especial interest. Darwin's plan to ameliorate the "tropic countries"—that
is, the countries along what he calls the "burning line" (I: 62, Canto I, l.
545)—was part of this larger project of "humanizing" torrid landscapes.

On the one hand, Darwin's reflections on weather reformation were
clearly continuous with British imperial aspirations and with the globalization of capitalism. Darwin's hope that weather reformation might eliminate disease, for example, was part of a narrowly nationalist colonial
biomedical discourse in which Britain justified its commercial involvement

in other countries with the claim that it was a British duty to help "cure" intrinsically diseased places. Darwin's image of reciprocal cooling in the tropics and warming in Britain was also at least in part an attempt to naturalize the project of British colonization. More generally, Darwin's view of nature as "inseparably bound up with novelty, fashion, and change" represented, as Bewell notes, a "new consumerist commercial vision of nature that would underpin Britain's emergence as an imperial nation," insofar as this "enlarged vision of nature is in keeping with that of a nation whose strength increasingly lay in its control and management of global natures."[28]

On the other hand, both Darwin's text itself and his specific weather reformation proposals emphasize the ways that the same networks that enabled the work of empire and capital could channel utopian aspirations less bound to a nationalist frame. As I have noted, both botanical gardens and books functioned as technologies of contiguity, for both brought samples or observations sufficiently close to one another that new relationships and possibilities could be imagined and fed back into the network. Darwin's *The Botanic Garden* sought to hybridize these separate technologies of contiguity by creating a *textual* "botanic garden" that would amplify the imaginative potential of a technology of contiguity—or, as Darwin put it in his Advertisement, "enlist Imagination under the banner of Science."[29] Thus, even if Darwin's weather reformation resonated with colonial and capitalistic projects, they were also premised on modes of international cooperation that operated in excess of narrowly colonial or capitalistic projects (or, at any rate, cannot be seamlessly aligned with either project). For Darwin, the process of altering all of the earth's weather was simply too big for one country to take on by itself, and it thus not only required the cooperation of multiple countries but also *enabled* such cooperation, precisely because weather was something that so clearly overflowed national boundaries. For Darwin, understanding existing weather patterns and reimagining new flows of wind forced one to imagine the *whole* of humanity as an active element within natural processes of transformation.

Shelley's Reformed Ecliptic

While Darwin's terraforming plans were monumental and global in scope, they pale in comparison to the transformation of nature outlined by Percy Bysshe Shelley in his first major poem, *Queen Mab*. This poem, like Darwin's *Botanic Garden*, is supplemented with an imposing number of notes, most quasi-scientific in character. However, where Darwin hoped that his poem would change praxis by popularizing scientific knowledge, Shelley sought in *Queen Mab* to enlist science in a radical philosophy of political and intellectual liberty.

The "Spirit" narrator of *Queen Mab* is given a dream tour of the past, present, and future of humanity. In the poem's final sections, which allows the Spirit a glimpse of the "sweet . . . scene" that the earth will become, Shelley developed the image of a terrestrial paradise, highlighted by the image of the lion "sheathing" his claws and becoming lamb-like in his nature.[30] Yet this transformation of predator-prey relationships seems to depend upon massive changes to the earth's surface, for humans must first "coalesce" with nature and "undertake regeneration's work," which involves a complete transformation of the planet (75 [Canto VI]). The polar caps are "unloosed" and the poles warmed (102 [Canto VIII]); the deserts of the world become wooded, habitable, and pacific (102–3 [Canto VIII]); and the formerly "illimitable plain" of the ocean is humanized, as "those [previously] lonely realms bright garden-isles [now] begem" (103–4 [Canto VIII]). As in the case of Darwin's account of terrestrial weather reformation, Shelley's narrator emphasizes that the transformation of the globe results in the destruction of "sick" places, enabling a utopia free of disease: "Health floats amid the gentle atmosphere" (104 [Canto VII]).

Perhaps most dramatic, even the axis of the earth will become straightened, so that the entire globe enjoys a temperate climate year round. In the period in which "man, with changeless nature coalescing, / Will undertake regeneration's work" the "ungenial poles" of the earth "no longer point / To the red and baleful sun / That faintly twinkles there" (75 [Canto VI]). Shelley's notes at the end of the book explain that the last two lines refer to

> the north polar star, to which the axis of the earth, in its present state of obliquity, points. It is exceedingly probable, from many considerations, that this obliquity will gradually diminish, until the equator coincides with the ecliptic: the nights and days will then become equal on the earth throughout the year, and probably the seasons also. (152)

In support of this claim, Shelley cited the work of authors such as the astronomer and mathematician Pierre-Simon Laplace and the astronomer Jean-Sylvain Bailly, and he cited as well the physiologist and philosopher Pierre Jean George Cabanis to argue that "there is no great extravagance in presuming that the progress of the perpendicularity of the poles may be as rapid as the progress of intellect; or that there should be a perfect identity between the moral and physical improvement of the human species"

(152–53).[31] Shelley thus offered what we might call a maximalist vision of terraforming, which involves not just transformations of specific terrestrial surfaces that channel and direct flows but a reorientation of the earthly sphere itself.

Yet if Shelley's vision of terraforming is more extensive than Darwin's, it is also more intensive, for in Shelley's account, terraforming reaches into humanity itself. In Shelley's note about the earth's ecliptic, it is not clear whether the climate changes that he describes are the *result* of human actions or if they occur independently and provide the foundation for human technical developments. This ambiguity is emphasized in Canto IX, which outlines a process in which

> human things were perfected, and earth,
> Even as a child beneath its mother's love,
> Was strengthened in all excellence, and grew
> Fairer and nobler with each passing year. (117 [Canto IX])

Shelley's phrasing suggests that perfection is not achieved through technocratic acts of willful agency but through modes that combine activity and passivity and that alter humans as much as the earth. Humans do not actively perfect their things, but rather, human things "are perfected." In the analogy that follows, the earth itself grows into excellence, though only because humans provide it with a maternal-like nurturing atmosphere. Both of these combinations of activity and passivity underscore the "coalescing" of humans and nature that occurs with the improvement of science and the progress of liberty. Human actions are in this sense not separate from nature but rather a portion of nature—a vector—through which nature improves itself. While this process eventuates in a state in which "every shape and mode of matter lends / Its force to the omnipotence of mind," such mental omnipotence is possible only when humans have so fully "coalesced" with nature that they have become completely transformed from the state in which they existed in Shelley's own time (235–36 [Canto VIII]).[32]

Shelley emphasized the implications of terraforming for human reformation by suggesting that this process would begin in his present with the transformation of the flows of food that pass through the individual human body. More specifically, terraforming begins when people switch from meat eating to a purely vegetable diet and abstain from alcohol consumption. Shelley suggested that these steps are more effective than any "mere reform of legislation" precisely because each individual functions as a material nexus through which various flows—flows of food and passion, for example—pass, and thus each individual is a point at which economics, politics, and nature can be reconfigured.[33] A reform of diet thus "strikes at the root of the evil" and has the potential to end colonialism, which Shelley described as the process by which "every corner of the globe is rifled" for commodities.[34]

Shelley's depiction of individual human beings as nexus points through which various flows pass pursues the principle of global flows even further than had Darwin. For Darwin, global flows happen around—and can potentially be controlled by—humans, yet for reasons that he does not explain, this world of flow seems not to reach into human beings themselves. For Shelley, by contrast, global flows do not simply pass over and around humans; they also move through them. (Or, as Shelley put it a few years later in "Mont Blanc," the "everlasting universe of things / Flows through the mind.")[35] As a consequence, every reformation of global flows is simultaneously a reformation of the flows that pass through, and determine the nature of, human beings. Timothy Morton suggests that in Shelley's early poetry, nature is presented as "an 'economy of the globe,' a homeostatic system of regulated flows, [that] has overwhelmed any final, arbitrating signifier to which it could be referred."[36] This description is correct in spirit, but Morton's emphasis on "homeostasis" conflicts with his basic point. We can better describe Shelley's understanding of global flows as oriented toward ecstasis, that is, self-transforming flows, rather than homeostatic standing waves.[37]

Where Darwin sought to direct global flows by creating a hybrid technology of contiguity—a textual botanic garden that placed side by side as many results of the sciences as possible—Shelley's more intensive understanding of flows required a more fundamental revisioning of textual form. As in the case of Darwin's poem, the notes of *Queen Mab* assemble various results of the sciences alongside one another. However, in *Queen Mab*, the perspective or view from which the results of science are contiguously aligned is neither quite a terrestrial view nor that "view from nowhere" that Thomas Nagel later diagnosed as the key to scientific objectivity. Rather, it is what I will describe as a gravitational view: A view of the earth as one of many spheres, but with the center of gravity positioned such that this view brings us back to our own globe.[38] Shelley's poem thus begins with an ascent upward through the earth's atmosphere ("The atmosphere in flaming sparkles flew") and outward through interstellar space, until

> Earth's distant orb appeared
> The smallest light that twinkles in the heaven;
> Whilst round the chariot's way
> Innumerable systems rolled,
> And countless spheres diffused
> An ever-varying glory. (13–14 [Canto I])

This is neither an indifferent survey of the universe as a whole nor a covetous view of other solar systems that humans might one day conquer. Rather, this is an image of an "interminable wilderness / Of worlds": a view, that is, of the earth disappearing into a potential infinity of many spheres, and thus a view from which one must ultimately turn away in order to refocus attention upon our terrestrial globe.[39]

For Shelley, writing long before the era of space travel, such a view is possible *only* from the perspective of poetry; it is a view that cannot be seen in fact but rather only by imagining our globe just at the point of its disappearing, perhaps forever, into a wilderness of other globes. Shelley's image is not original, for it calls to mind Anna Letitia Barbauld's closing image in "A Summer Evening's Meditation" (1773), in which the narrator is impelled through our solar system

> To the dread confines of eternal night,
> To solitudes of vast unpeopled space,
> The desarts of creation, wide and wild;
> Where embryo systems and unkindled suns
> Sleep in the womb of chaos[.][40]

Barbauld's image of "solitudes of vast unpeopled space" works in the service of the Christian goal of refocusing the reader back on this earth (this "mansion fair and spacious for its guest, / And full replete with wonders"; 138) and, even more important, encouraging the reader to develop a practice of patience, by means of which the reader can, "content and grateful, wait th' appointed time" when he or she makes the transition to that eternal afterlife in which "the glories of the world unknown" shall be revealed.[41] Shelley's "wilderness of worlds" also encourages us to value properly the one sphere on which we live and from which—for Shelley as much as for Malthus—no emigration is possible. However, where Barbauld's image of infinite worlds orients us toward the afterlife, Shelley's wilderness of worlds focuses attention on the future of this earth. For Shelley, this refocusing requires both the images of the poetry and the sciences referenced in the notes. In the case of *Queen Mab*, the genre of philosophical poem does not simply assemble the results of the sciences alongside one another in order to amplify the role of imagination in science. Rather, this philosophical poem brings the results of the sciences back to earth, which allows us to orient the project of terrestrial terra-forming toward values such as the reduction of violence and suffering and the elimination of political and economic inequity. The Shelleyan philo-sophical poem treats the work of science itself as a global flow, and it seeks to locate—and ultimately, to direct—the orthogonal forces upon which science and technology depend.

Globalization, Provincialization, and the Anthropocene

Darwin's image of an international fleet of iceberg-bearing ships and Shel-ley's image of a reformed planetary ecliptic help us approach productively a recent debate about the relationship of aspirations for human freedom to the limits on freedom that seem inherent in the concept of the Anthropo-cene. Within the humanities, the historian of India Dipesh Chakrabarty's essay "The Climate of History: Four Theses" has focused this debate, in part because this essay seemed to many critics to reject the project of human freedom for which Chakrabarty had strenuously advocated in his earlier postcolonial theory. In his influential *Provincializing Europe: Post-colonial Thought and Historical Difference* (2000), for example, Chakrabarty argued that the image of historical progress leading to freedom—an image common to the European liberal and Marxist traditions and also to most postcolonial movements and theory—must be retained, for only this image allows us to orient ourselves toward the telos of global justice. Yet this image of history also implied that areas of our globe outside of Europe, and everyday practices that do not conform to European bourgeois patterns of behavior, are archaic, prehistorical, or anachronistic. For liberal theorists, for example, this image of history justified, in the name of freedom, both enslaving populations of other areas of the globe or treating these as chil-dren who must be tutored in the European model of liberal citizenship.[42] Though one might see such liberal imperialism as simply bad faith—that is, as providing ideological cover for various forms of domination— Chakrabarty stressed that precisely this same image of history unfolding from Europe also underwrites most Marxist accounts of the possibilities for human freedom.[43] Chakrabarty's critique was not intended to dismiss Marxist (or liberal) aspirations for global human freedom but to link their orientation toward global justice to other ways of understanding human social relations that do not conform to the bourgeois/liberal model. He thus argued in favor of another mode of writing history, which both stresses the extent to which local and indigenous patterns of life cannot be judged within the terms of progressive history (from which perspective they will always appear anachronistic) and contests and decentralizes the liberal-Marxist progressive model of time.

Roughly a decade after the publication of *Provincializing Europe*, Chakrabarty published "The Climate of History: Four Theses." Focusing on the implications of global warming for history, this essay seemed to undo—or at least change significantly—the aspirations for freedom and justice that were the point of *Provincializing Europe*. Chakrabarty's basic claim in "The Climate of History" is simple: Given the "current planetary crisis of climate change or global warming," the fate of humanity likely depends on renouncing many of those aspirations for freedom that are part of both the liberal and Marxist traditions (197). The problem, Chakrabarty contends, is that "the mansion of modern freedoms stands on an ever-expanding base of fossil-fuel use," for "most of our freedoms so far have been energy-intensive" (208). Though this does not mean that "analytic frameworks engaging questions of freedom by way of critiques of capitalist globalization" are therefore "obsolete," "these critiques do not give us an adequate hold on human history once we accept that the crisis of climate change is here with us and may exist as part of this planet for much longer than capitalism or long after capitalism has undergone many more historic mutations" (212). Chakrabarty suggests that we can address the fact that the collective actions of the human species now constitute a planetary force only by framing aspirations for freedom *within* those limits that are part and parcel of being a biological species that can affect the entire globe (212–20).

For postcolonial scholars such as Ian Baucom, "The Climate of History" does not so much reframe postcolonial theory and its concept of freedom as abandon the latter. "The Climate of History," Baucom claims, no longer "orient[s] us toward a future measured against the promise of freedom," as was the case for *Provincializing Europe*, but instead "direct[s] us to (and desperately against) a future marked by the threat of extinction."[44] In his response, Chakrabarty in essence grants Baucom's point, noting that where he had earlier stressed the need to interrupt the premise of progressive history by engaging the experiences of groups understood within progressive history as anachronistic, the ecological limits inherent in species existence that are underscored by the concept of the Anthropocene *cannot* be experienced and hence can only be included in our considerations as scientific statements of limits.[45] Chakrabarty sees the collective activity of humans in the age of the Anthropocene as leading us to collective extinction and hence—if we wish to avoid that fate—necessitating collectively shared limits on what counts as freedom. Baucom, by contrast, sees Chakrabarty's replacement of "freedom" with "survival" as the means by which Chakrabarty falls prey to the same universalizing move that he criticized in his earlier work. Baucom's solution is to continue the project of *Provincializing Europe* by retaining the historical telos of freedom but ensuring that the content of freedom remains open *both* to that "ontological plurality of the human" stressed by Chakrabarty in his earlier work (which ensures that the Eurocentric version of the history of freedom will always be "extensively interrupt[ed] and modif[ied]" by the experiences and knowledges of non-European peoples) *and* "the post-natural actors, agents, and actants of cyclones, heatwaves, and melting ice" characteristic of the Anthropocene (139).[46]

Though it is tempting to take sides here by opting either for survival or freedom, Shelley's approach to climate change offers us another path, one that brackets the question of survival from discussions of the Anthropocene. In our era of global warming and species extinction, what is perhaps most striking about Shelley's terraforming proposal is the *optimism* that underwrites it. Shelley valorizes events, such as the melting of the polar ice caps, that we now intensely fear. It is certainly possible that this optimism was grounded in a massively deficient understanding of the nature of global ecological processes. Though Shelley, like the other Romantic-era authors I have considered, understood global processes such as weather in terms of flows, their understandings of these were coarse in comparison with our contemporary knowledge of the interdependencies and feedback loops intrinsic to global ecological dynamics. Given these deficiencies of Shelley's knowledge, we might be tempted to see his poem as an early instance of the technocratic belief that, just as humans could build ships to guide them across the oceans, they could also successfully engineer the entire global environment.[47] If survival—whether of humans as a species or of humans in combination with their institutions—is absent from Shelley's (and Darwin's) accounts, it may seem that this is because neither took seriously enough the consequences of human action becoming a geological force.

Yet such a critique does not fit Shelley well, for it misses the ecstasis and exuberance that mark his vision of transforming humans merging with transforming nature and that better explain the absence of survival from his account. For Shelley, the becoming elemental of humans is part of a process of "perfection." Yet perfecting names, paradoxically, a process of employing the ruins and wreckage of human history as elements for recuperation. In the just future to come, the earth's inhabitants are surrounded by the wreckage of past human injustice, such as ruined castles and dungeons, yet

> These ruins soon left not a wreck behind:
> Their elements, wide-scattered o'er the globe,
> To happier shapes were moulded, and became
> Ministrant to all blissful impulses:
> Thus human things were perfected, and earth,
> Even as a child beneath its mother's love,
> Was strengthened in all excellence, and grew
> Fairer and nobler with each passing year. (117 [Book 9])

This is not a technocratic vision of humans becoming elemental by enslaving the blind forces of an external nature to their will, nor is it a survival-oriented project of protecting humans from a threatening earth. Rather, perfecting names for Shelley the redemption of those ruins that have emerged from human history.[48]

For Shelley, the project of redemptive perfection begins in the experience of sympathetic joy. Following Spinoza, Shelley understood joy as a modality of love that is both other oriented and transformative of the self. More specifically, joy signals the expansive transformation of the self by means of elements that are common to the self and something beyond it—or, as Marjorie Levinson puts it, "the awareness of becoming joined to another body harmonious with one's own."[49] For Shelley, the redemption of past injustices requires creation of a common element that enables such transformation. In *Queen Mab*, for example, after surveying past and present injustices in the first parts of the poem, the Spirit looks toward a redemptive future, which enables the experience of joy:

> Joy to the Spirit came.
> Through the wide rent in Time's eternal veil,
> Hope was seen beaming through the mists of fear;
> Earth was no longer hell;
> Love, freedom, health had given
> Their ripeness to the manhood of its prime,
> And all its pulses beat
> Symphonious to the planetary spheres.[50] (99–100 [Book 8])

Perfection names for Shelley, as for Spinoza, a dynamic process of humans and earth moving together into a common future, and this is a movement that will lead to—or just is—human freedom. For Shelley, such a movement can neither be guided by advanced Europeans who know what freedom really is, nor does it require that humans compromise their pursuit of freedom in the face of the threat of extinction. Rather, it names the joyful process of redeeming the ruins of history by resolving these into their elements and molding these elements into "happier shapes."

From One to Many Spheres

I flesh out this Shelleyan vision of a joyful Anthropocene in what follows, but to clarify why such a vision is even more necessary now than in Shelley's time, I return to another of his images, that of the earth amid an interminable wilderness of worlds. This image helps us locate a development in global thinking that is registered by neither Chakrabarty nor his critics. Chakrabarty noted that "The Climate of History" originated in his recognition that the kind of economic and political globalization addressed in *Provincializing Europe* had little to say about the different kind of globalization revealed by climate change and global warming. Yet we now live in an era that, perhaps paradoxically, seems to be distancing itself from concepts of globalization altogether. Such a claim may seem counterintuitive, since we tend to think of the present as the real era of globalization, that is, the period in which processes of global networking and economic interconnectedness that were only just beginning in the eighteenth century have finally reached their full intensity and force. Yet many of these processes of contemporary globalization are encouraged by a *waning* of the belief that we must understand flows in terms of the logic of a globe from which no emigration is possible.

Hannah Arendt had already pinpointed the peculiar contemporary status of globalization in her discussion, at the start of *The Human Condition* (1958), of the implications of the 1957 launch of the Sputnik satellite, an "event, second in importance to no other, not even to the splitting of the atom."[51] What this event signified, according to Arendt, was not only a literal but, more importantly, an existential movement away from the earth, which latter she defined as that "habitat in which [humans] can move and breathe without effort and artifice" (and hence, "the very quintessence of the human condition") (2). Arendt noted that many observers expressed enthusiasm about this movement of Sputnik into outer space, seeing in it a "first step toward escape from man's imprisonment to the earth," and she argued that this movement may have signaled the end—or at least a massive transformation—of the human condition (1). "The most radical change in the human condition we can imagine," Arendt wrote,

would be an emigration of men from the earth to some other planet. Such an
event, no longer totally impossible, would imply that man would have to live
under man-made conditions, radically different from those the earth offers
him. Neither labor nor work nor action nor, indeed, thought as we know it
would then make sense any longer.

Though these "hypothetical wanderers from the earth" would still be
human beings, their human condition would differ radically from that of
those living on the earth upon which humans first emerged (9).

Whether or not one adopts the rest of Arendt's analysis, the Sputnik
launch seems indeed to have marked the end of the Romantic understanding of globalization, for from 1957 on, "flow" no longer had to be understood as limited to the one sphere upon which we currently live. The earth
is now one sphere of many, though in a decidedly un-Shelleyan sense.
Earth may be the one sphere on which we currently happen to live, but it
is also one from which some *can* emigrate, at least in principle, should we
embrace the technological ability to "escape from [our] imprisonment to
the earth." When Earth is no longer understood as a globe from which, as
Malthus put it, "emigration would of course be excluded," this alters how
the global flows of the earth are understood.

The contest that has been waged since the 1970s over the ecological
implications of those population-flow dynamics that Malthus sought to
identify and describe exemplifies this transformation of the meaning of
terrestrial global flows. As I noted in Chapter 2, the politics of Malthusianism have always been difficult to pin down. Though many Romantics saw
Malthus's book as propaganda for an antidemocratic, class-stratified status
quo, ecologically oriented thinkers of the 1960s and 1970s saw in it a
resource for critiquing the liberal commitment to economic growth. The
famous 1972 Club of Rome report, for example, which argued that the
dominant Fordist model of manufacture was producing ecological and
social crises, grounded its claims in rigorously Malthusian logic.[52] As
Melinda Cooper notes, the report's authors suggested that "the exponential
growth of population and industry could not continue indefinitely without
running up against the limits inherent in the other variables under study—
namely, agricultural production, energy supplies, and pollution" (16). In
other words, there are limits to economic and industrial growth precisely
because "the earth is finite."[53]

As I also noted in Chapter 2, what is perhaps more surprising than this
(vaguely) leftist embrace of Malthusianism was the neoliberal response,
which involved a rejection of this global understanding of flows and

ultimately a rejection of the very concept of globalization. "Postindustrial" thinkers such as Daniel Bell argued that this Malthusian-cum-ecological reasoning assumed a "closed system"—the Malthusian sphere from which no emigration was possible—whereas new technological advancements contested precisely this assumption. Julian Simon's *The Ultimate Resource* (1981, 1996), for example—a book supported in the 1980s and 1990s by politicians such as US president Ronald Reagan—pointed to the moon as a place from which resources could be derived.[54] The authors implied that the mere fact that we could escape the earth in order to mine moon metals allowed us to rethink our relationship to our own globe by realizing that all Malthusian "limits" were in fact illusory:

> Each epoch has seen a shift in the bounds of the relevant resource system. Each time, the old ideas about "limits," and the calculation of "finite resources" within the bounds, were thereby falsified. Now we have begun to explore the sea, which contains amounts of metallic and perhaps energy resources that dwarf the deposits we know about on land. And we have begun to explore the moon. Why shouldn't the boundaries of the system from which we derive resources continue to expand in such directions, just as they have expanded in the past?[55]

The passage exemplifies the change in global thinking that the Sputnik satellite launch introduced: As soon as humans could reach another sphere, our coordinates and assumptions for understanding the flows that occur upon our sphere were thrown into flux. Flows are no longer necessarily restricted to globes but can now occur simultaneously on and between globes.

Shelley's mode of global thinking harbors resources for countering the centrifugal force of this neoliberal commitment to thinking "beyond globalization." Shelley's vision of planetary adjustments and a plurality of globes ties globalization to a perspective of the earth as one of many spheres, yet his many-sphere image emphasizes that contemporary visions of an unbounded "open" system—a universe of interminable globes and a lack of any ultimate limits—is ultimately a vision of unconquerable, irredeemable wilderness, rather than of plenitude. For Shelley, there are indeed other spheres, but their function is to focus us back on this one earth upon which we can live. Shelley's image thus suggests that the criterion for determining our lived relationship to questions of limits is not that of truth—that is, the (ultimately undecidable) question of whether current limits are "really" final limits or not—but rather a linkage of freedom, justice, and beauty, that is, the question of the kind of life we want to live on the one earth upon which we have evolved.

Neoliberalism, Seeds, and Robinson's *Aurora*

We can begin to clarify what a Shelleyan version of the Anthropocene in the age of neoliberalism might mean by turning to the contemporary (and arguably neo-Romantic) science fiction of Kim Stanley Robinson, in particular his novel *Aurora* (2015). *Aurora* takes up all of the topics I have discussed in this chapter, including the operation of untethering, the possibility of emigration from Earth, the question of flows across bounded surfaces, network construction, and the relationship of our earthly globe to what Arendt called the human condition. Reading Robinson's novel through the lens of Shelley's image of a wilderness of worlds also helps us understand more fully the extent to which recent neoliberal efforts to overcome limits transform the operation of untethering into a new principle, by reimagining human reproduction and relations in terms of plant modes of dissemination such as seeds and spores.

Most of Robinson's novels fall within the subgenre of so-called hard science fiction, which draws heavily on the work of—and seeks to restrict itself to the boundaries of the possible established by—existing communities of scientists, such as physicists, biologists, geologists, and climatologists. Much of Robinson's work has focused on the question of how human political, social, and biological relations would be transformed if, indeed, it became possible for some people to emigrate to another nearby sphere, such as Mars, while the vast majority of the human population remains on a globally warming Earth.[56] Though *Aurora* also takes up these questions, it departs significantly from Robinson's earlier novels about emigration from our globe, in which humans ventured outward toward different parts of our solar system but remained tethered to the umbilical center of Earth through political, scientific, and economic relations. *Aurora*, by contrast, considers emigration to a sphere so far away—one of the potentially habitable planets circling Tau Ceti, a real star system located twelve light-years from Earth—that the humans living there could no longer remain connected to Earth in any meaningful way. Though the spaceship—or, perhaps more accurately, ark—represented in *Aurora* travels at up to one-tenth the speed of light, the vast distance between Earth and Tau Ceti means that it still takes 160 years for the ship to reach its destination, and, once it has arrived, radio transmissions traveling at the speed of light from Earth nevertheless take twelve years to reach them. The distance between Earth and Tau Ceti required Robinson to cast this story within the science-fiction subgenre of the "generation ship," which is based on the premise that multiple generations of humans would have to live and die within the confines of a spaceship before it could reach its destination.[57]

What is perhaps most striking about Robinson's version of the generation-ship novel is that *Aurora* ultimately narrates a *failed* attempt to emigrate from Earth. The novel begins 160 years after the original generation of travelers departed from Earth, as the roughly 2,100 descendants of that original crew are arriving at Tau Ceti.[58] Though the ship is immense and contains twenty-four different biomes, the travelers have encountered constant problems trying to keep the biomes and their plant and animal inhabitants (including humans) healthy. This is in part because bacteria and viruses mutate more quickly than other forms of life on the ship and in part because of the extraordinary difficulty of taking into account every ecological cycle upon which living beings depend.[59] It is, as a consequence, not clear that humans will be able to survive on the ship much longer. Yet when a contingent from the ship tries to establish a base on one of the planets of Tau Ceti, many of the landing party are killed by something indigenous to the planet's soil—perhaps a new kind of prion, but in any case, something so different from earthly categories of threat that it is not even clear what it is or how to combat it. The inhabitants of *Aurora* are then faced with a hard choice: continue on toward yet another (possibly) inhabitable planet within the Tau Ceti system or admit failure and head back to Earth, which would mean seven more generations of humans living out their existence on the ship before it arrives. This choice provokes a civil war, which results in the ship—which was designed to be split in two—dividing, one half headed outward in search of another inhabitable planet and the other half returning to Earth. The narrative follows only those who return to Earth. Unhappily for this group, the ship's ecology seems to be failing for good, rendering it unlikely that another seven generations of humans could survive until the ship reaches Earth. Luckily—albeit also as something of a narrative deus ex machina—the ship receives radio transmissions from Earth that document advances in suspended-animation techniques, with the result that the ship's denizens are able to freeze themselves for the duration of the return voyage.

Yet the ship's inhabitants "return" to an Earth that they themselves have never known and that has largely forgotten about the original voyage outward of these emigrants. Nor are the various Earth governments particularly interested in these survivors. However, the latter are at one point asked to present their thoughts at an international forum devoted to the possibility of sending out new ships to even more distant star systems. In response to this proposal, Aram, one of the survivors, delivers a speech with an Arendtian ring:

> No starship voyage will work. . . . This is an idea some of you have, which ignores the biological realities of the situation. We from Tau Ceti know this better than anyone. There are ecological, biological, sociological, and psychological problems that can never be solved to make this idea work. The physical problems of propulsion have captured your fancy, and perhaps these problems can be solved, but they are the easy ones. The biological problems cannot be solved. And no matter how much you want to ignore them, they will exist for the people you send out inside these vehicles. . . . The bottom line is the biomes you can propel at the speeds needed to cross such great distances are too small to hold viable ecologies. The distances between here and any truly habitable planets are too great. And the differences between other planets and Earth are too great. Other planets are either alive or dead. Living planets are alive with their own indigenous life, and dead planets can't be terraformed quickly enough for the colonizing population to survive the time in enclosure. Only a true Earth twin not yet occupied by life would allow this plan to work, and these may exist somewhere, the galaxy after all is big, but they are too far away from us. Viable planets, if they exist, are simply *too—far—away.* (459–60)

Where Arendt had granted the possibility of humans emigrating to another planet, Aram argues that, for practical reasons, this is not in fact possible. While both *Aurora* and Robinson's other novels about terraforming suggest that humans can live on other spheres within our solar system, humans in that case remain in a fundamental sense tethered to Earth, the globe of human origin. But to travel from Earth to another inhabitable planet, humans would require a spaceship nearly the size of Earth itself, for anything smaller will not allow for biomes with "viable ecologies." As a consequence, Aram concludes—and one senses that this is also Robinson's conclusion—that it makes no sense to try to send humans to other stars. To return to Shelley's *Queen Mab*, though "the countless spheres" of the universe present an image of "ever-varying glory," any attempt to reach other planets transforms them into a wilderness.

The discussion's moderator objects to Aram's statement, however, and his counterargument reveals the inhuman logic that stands behind aspirations to colonize the cosmic wilderness. The moderator contends that no general conclusions should be drawn from the individual case of Aram and his fellow survivors, and one ought instead to view this from a broader perspective:

There are really no physical impediments to moving out into the cosmos.
So eventually it will happen, because we are going to keep trying. It's an
evolutionary urge, a biological imperative, something like reproduction
itself. Possibly it may resemble something like a dandelion or a thistle releasing its seeds to the winds, so that most of the seeds will float away and die.
But a certain percentage will take hold and grow. Even if it's only one
percent, that's success! And that's how it will be with us— (460)

Before the moderator can finish his sentence, Freya, the novel's main protagonist, runs up to the moderator and begins to pummel him with her
fists (461). Later, when Freya has calmed down and Aram asks her to explain
her actions, she stresses the latter's comparison of human beings to seeds or
spores:

It isn't just foolish, it's sick. Did you hear what he said? Dandelion seeds?
Ninety-nine percent sent out to die, as part of the plan? Die a miserable
death they can't prevent, children and animals and ship and all, and all
for a stupid idea someone has, a dream? Why? Why have that dream? (462)

Through her act of spontaneous violence, Freya seeks to oppose the
extraordinary transformation of the operation of untethering from the
earth that the moderator proposes. In Malthus's foundational images of
populations covering the globe, vegetation played a paradigmatic role, for
he exemplified population growth with the image of the earth being encircled by flows of fennel and urged his readers to imagine human populations
as motivated by this same basic drive. Malthus also claimed that population
questions should not be considered from the perspective of individuals but
rather with a view of the species as a whole: "In reasoning upon this subject, it is evident that we ought to consider chiefly the mass of mankind
and not individual instances."[60] Yet Malthus also ultimately undid his initial identification of human and plant populations, both by stressing the
orthogonal drag of the less intensive growth of human food sources on
human population growth and (in later editions of his text) by noting
human capacities of "moral restraint" that enabled humans consciously to
moderate human population growth.[61]

In *Aurora*, by contrast, the moderator pushes even further the operation
of untethering that underwrote Malthus's opening image of vegetative
reproduction. For both Malthus and the moderator in *Aurora*, humans and
plants have a blind reproductive drive that propels them to cover every
available territory. However, the moderator suggests that human reproduction should be consciously and actively reconceived and modeled on the

reproductive strategies of plants such as dandelions, which cast off as many seeds as possible so that at least one or two will survive. As a distant descendant of one of the original emigrants to leave Earth several centuries earlier, Freya is well aware that this transplantation of human reproduction and survival into a plant model can never be restricted to the choices of those individuals who choose to leave Earth but of necessity also forces all their progeny into this same mold.

As both Freya's initial anger and subsequent incomprehension—"Why have that dream?"—underscore, the moderator's dream of transforming humans into seeds is ultimately incoherent. Though dandelions and thistles indeed release "seeds to the winds, so that most of the seeds will float away and die," plant species do so *within* the flows and orthogonal drag of the earth; neither dandelions nor thistles send seeds into space. This vegetable mode of propagation only makes sense, in other words, as a kind of flow around a globe. Though the moderator partially acknowledges this through his use of analogy—the movement of humans to the stars would only be "like" the thistle's release of seeds—it is unclear to Freya (and likely to many of Robinson's readers) what kind of human goal would be served by this transformation of human reproduction. Even if 1 percent of those human "seeds" survive, they will remain effectively isolated, even in terms of communication, by the vast interstellar distances between Earth and the 1 percent who survive. This is no longer colonization, which presumes some communication between origin and colony, but simply mute dissemination. It is for this reason that sending humans beyond the solar system turns what Shelley described as the "ever-varying glory" of the stars into an "interminable wilderness / Of worlds." And hence Freya's question: Why have that dream at all?

Building Beaches: Joy and the Allegory of the Tube

Treating the plot of *Aurora* as an allegory for contemporary neoliberalism helps us expand on the connection between ecstatic joy and the joint transformation of humans and the earth that Shelley proposed in *Queen Mab*. Reading *Aurora* allegorically suggests that the novel describes not the future of contemporary neoliberalism—for example, some stage that occurs after the neoliberal desire to mine the resources of other planets in our solar system has been realized—but rather our present. The parallels between *Aurora* and the present are illuminated by the fact that the spaceship's inhabitants occupy a tube, not a globe; that is, they live within a container, not upon a sphere. The premise of this enclosed form of living is that it would enable complete control of the environment, in contrast to the relative lack of such control that characterizes human existence lived on the surface of the earth. As the urban historian Chris Otter has noted, tube-living has in fact become the current state of affairs for a sizable minority of humans, for the development of a "technosphere" of infrastructure and devices

> allows humans to progressively abandon a largely outdoor existence, and to retreat into increasingly sealed, climate-controlled spaces. . . . Air conditioning, for example, has transformed American housing, energy use and demography. . . . Air conditioning has facilitated a "great enfolding" of humans. One recent study found that Americans spent only 7.6 per cent of their time outdoors. . . . The technosphere is a new phase in the history of human niche-construction. It is ruthlessly cleansed, with sanitized surfaces, vacuum cleaners, disinfectants and antibacterial soaps.[62]

The great enfolding enabled by devices such as air conditioning and antibacterial cleansing has created vast *Aurora*-like tubes on Earth. Though Erasmus Darwin had hoped to create a fully air-conditioned globe, it has turned out that his dream could only be realized by enfolding a significant amount of human life on Earth within interconnected tubes, which allow humans to move from air-conditioned, sterilized houses to air-conditioned, sterilized cars, to air-conditioned, sterilized work spaces, and then back again to houses.[63] However, just as the tube-existence of the *Aurora*'s inhabitants is ultimately unviable within Robinson's novel, so too, it turns out, is our actual mode of living within tubes on Earth, as the fact of global warming makes clear.

The return of *Aurora* to Earth is thus an allegory not only for our need to exit contemporary tube-life and step into the exposed and far less physically comfortable position of globe dweller but also for the necessity of taking the elements of this ruined earth and molding these into "happier shapes." For Freya, who has never known anything but tube-life, the movement out of her container is literally overwhelming, and for many months she is unable to bear the vastness of the earthly horizon and sky and must remain in enclosed spaces. Moreover, when Freya is finally able to emerge from her Earth-tube, it is onto a globe that has been massively damaged by global warming. Freya is only able to begin to live on the earth without fear of its vast spaces after a conversion experience that follows the path of joy. On Freya's Earth, "there are no beaches," for

Sea level rose twenty-four meters in the twenty-second and twenty-third
centuries of the common era, because of processes they began in the twenty-
first century that they couldn't later reverse; and in that rise, all of Earth's
beaches drowned. Nothing they have done since to chill Earth's climate has
done much to bring sea level back down; that will take a few thousand more
years. (468)

Freya and other survivors of *Aurora* join a group that creates new beaches.
This group uses technology to reform elements of the ruined earth—in this
case, the relationship of water and sand—into happier shapes. They do so
not for the sake of survival or to align the globe with the climatic ideal of a
specific geographic region, as in Darwin's desire to make the entire world
conform to the standard of a comfortable British summer day. Rather, they
do so to enable an affective connection to that "lifeway that went right back
to the beginning of the species in south and east Africa, where the earliest
humans were often intimately involved with the sea" (468).

This form of beach building draws out the implications of the strange
Platonic-Spinozan allegory of the tube that results from reading Robinson
through Shelley. In Plato's allegory of the cave, Shelley's spiritual journey
in *Queen Mab*, and Robinson's allegory of the tube, the achievement of
earthly justice requires emergence from an enclosed space, and this emergence enables a new, redemptive vision. Yet in Shelley's *Queen Mab*, this
vision does not move the Spirit away from the earth and toward the eternal
Forms but instead requires a reenvisioning of the earth and its polar
coordinates:

man, with changeless nature coalescing,
Will undertake regeneration's work,
When its ungenial poles no longer point
        To the red and baleful sun
        That faintly twinkles there.

In the late eighteenth and early nineteenth century, "regeneration" was
a theological term that meant "rebirth," or "the act of being born again by
a spiritual birth, or becoming a child of God."[64] For Shelley and Robinson,
regeneration requires an embodied realignment of the relationship between
human bodies and the earth's polarities, but this itself requires a reorientation toward the injustices of the past and possibilities of a just future. To
emphasize the coalescence of humans and nature, Shelley describes this as
a literal reorientation of the earth's poles in their relationship to the sun,
while Robinson stresses the "subjective" side of this reorientation through

his representation of Freya's hard-won ability to endure the earth's immensity. Shelley's Spirit and Robinson's Freya orient themselves toward the forms of Beauty, Truth, and the Good, but such a project is understood as reparative rather than utopian, that is, a matter of taking relationships among humans and between humans and nonhuman nature both as they have been and as they are now, and seeking to redeem these through joy.

Aurora suggests that there are two sides, or slopes, to this kind of reparative beach building. On the one hand, the rebuilt beach is a scene of corporeal instruction, through which one learns to locate elements common to one's own bodily capacities and the forces of waves and sand, which in turn enables a reorientation toward the earth. The novel illustrates this slope of the beach in its extended final scene, in which Freya nearly drowns in the waves close to the shore before learning how to coordinate her body sufficiently to the rhythm of the waves so that she can crawl back out of the water.[65] Yet even as the literal slope of the rebuilt beach is oriented toward the individual human body, the project of creating a beach commons is also "sloped" toward a long history of historical and even prehistorical human ancestors. This latter, more metaphorical slope of the beach is the surface along which questions of justice are engaged. As I have noted, Chakrabarty's critics worry that the concept of the Anthropocene unjustly blames "humans in general" for our current situation and thus encourages solutions to global warming that are unjust to those already suffering the most from the specific kinds of human relations—capitalist, racist, and sexist—which are responsible for these problems. Freya's project of beach building orients itself toward these questions of justice by creating a contemporary commons, rather than an enclosed beachfront property, that also forms a commons with past patterns of human existence. This is not a commons with all past forms of human existence; as the narrator notes, the rebuilt beach makes common cause only with "the joy of the relatively few humans who were lucky enough to live on the strand" (469).[66] It is, rather, a very specific commons that aims to mold specific ruins into specific happier forms. Yet the rebuilt beach enables joy *because* it is both a corporeal commons established at the interface of water and sand and a transhistorical commons that connects with the long history of human beach dwellers.

Conclusion

The project of rebuilding beaches on a massively damaged Earth may seem like an unhelpfully playful, even capricious, image with which to end a discussion of how to respond to the overwhelming threats of the

Anthropocene. As Bill McKibben has noted, for all intents and purposes we no longer live on that Earth that was inhabited by humans for the last two million years; instead, we now live on what McKibben calls, to capture its strangeness, "Eaarth."[67] Where for the last ten thousand years, Earth had a stable average temperature range warm enough that "the ice sheets retreated from the centers of our continents so we could grow grain, but cold enough that mountain glaciers provided drinking and irrigation water to those plains and valleys year-round," Eaarth is, by contrast, growing much hotter and is characterized by increasingly acidic oceans; a greater number of huge, unpredictable storms; more and more disease-bearing ticks, mosquitoes, and other pests; declining grain yields; and expansion of areas of the globe that are simply uninhabitable for humans. Lost beaches, it seems, will be the least of our worries as we determine whether large and complex human cities and societies are actually compatible with Eaarth.

Yet the problem with McKibben's description, however accurate it may be, is that—like Chakrabarty's approach—it orients us toward questions of survival and its associated logics of defense, limits on freedom, violence, and immunity. Robinson's description of beach building, by contrast, turns us toward Shelley's understanding of what it can mean for humans to coalesce joyfully with nature and encourages us to imagine and invent other projects that would enable the kind of serious play involved in reparative beach building, that is, the kind of play that produces joy *because* it combines an individual scene of instruction with transhistorical projects of justice and redemption. These projects keep the injustices of racism, sexism, and capitalist exploitation in the foreground, for it is only by acknowledging long and multiple histories of injustice that we can approach the ruins of history *as* elements—that is, starting points that cannot be dissolved away into the abstraction of "the human"—which can then be remolded into happier shapes.[68] The projects that McKibben champions, such as a turn toward local food and energy production, might indeed be other examples of this approach.[69] However, for them to be so, "the local" must then name a site of both individual corporeal instruction and historical redemption. This would mean, for example, keeping in the forefront, and seeking to redeem, the long history of injustices associated with each locality within which new practices of "the local" are to be developed. (To take an example from my own locale, Trees Durham is a nonprofit organization that seeks to readdress a long historical relationship between city-sponsored tree planting and racial redlining practices in Durham, North Carolina, by guiding future city-sponsored tree-planting efforts.[70]) Though it seems to me unlikely that all of the more contentious projects advocated by Stewart Brand, which range from increasing urbanization to more nuclear power to transgenic crops and even to geoengineering projects such as placing sulfates in the stratosphere, could participate in the coalescence of humans and nature to which Shelley pointed, even these extreme possibilities should not be rejected out of hand.[71] However, the burden should be on those who advocate these measures to show how, concretely, such projects engender joy by engaging the two slopes of individual instruction and historical redemption exemplified in Robinson's project of beach building.

5.   Liberalism and the Concept of the Collective Experiment

> The reality of the public realm relies on the simultaneous presence of innumerable perspectives and aspects in which the common world presents itself and for which no common measurement or denominator can ever be devised. . . . Being seen and being heard by others derive their significance from the fact that everybody sees and hears from a different position.
> —Hannah Arendt, *The Human Condition*

> The game of liberalism . . . means acting so that reality develops, goes its way, and follows its own course according to the laws, principles, and mechanisms of reality itself. . . . More precisely and particularly, freedom is nothing else but the correlative of the deployment of apparatuses of security. An apparatus of security . . . cannot operate well except on condition that it is given freedom, in the modern sense [the word] acquires in the eighteenth century: no longer the exemptions and privileges attached to a person, but the possibility of movement, change of place, and processes of circulation of both people and things.
> —Michel Foucault, *Security, Territory, Population*

In his classic text *On Liberty* (1859), John Stuart Mill suggested that liberalism was intimately connected with what he called "experiments in living."[1] Mill meant that individuals, and voluntary associations of individuals, ought to be allowed to live as each saw fit, provided that their experiments did not diminish the liberty of others to live as they saw fit. The state's purpose was to protect and facilitate such individual experiments in living, including in areas such as the education of children: "An education established and controlled by the State should only exist, if it exist at all, as one among many competing experiments, carried on for the purpose of example and stimulus, to keep the others [i.e., non–State forms of education] up to a certain standard of excellence" (302).

Mill's claim about the fundamental importance of experiments in living for both individuals and society as a whole was based on several premises. These included the principles that individuals differed from one another and that it was difficult (and generally impossible) to prove that one way of life was best for everyone. Experiments in living had a quasi-scientific social function for Mill, for they allowed members of a particular society, as well as the long-term developmental process he called "civilization," to explore advantages and disadvantages of many forms of life. Both government and social opinion, by contrast, tended to produce uniformity. As a consequence, the best role for the state, beyond enforcing laws that enabled experiments in living, was to serve as what Mill called a "central depository" for the many experiments in living that he hoped contemporary society would engender:

> Government operations tend to be everywhere alike. With individuals and voluntary associations, on the contrary, there are varied experiments, and endless diversity of experience. What the State can usefully do, is to make itself a central depository, and active circulator and diffuser, of the experience resulting from many trials. Its business is to enable each experimentalist to benefit by the experiments of others; instead of tolerating no experiments but its own. (306)

The laissez-faire, or "letting be," of Mill's version of liberalism was thus less a matter of facilitating the operations of the free market than of enabling experiments in living, for these latter—at least when their results were collected by a central state repository—would make possible scientifically oriented progress in collective forms of living.

Though Mill's depiction of the relationship between liberty and experiments in living is a classic statement of Victorian liberalism, there are striking resonances between his account and our contemporary understandings of what I will call "collective experiments." Mill's advocacy of social diversity and individual freedom in choosing how one wishes to live, for example, has been echoed since the 1960s in widespread efforts to encourage respect and legal protection for a wide variety of class, gender, racial, religious, and medical forms of diversity. The quasi-scientific rationale of Mill's defense of experiments in living—his claim that many such experiments enabled "society" to observe and choose the best among these—is also echoed in the efforts of neoliberal thinkers and policy makers to create population-wide health experiments. In the 1970s, for example, Louis Lasagna, head of the Center for the Study of Drug Development (CSDD), a neoliberal think tank focused on the pharmaceutical industry, claimed that clinical trials were too "artificial" to ascertain the efficacy of pharmaceutical drugs. He argued for what he described as a more "naturalistic" method of testing experimental drugs, namely, allowing pharmaceutical companies to sell minimally tested potential drugs to anyone who wished to take them and carefully monitoring the results of this collective experiment.[2] Though Lasagna's proposal did not, in the short term, persuade the Food and Drug Administration (FDA), something like his vision of medicine has nevertheless become standard practice, partly from a relaxation of standards around experimental drugs, partly through the triumph of the consumer model of the patient-doctor relationship, and partly as a consequence of the development of data- and tissue-gathering protocols that now track large swathes of national populations.[3] We cannot know whether Mill would have condoned these health-oriented experiments in living, but it seems fair to say that these latter are implied by the quasi-scientific logic of experimentation outlined in *On Liberty*.

The links among health, population, and experiments in living so important in our own moment also provide a lens that allows us to understand better earlier interest in collective experiments. In the early eighteenth century, for example, the Scottish physician John Arbuthnot contended that opponents of the newly introduced method of smallpox inoculation sought to prevent a widespread, and necessary, collective experiment from occurring. He noted that though one published criticism "pretends to be an Admonition to Physicians not to meddle in this Practice of *Inoculation*, 'till they are better ascertain'd, by experience, of the Success of it," that publication was actually a "most warm Dissuasive, not only to Physicians, but to all Sorts of people, not to practice [inoculation] at all" and thus amounted to an effort "to deprive [people] of all Possibility of coming by Experience."[4] Drawing on what would become a central principle of subsequent liberalisms, Arbuthnot contended that each individual is the best judge of his or her interests, and therefore each individual should decide whether or not to be inoculated against smallpox (3). Allowing individuals to choose would enable precisely that collective "experience" that could determine the efficacy of inoculation as a medical practice. Arbuthnot thus praised physicians who, "from their Disinterestedness and Innate Love to Mankind, are willing, that an Experiment should go on, which, in Proportion to the Extensiveness of the Practice, must necessarily diminish the Mortality of the *Smallpox* in general" (39–40). Arbuthnot's early-eighteenth-century connection between individual choice and a larger social health experiment suggests that Mill, in valorizing experiments in living, was himself further developing an implicit—though, I will argue, essential—connection between liberalism and the concept of collective experimentation.

My goal here is to document the centrality of the concept of collective experimentation to the history of liberalism. I argue that liberalism has been, since its eighteenth-century origins, intrinsically bound to the concept of collective experimentation. This concept has served as a conceptual matrix for bringing together five elements—populations, institutions for data gathering, individual rights, progress, and what I will call "test subjects"—that have been central to subsequent forms of liberalism, and different versions of liberalism can be distinguished by the ways that they combine these elements. Understanding collective experimentation as central to the theory and practice of liberalism illuminates the close link between liberalism and biopolitics, for the purpose of collective experimentation is, in all versions of liberalism, to maximize human capacities in ways only possible at the scale of populations. However, to judge by the examples I discuss here, this approach is either equivalent to, or easily slips into, an immunitary logic that necessarily exempts test subjects from both the freedoms promoted by liberalism and the benefits that purportedly result from collective experimentation.

I develop five case studies, each focused on a specific author: John Arbuthnot, Edmund Burke, John Stuart Mill, Friedrich Hayek, and Ulrich Beck. I begin with Arbuthnot's early-eighteenth-century case for continuing the collective experiment of smallpox inoculation, noting that though his advocacy was not itself a brief for liberalism, he established the matrix of elements—populations, institutions for data gathering, individual rights, progress, and test subjects—that subsequent liberalisms would reconfigure in various ways. The second section focuses on Edmund Burke's late-eighteenth-century reconfiguration of these elements. Burke argued against conscious experimentation in politics but also defended a mode of collective experimentation that purportedly occurred over longer time-scales and resulted in durable traditions and institutions. Burke thus focused attention on what we might call the institutional preconditions for conscious collective experimentation. I then turn to Mill, arguing that he took seriously Burke's stress on institutional preconditions for collective experimentation but sought to locate these in a new legal order, rather than in the traditional institutions of church and state favored by Burke. The fourth section focuses on Friedrich Hayek's reconfiguration of liberalism in the 1940s and 1950s, noting that his *neo*liberalism, as it came to be called, combines Burke's and Mill's approaches through the argument that optimized collective experimentation requires subordinating all institutions to the requirements of one traditional institution, namely, the market. (Lasagna's pharmaceutical experiments in living take these neoliberal premises as their starting point.) In the fifth section, I take up a final reconfiguration of the matrix of collective experimentation, namely, Ulrich Beck's concept of the "risk society." I end with Beck's account because he exposes the importance of test subjects to contemporary modes of collective experimentation in a way that potentially begins to move the latter concept away from liberalism.

Smallpox, Collective Experiments, and the Wealth of Nations

Something of a polymath, John Arbuthnot published important mathematical texts on probability, was appointed physician to Queen Anne, was a member of the Royal Society, and was a Scriblerian, along with other eighteenth-century British literary luminaries such as Jonathan Swift and Alexander Pope. In 1719, he became a vocal advocate of the newly introduced method of smallpox inoculation.[5] Arbuthnot's admonition to his readers to understand smallpox inoculation as "an Experiment [which] should go on" appeared in a short text entitled *Mr. Maitland's Account of Inoculating the Smallpox Vindicated, from Dr. Wagstaffe's Misrepresentations of That Practice, with Some Remarks on Mr. Massey's Sermon* (2). Arbuthnot was not alone among early-eighteenth-century authors in referring to smallpox inoculation as a valuable "experiment" upon the social body. His contemporaries Cotton Mather and Thomas Nettleton, for example, also described smallpox inoculation in this way.[6] However, Arbuthnot's text illuminates the forging of a matrix that linked the concepts of the collective experiment, progress, the health and power of the state, and individual rights. Though Arbuthnot's own interpretation of this matrix was only ambiguously liberal, his account helps us understand why subsequent liberals, from Mill to Hayek to Lasagna, have continued to return to, and reinterpret, this same constellation of concepts.

As suggested by its title, Arbuthnot's text was a point-by-point refutation of an earlier public letter by a Dr. Wagstaffe and a published sermon by a Mr. Massey, who each had levied various arguments against the utility and morality of inoculation. Arbuthnot contended that Dr. Wagstaffe's arguments against the new practice were also arguments against the introduction of *any* new practice into medicine and thus opposed the very possibility of progress in medicine. Arbuthnot claimed that medical practice has been, since its origins, bound up with collective experiments, and noted that it "is . . . strange to forbid the Practice, 'till that is determin'd, which can only be found out by Practice. According to this Principle, it had been impossible ever to have found out any Thing in Medicine" (11).[7] Arbuthnot noted that though it might be tempting to set, as a minimal criterion, agreement among physicians concerning the efficacy of a new practice, such unanimity was never found among physicians (15). For

Arbuthnot, the possibility of progress in medicine depended upon collective experiments, the results of which were initially and necessarily uncertain. The practice of inoculation was simply the most recent of such experiments.

To describe the progress of medicine as the telos of collective experiments was, however, to raise questions about the telos of the progress of medicine. Who, or what, benefits from medical progress? Arbuthnot was less clear on this point. He suggested that the first, and most obvious, beneficiary of smallpox inoculation was the specific individual who was inoculated. Given that smallpox inoculation only worked if it occurred before an individual contracted smallpox naturally, decisions about whether to inoculate would fall to parents, who made this decision on the basis of parental love. The second beneficiary of smallpox inoculation was "Mankind." This collective beneficiary was also connected to the collective experiment through a relationship of love, namely, the "innate Love of Mankind" of physicians who wish to "diminish the Mortality of the *Small Pox* in general" (39, 40).

There was also a third beneficiary of the collective experiment, situated between the individual and "Mankind": the state. After noting his belief that inoculation in Britain contributed to "the *Publick Good*" (1), Arbuthnot subsequently clarified the connection between inoculation and public good along the political-arithmetical lines earlier developed by authors such as John Graunt and William Petty.[8] Arbuthnot proposed that if, as suggested by the Bills of Mortality, smallpox kills one in ten people who have not been inoculated (19) but kills only one in one hundred people who have been inoculated (21), then if smallpox inoculation were "obtain'd universally [it] would save to the City of *London* at least 1500 people yearly" (21).[9] Inoculation contributed to the public good by increasing what Arbuthnot described as "yearly recruits" for London, with the London population functioning as synecdoche of the entire British population (19).

As I have noted in previous chapters, this political-arithmetical link between population and national public good was not unusual in the late seventeenth and early eighteenth century, and it could as easily promote a mercantilist as a liberal position. Yet Arbuthnot made explicit reference to the proper limits posed to state power by individual "rights" in a way decidedly absent from the work of political arithmeticians such as William Petty or John Graunt. Arbuthnot acknowledged that if the strength of the state was the only telos of the collective experiment of smallpox inoculation, it might seem more prudent to legislate compulsory inoculation. Arbuthnot asserted, though, that the telos of national strength was limited by the "natural Rights of Mankind." Contrasting the "prudence" of the statesman with the "rights" of individuals, Arbuthnot noted that

> if Prudence only were to be consulted, it would perhaps be much more the Duty of the Legislature to order, than to forbid this Practice . . . [for] they would, by this Method, diminish the Mortality, and encrease the Number of their People; and the Magistrate is forc'd often upon more arbitrary Proceedings in any Pestilence: But as that would seem too great an Encroachment upon the natural Rights of Mankind, I should not approve of it. But on the other Hand, it would be a most Tyrannical Encroachment upon the same Rights, to debar Mankind from the lawful Means of securing themselves from the Fear and Danger of so terrible a Plague. (35)

While Arbuthnot accepted that legislators could compel medical measures during states of emergency caused by "Pestilence," he argued that natural rights prevented such legislative demands from being introduced during the normal course of affairs. Moreover, these natural rights of individuals required that individuals be allowed to choose to participate in collective experiments such as smallpox inoculation.[10]

The collective experiment of smallpox inoculation was thus purportedly an instance in which the progress of medicine, the needs of state, and the natural rights of individuals virtuously converged. Yet if smallpox inoculation was an experiment to determine whether this procedure would strengthen the state, how and where could the results of this experiment be gathered and judged?

Arbuthnot provided relatively little detail regarding this point. Implicit in the form of Arbuthnot's book was the premise that the republic of letters established by texts such as his, as well as by natural-philosophical journals such as the *Philosophical Transactions of the Royal Society*, would serve as one means of gathering information about, and assessing the outcome of, this experiment in Britain.[11] Arbuthnot further suggested that lack of any proof that this experiment was *not* working in other countries also provided a basis for belief in its efficacy. For example, with respect to the question, posed by Wagstaffe, of whether those inoculated against smallpox could later contract the disease a second time, Arbuthnot contended that

> this Practice of *Inoculation* has been continu'd for many Years in several Countries; if the *Inoculated* had been subject to catch the *Small Pox* a second Time, something of this Kind must have happen'd; and a very few Instances of this, must have put an End to the whole Practice: For can any one imagine, that

People in their Senses would have continu'd a troublesome Experiment,
which was not effectual for the Purpose for which it was design'd? (13)

Arbuthnot implied that, in the absence of widespread knowledge that a technique such as smallpox inoculation does *not* work, the experiment should be allowed to continue. At the same time, it is not easy to see how Arbuthnot's principle of relying on the experience of other countries should function in the case of entirely novel techniques.

As smallpox inoculation became more common in Britain during the eighteenth century, commentators continued to stress the importance of continuing this experiment, while also aligning it ever more closely with a liberal—that is, a commerce- and trade-oriented—vision of public good. Mid-eighteenth-century advocates of inoculation sought to extend this practice from the aristocratic and middling classes to the laboring class and the poor by means of hospitals, and they often defended their proposals by pointing to the effects of smallpox on commerce, rather than simply national population size. In a 1763 sermon dedicated to smallpox hospitals for the poor, John Green drew on Arbuthnot's principle by noting that "the certain experience of near half a century has manifested [inoculation's] success in various nations."[12] But where Arbuthnot presented the dangers of smallpox in terms of its (literal) decimation of the national population, Green underscored the effects of smallpox on commerce. Green contended that, since "men cannot flee from place to place, to avoid the danger of infection," epidemics ensure that "multitudes will soon be reduced to poverty, manufactures will be stopt, [and] commerce will stand still" (17). Green's stress on trade suggested that the collective experiment of inoculation would not be concluded until the national populations of all countries engaged in international trade had been inoculated.[13]

While Arbuthnot's demand that the collective experiment of smallpox inoculation continue was not itself necessarily liberal, it brought together and illuminates a series of considerations that reappear in the subsequent discussions of collective experimentation I consider in this chapter. First, Arbuthnot underscored the link between collective experimentation and large collectives (that is, populations). For Arbuthnot, the success or failure of this collective experiment could only be determined at the level of experience of thousands of individuals or even whole countries. Second, Arbuthnot's account highlights the difficulty of collecting the results of the collective experiment, since the regularities that it can reveal are distributed across large numbers of people. Third, and perhaps most important for the next section of this chapter, Arbuthnot's text illuminates (no doubt despite Arbuthnot's intentions) the ambivalent relationship of collective experimentation to individual freedom, choice, and rights. Arbuthnot implied that individual rights were a key facilitator of the collective experiment of smallpox inoculation, suggesting that this collective experiment could take place only if each individual was free to choose whether to be inoculated. Yet Arbuthnot also noted that parents must decide for their children, which complicates the question of individual freedom. Arbuthnot also chose not to comment on the fact that some of the earliest successful British inoculations to which he referred in his text—those on prisoners at Newgate prison—were "freely chosen" only in the very strained sense that prisoners were offered release from jail should they agree to be the initial test subjects (and, of course, survived the effects of this new technique).[14] As Genevieve Miller notes, these quasi-forced smallpox inoculations were vital for establishing for the British royalty that inoculation worked, and the example of the British royalty choosing inoculation was itself key to its subsequent more widespread adoption in Britain.[15] It would thus seem that the period during which individuals chose freely was in fact a *second* stage that followed a period of heavily constrained choice for a smaller number of test subjects. Or, to put this another way, the happy convergence of individual rights, choice, and collective experimentation to which Arbuthnot pointed was dependent, in a way he did not acknowledge, upon an enabling set of conditions that involved collective experimentation *without* individual rights and free choice.

This latter fact opens up a whole new set of questions that were taken up by subsequent theorists of liberalism. First and foremost was the question of the preconditions of that form of collective experimentation that was linked to individual choice and rights. Arbuthnot's stress on rights and individual choice in the case of smallpox inoculation also raised the question of whether there might be collective experiments that could *only* be pursued if a very large percentage of a population were involved and, if so, whether this might require involving individuals in collective experimentation regardless of their explicit desire to participate. Finally, might there be collective experiments that involve variables that are more or less unrelated to individual choice?

Burke and Experimental Politics

Between Arbuthnot's early-eighteenth-century text and the mid-century sermons extolling smallpox hospitals and inoculation, smallpox inoculation had become a thread for twining together the collective experiment with what eventually become known as liberalism. The concept of rights was especially important in this twining operation, in part because it obscured the relationship between the collective experiment of rights-bearing citizens who made informed choices and the preceding stage of experimentation on those who did not possess those same rights. This latter group included both the original test subjects (Newgate prisoners and slaves in the colonies) but also the normal recipients of smallpox inoculation, namely, children. Disregarding this actual complexity of the relationships among collective experimentation, rights, and choice, mid-century advocates of smallpox inoculation instead contended that smallpox inoculation was a collective experiment that respected individual rights *and* benefited the state by eliminating potential hindrances to trade. Yet this emphasis on the joys of unimpeded trade relied upon another form of obscurity. Arbuthnot and mid-century commentators suggested that parental love was the key sentiment that encouraged individuals to participate in the collective experiment of smallpox inoculation. However, these motivating affects were then conflated with love for the state and the allied worry about the interruption of trade, in the sense that commentators suggested that the success of the collective experiment was *not* to be judged at the level of the individual (for example, the results of inoculation for specific parents and their children) but only at the level of the effect on the state and trade.

The French Revolution provided Edmund Burke with the opportunity both to recognize these conflations and confusions that were essential to the version of the collective experiment that emerged through the link of smallpox inoculation and a liberal emphasis on trade and to reconfigure the relationship of the collective experiment to liberalism. Burke's reflections on collective experimentation also help us understand better the "mode" of his political thought. Burke's political theory is notoriously difficult to categorize, for he has been described convincingly both as a key liberal theorist and as the central advocate for a conservative philosophy that is understood as distinct from liberalism.[16] Burke's engagement with the topic of collective experimentation clarifies this apparent conflict between his liberalism and conservatism, revealing Burke as a theorist of a mode of liberalism that stressed the importance of enduring institutions for the very possibility of collective experimentation.

Burke developed the connection of collective experimentation to the "conservative" aspect of his philosophy in *Reflections on the Revolution in France* (1790). Concerned about, as Burke suggested in his book's subtitle, "the proceedings in certain societies in London" that explicitly supported the revolution in France, Burke criticized the French revolutionaries by contrasting two different concepts of political experimentation. Burke held that the health and wealth of the state depended on supporting time-honored traditions, and hence he opposed *consciously directed* political "experiments":

> We wished at the period of the Revolution, and do now wish, to derive all we possess as *an inheritance from our forefathers*. Upon that body and stock of inheritance we have taken care not to inoculate any cyon [i.e., scion; graft] alien to the nature of the original plant. All the reformations we have hitherto made, have proceeded upon the principle of reverence to antiquity; and I hope, nay I am persuaded, that all those which possibly may be made hereafter, will be carefully formed upon analogical precedent, authority, and example.[17]

Burke's stress on the importance of reverence for precedent, authority, and earlier example (that is, "the inheritance from our forefathers") led him to castigate the leaders of the French Revolution as political experimentalists who desired that "the whole fabric [of the French government] should be at once pulled down, and the area cleared for the erection of a theoretic, experimental edifice in its place" (188). Drawing on the language of parental affection that authors such as Arbuthnot had linked to the collective experiment, Burke suggested that the French political experimentalists had "nothing of the tender parental solicitude which fears to cut up the infant for the sake of an experiment" (245). However, Burke's animus against explicit innovations in politics also extended to the concept of the "rights of man," which he contended were simply weapons used by political experimentalists to disregard the wisdom embodied in traditions and institutions (95, 128).[18]

At the same time, though, Burke's defense of tradition pointed to a virtuous mode of *unconscious* political experimentalism. Though Burke opposed the "erection of a theoretic, experimental edifice" in political affairs, he nevertheless asserted that the "science of constructing a commonwealth, or renovating it, or reforming it" was an "experimental science" (90). However, this experimental science of governing is,

> like every other experimental science, not to be taught *a priori*. Nor is it a short experience that can instruct us in that practical science; because the real effects of moral causes are not always immediate; but that which in the first instance is prejudicial may be excellent in its remoter operation; and its

excellence may arise even from the ill effects it produces in the beginning.
. . . The science of government being therefore so practical in itself, and
intended for such practical purposes, a matter which requires experience,
and even more experience than any person can gain in his whole life, however sagacious and observing he may be, it is with infinite caution that any
man ought to venture upon pulling down an edifice which has answered in
any tolerable degree for ages the common purposes of society, or on building
it up again, without having models and patterns of approved utility before his
eyes. (90)

For Burke, traditions and long-standing institutions were the *results* of
successful experiments, and it was through these that the experience of the
past was stored and made available for the present.

Burke acknowledged that traditional institutions changed over time but
stressed that they did so at a pace generally imperceptible to members of
any given generation. While this was frustrating for those who wished for
faster change, Burke argued that only this mode of tradition-reverencing
experimentalism could facilitate progress, for only imperceptibly slow
change linked generations with one another and hence enabled the *extension* of collective experience, rather than forcing each generation of legislators to start anew. The French mode of political experimentalism, by
contrast, destroyed that continuity upon which the possibility of progress
depended:

By this unprincipled facility of changing the state as often, and as much,
and in as many ways as there are floating fancies or fashions, the whole
chain and continuity of the commonwealth would be broken. No one generation could link with the other. Men would become little better than the
flies of a summer. (141)

Burke's image of humans reduced to seasonal flies is powerful not only
because it demotes humans to a lower order of animality (insects) but also
because he uses this animal image to underscore that at stake is the survival
of a properly human relationship to political experimentation. Burke's
image suggests that the human species would, like seasonal flies, no doubt
survive even if humans perpetually destroyed and recreated their political
and social institutions. Yet he suggests that such mere survival would
destroy the possibility of *experimentation*, since no knowledge would be
transmitted from one "season" to the next.

Burke's claim that social experimentalism required the stability that
could only be provided by traditional institutions, such as the church and

a hierarchical social order, was not a rejection of liberalism but rather set limits on individual rights and freedom and reconceived the temporality of the liberal principle of progress. Arbuthnot had focused on a specific collective experiment, smallpox inoculation, and had suggested that this experiment not only proceeded without coming into conflict with individual rights but in fact depended upon individual choice. Burke, by contrast, argued that many collective experiments—namely, those undirected and unconscious collective experiments that resulted in traditional institutions—were not of this kind. This latter mode of collective experimentation was *not* dependent upon individual rights and choice and was in fact threatened by overly formalized conceptions of "the rights of man." Burke did not oppose the kind of collective experimentation and medical progress described by Arbuthnot but sought to establish the larger social-experimental frame within which experiments such as smallpox inoculation operated. Burke argued that without the social stability provided by the unconscious collective experimental mode that produced traditional institutions, social life would be so chaotic that more specific, consciously constructed experiments of the sort described by Arbuthnot would be impossible. Burke thus ended up with a version of liberalism in which collective experimentation was vital but within which individual liberties were not absolute. Rather, individual liberties met a necessary limit whenever they threatened the social stability that was the precondition of the progress that collective experimentation enabled. From this perspective, Burke employed the concept of the collective experiment to develop a mode of liberalism, or "*free government*," that "temper[s]" the "opposite elements of liberty and restraint in one consistent work" (353).[19]

Mill, Experiments in Living, and the Progress of Knowledge

Burke, then, distinguished between two modes of collective political experimentation: a more primary and unconscious mode that resulted in tradition and long-lived institutions and a conscious mode of political experimentation that tended, because it did not respect the constraints of that former mode of unconscious experimentation, to destroy the very possibility of social progress and civilization. Burke's distinction was a powerful way of highlighting what I have described as the institutional preconditions for the kind of collective experimentation and medical progress described by earlier authors such as Arbuthnot. Burke stressed that "experimentation" could be distinguished from mere change only if the results of experimentation could be gathered and stored in an institutional order. Though institutions such as the Royal Society and correspondence among physicians could facilitate the collation and assessment of the results of a collective experiment like smallpox inoculation, Burke implied that these institutions depended upon the more primary institutions of church and state that had emerged from the unconscious mode of collective experimentation. While subsequent nineteenth-century liberal commentators disagreed with Burke about the implications of this position for specific aspects of church and state, they did not, in general, dispute the logic of Burke's distinction.[20]

John Stuart Mill was an exception to this rule and was so precisely because he took seriously Burke's stress on the importance of institutions that enabled social stability and collective memory. Where Burke concluded that only traditional institutions could serve those functions, Mill located a new source of social order and collective memory by reconceptualizing the relationship of collective experimentation to individual rights, knowledge, and progress.

In *On Liberty*, Mill presented institutions with the power to punish and discipline individuals—which included the state but also social opinion—as impediments to the progress of knowledge.[21] Because Mill believed that each individual was a potential site of an experiment in living, and because he believed that such experiments, if properly documented, would allow other individuals to locate their own paths to happiness, institutions that produced conformity hindered the discovery of truth. Mill believed that while the institutions of mid-nineteenth-century British society had the capacity, unprecedented in human history, to enable widespread experiments in living, in actual fact those institutions threatened "a despotism of society over the individual, surpassing anything contemplated in the political ideal of the most rigid disciplinarian among the ancient philosophers" (227).[22] His text thus aimed to illuminate a principle that, if followed consistently, would liberate the experimental potential of contemporary British institutions by limiting their ability to enforce conformity.[23]

Mill's "simple principle" was the following:

That principle is, that the sole end for which mankind are warranted, individually or collectively, in interfering with the liberty of action of any of their number, is self-protection. That the only purpose for which power can be rightfully exercised over any member of a civilized community, against his will, is to prevent harm to others. His own good, either physical or moral, is not a sufficient warrant. He cannot rightfully be compelled to do or forbear because it will be better for him to do so, because it will make him happier, because, in the opinions of others, to do so would be wise, or even right. (223–24)

Such a principle would enable individual experiments in living, since each individual would be free from judicial restraint and bullying by others and so could freely determine how he or she wished to live. Facilitating these individual experiments in living would then lead to greater general happiness, for each individual could design his or her own experiment in living and, hence, develop his or her own "character," by taking into account the virtues and problems of earlier experiments in living.[24] Mill wrote:

> As it is useful that while mankind are imperfect there should be different opinions, so is it that there should be different experiments of living; that free scope should be given to varieties of character, short of injury to others; and that the worth of different modes of life should be proved practically, when any one thinks fit to try them.[25]

While Mill, like Burke, opposed the idea that individuals have "abstract right[s]" in any meaningful sense of the term, he asserted that the "permanent interests of man as a progressive being" meant that each individual ought to be treated as having a right "to act, in things indifferent, as seems good to his own judgment and inclinations" (224; 271n).[26] Only in this way could society as a whole become experimental.

Yet what ultimate purpose was served by society becoming more experimental? Mill suggested that an experimental society enabled social progress and that progress was the vocation of human beings (that is, humans are a "progressive being" [224]). Though Mill did not define the axis or axes along which progress was to be measured, he suggested two separate criteria: increase in knowledge and in individuality.[27] Mill implied that the human species progresses to the extent that its knowledge of the natural and social worlds expands. He contended, for example, that progress "ought to superadd . . . one partial and incomplete truth" to other partial and incomplete truths (252).[28] Mill also suggested that progress in knowledge was a means to the end of greater individuality (that is, a greater capacity of each individual to determine the character and trajectory of his or her own life). From this perspective, he worried that his British contemporaries were embracing the progress of knowledge yet rejecting an increase in individuality:

> We continually make new inventions in mechanical things, and keep them until they are again superseded by better; we are eager for improvement in politics, in education, even in morals, though in this last our idea of

improvement chiefly consists in persuading or forcing other people to be as good as ourselves. It is not progress that we object to; on the contrary, we flatter ourselves that we are the most progressive people who ever lived. It is individuality that we war against: we should think we had done wonders if we had made ourselves all alike; forgetting that the unlikeness of one person to another is generally the first thing which draws the attention of either to the imperfection of his own type, and the superiority of another, or the possibility, by combining the advantages of both, of producing something better than either. (273)

For Mill, the point of rendering society more fully experimental was to ensure that the progress of knowledge remained tightly linked to an increase in individuality.

Mill's reconfiguration of the concept of collective experimentation brought the question of human diversity to the fore in a way not evident in either Arbuthnot's or Burke's accounts. While Arbuthnot's statistical approach to smallpox and inoculation implied biological differences that accounted for differing human responses to smallpox and inoculation, this was not a point upon which Arbuthnot dwelled. Moreover, he suggested that, in the register of conscious decision making, every rational person would judge and act in the same way. Burke, for his part, connected collective experimentation to the hierarchical differences of power and privilege that were part of the British tradition but acknowledged that these did not correspond to innate differences among actual individuals. For Mill, by contrast, individual diversity was central to his understanding of collective experimentation, for he contended that it is *because* individuals differ from one another that, absent external compulsion to conform, they will arrive at different opinions and choose different modes of life and hence make an experimental society possible. Differences among individuals were, for Mill, the motor of collective experimentation and progress.

Mill accounted for the relationship of individual differences to social progress in two different—and arguably incompatible—ways. On the one hand, he aligned his account with a German Romantic understanding of the importance of human diversity. Mill signaled the Romantic origins of his understanding of individual diversity in his epigraph, drawn from Wilhelm von Humboldt's posthumously published text *The Limits of State Action*, in which Humboldt contended that "the grand, leading principle, towards which every argument unfolded in these pages directly converges, is the absolute and essential importance of human development in its richest diversity."[29] Mill cited Humboldt at more length in the section of *On*

Liberty entitled "Of Individuality," drawing out there the importance for Mill of Humboldt's approach to diversity:

> Few persons, out of Germany, even comprehend the meaning of the doctrine which Wilhelm Von Humboldt, so eminent both as a *savant* and as a politician, made the text of a treatise—that "the end of man, or that which is prescribed by the eternal or immutable dictates of reason, and not suggested by vague and transient desires, is the highest and most harmonious development of his powers to a complete and consistent whole;" that, therefore, the object "towards which every human being must ceaselessly direct his efforts, and on which especially those who design to influence their fellow-men must ever keep their eyes, is the individuality of power and development;" that for this there are two requisites, "freedom, and a variety of situations;" and that from the union of these arise "individual vigour and manifold diversity," which combine themselves in "originality."[30]

For Humboldt, the full expression of unique individuality and diversity necessitated restrictions on state power, and Mill drew on Humboldt's claims in order to support his own attack on the state's tendency to produce conformity.

While this connection between human diversity and limits on state power was specific to Humboldt, similar positive evaluations of human diversity were shared by many German proto-Romantic and Romantic authors. Humboldt's stress on the unique nature of each individual echoed Johann Gottfried Herder's earlier assertion that

> it is, I think, the most flat-footed opinion that ever entered a superficial head that all human souls are alike, that they all come into the world as flat, empty tablets. No two grains of sand are like each other, let alone such rich germs and abysses of forces as two human souls—or I have no grasp at all of the term "human soul."[31]

There were multiple sources for late-eighteenth-century German interest in human diversity, including Gottfried Leibniz's theory of individuals as unique monads each striving toward their own perfection, or *entelechy*. Leibniz's philosophy remained a key reference point in German philosophy throughout the eighteenth century, and it was important for Humboldt personally as he was composing *The Limits of State Action.*[32]

However, Pietism also promoted interest in individual uniqueness in Germany, France, and England, and the explicitly religious orientation of this movement helps us understand better both the quasi-religious resonance of claims about individual difference and uniqueness in authors such as Humboldt and Mill and their accounts of the relation of individual difference to communal bonds. For Pietists, individual uniqueness was not an end in itself but instead the individual's starting point for strengthening his or her bonds to others in the form of conversation and exchange. As their name suggested, late-seventeenth- and eighteenth-century Pietists stressed the inner experience of religion, in contrast to what they saw as the orthodox Lutheran and Calvinist overemphasis on doctrine. Inner experience took the form of thoughts, feelings, emotions, and even sexual desire. The Pietist's task was to link these inner experiences into a progressive narrative, that is, to enable self-development, or *Bildung*, on the basis of those experiences. Yet for Pietists, *Bildung* could not be pursued alone, but only in conversation with the different, unique members of his or her congregation.[33] Pietists emphasized conversation because they believed that God revealed himself only partially through each individual (and, in a more general sense, revealed himself only partially in each of the different world religions). Exploration of the diversity among individuals in a congregation through conversation was a means by which an individual could facilitate his or her *Bildung*. Equally important, it was the means by which the community could more fully reveal and manifest God's presence on earth.

The link between knowledge and freedom that both Humboldt and Mill connected to the full expression of unique individuality derives much of its affective force from this basic Pietist religious schema. Earlier liberal political economists such as Hume and Smith had suggested that a remote, providential "invisible hand" would connect the dispersed activities of separate individuals if each sought only to better his or her own condition. Humboldt and Mill, by contrast, stressed both that individuals must *perfect* themselves, rather than simply pursuing self-interest, and that individuals can do so only by engaging one another actively through self-chosen communal ties. The point of limiting the state, for both Humboldt and Mill, was to enable those communal ties, which would in turn enable something far more chiliastic than simply the utilitarian increase in collective happiness.

Yet even as Mill's text suggested, through its references to Humboldt, a quasi-religious reverence for each individual, Mill also drew on the schema of "genius" to suggest a less inclusive way that individual experiments in living contributed to human progress. Though Mill wished every individual to conduct his or her own experiments in living freely by consulting the results of past experiments of others, he also noted that

there are but few persons, in comparison with the whole of mankind, whose experiments, if adopted by others, would be likely to be any improvement on established practice. But these few are the salt of the earth; without them, human life would become a stagnant pool. Not only is it they who introduce good things which did not before exist; it is they who keep the life in those which already existed.[34]

This suggested that each individual should express him- or herself not precisely for his or her own sake but in order to create a milieu within which a small number of geniuses could arise.[35] It is only the latter, Mill suggested, who actually enable collective progress. Yet since one cannot predict when and where such geniuses will arise, each individual must express his or her individuality, so that the few geniuses will not be lost. Or, as Mill put it, while it is true that "persons of genius . . . are, and are always likely to be, a small minority . . . in order to have them, it is necessary to preserve the soil in which they grow. Genius can only breathe freely in an *atmosphere* of freedom" (267).

Mill's account of liberalism thus drew on two different rationales for explaining why unique individuality must be protected and encouraged to express itself. On the one hand, each individual was presented as, if not quite an end in itself, at least a vital part of a collective process of progress. On the other hand, Mill's distinction between a small number of flower-like geniuses and the rest of us, who function as nutrient-rich soil for those flowers, suggested that most unique individuals contribute relatively little to collective progress. (Or, more specifically, contributed nothing uniquely individual to collective progress.) These two rationales pointed toward two different, arguably even opposed, versions of liberalism. The first rationale implied a liberalism in which collective progress required that social institutions be subordinated to the end of encouraging each individual to pursue his or her own experiments in living. The second rationale pointed, by contrast, to a liberalism in which collective progress required that social institutions be subordinated to the end of producing a small number of geniuses. While Mill seems to have seen no conflict between these two visions of liberalism, subsequent neoliberal commentators recognized both the difference and conflict between these visions of liberalism and opted emphatically for the latter.

Hayek and the Neoliberal Reconfiguration of Markets as Information Processors

Where Mill's concept of a liberal society committed to experiments in living sought to balance a Romantic stress on the importance of each individual with a more competition-oriented understanding of society as the means for creating an "atmosphere" within which geniuses could emerge, the more recent neoliberal schema of population-wide medical experiments that I noted at the start of this chapter privileges the latter, competitive schema. For a neoliberal like Lasagna, selling experimental drugs to anyone who felt that he or she might benefit from them would purportedly reveal, by means of a "naturalistic" method of drug discovery, which drugs actually work for a given disease. Yet such a method would also necessarily produce "losers" who opted to take would-be cures that did not work. At first blush, this grim neoliberal vision of collective benefit produced through population-level experiments seems different in spirit from Mill's rosier image of a liberal society progressing collectively through experiments in living. Yet the grimness of the neoliberal vision is the result of neoliberals considering in more depth than Mill the informatic question of how, precisely, results of individual experiments could be collected and compared. It is also the result of their willingness to return to the biopolitical dimension of collective experiments evident in Arbuthnot's advocacy for small-pox inoculation. Taking seriously these two dimensions of experiments has led neoliberals to hybridize the Burkean and Millian approaches to collective experimentation and liberalism, with "the market" and law now understood as the only time-proven institutions of tradition that can enable collective progress.

Mill suggested that the state should serve as a "grand repository" of different individual experiments in living, with individuals presumably able to consult this repository as each planned his or her own experiment in living. However, neoliberals have been convinced, since the origins of this movement in the 1930s, that the devil lies in the details of how, precisely, information from a diversity of individual perspectives is gathered and coordinated. They concluded, *pace* Mill, that the state is the *least* likely facilitator of such an endeavor and argued instead that "the market"—that is, capitalist economic relations—is the most efficient possible coordinator of the vast diversity of experiments in living that modern societies enable.

Friedrich Hayek's contributions to the so-called planning debate of the 1920s and 1930s were central to the neoliberal effort to rethink Mill's interest in individual diversity and experiments in living from an informatic perspective. By the early twentieth century, many European countries, as well as the United States, had turned to centralized government planning in order to address social concerns such as healthcare, workers' compensation, and public works programs in cases of high unemployment, and this approach intensified during both the First and Second World Wars.

However, economists associated with the Austrian School, including Ludwig von Mises and Friedrich Hayek, mounted the argument that government planning was intrinsically flawed, at least when it impinged on economic matters.[36]

Hayek made this argument in two ways. On the one hand, he argued that central planning required a unanimity about ethical values that was not in fact possible. Hayek noted that, though advocates of centralized planning often referred to their ultimate goal in terms of an abstract "'common good,' 'general welfare,' or the 'general interest,'" any concrete act of planning demanded specific choices about which resources to use, and in which ways.[37] Such choices necessarily involved tradeoffs between different values. Hence, "to direct all our activities according to a single plan presupposes that every one of our needs is given its rank in an order of values which must be complete enough to make it possible to decide among all the different courses which the planner has to choose" (101). For example, when planners

> have to choose between higher wages for nurses or doctors and more exten-
> sive services for the sick, more milk for children and better wages for agricul-
> tural workers, or between employment for the unemployed or better wages
> for those already employed, nothing short of a complete system of values in
> which every want of every person or group has a definite place is necessary
> to provide an answer. (116)

The problem, Hayek contended, was that "not only do we not possess such an all-inclusive scale of values" but "it would be impossible for any mind to comprehend the infinite variety of different needs of different people which compete for the available resources and to attach a definite weight to each" (102). As a consequence, what in fact happens under centralized planning is that the specific individual values of those in charge of planning are imposed upon everyone. Since a planning society requires individuals to "conform to the standards which the planning authority must fix," the vast majority of individual differences will necessarily be disregarded, in the sense that "the diversity of human capacities and inclinations" will be "reduced" to "a few categories of readily interchangeable units" that "deliberately . . . disregard minor personal differences" (130).

Where this first argument points to a paradox intrinsic to the idea of social planning, Hayek's second argument underscored what he saw as the strength of liberal economic relations in resolving this problem of competing values. Hayek contended that a government simply cannot gather the information it would need in order to plan economic activities—information about, say, raw materials, production costs, and consumer preferences—because this information can never be brought together at a single point; rather, it exists "solely as . . . dispersed bits of incomplete and frequently contradictory knowledge which all the separate individuals [of an economy] possess."[38] Hayek contended that each individual is situated in, and has the most knowledge of, his or her own particular "time and place" and his or her "local conditions" (521, 522). As a consequence, "practically every individual has some advantage over all others in that he possesses unique information of which beneficial use might be made, but of which use can be made only if the decisions depending on it are left to him or are made with his active cooperation" (521–22). Hayek argued that this distributed knowledge is especially important in the context of changing economic conditions, such as rising or falling production costs or changes in availability of raw materials (523). He argued that the only possibility of "planning" in such a state of distributed knowledge is to enable economic competition, for "competition . . . means decentralized planning by many separate persons" (521).[39] For Hayek, the "price system" of capitalist competition functions as the mechanism by which distributed individual perspectives are brought together and by which economic problems are thus "solved" (525).[40] Or, as Hayek put it, "the whole acts as one market, not because any of its members survey the whole field, but because their limited individual fields of vision sufficiently overlap so that through many intermediaries the relevant information is communicated to all" (526). Where centralized government planning necessarily ignores most of the diverse perspectives of individuals in favor of the specific perspectives of those in charge of planning, a competitive market synthesizes all of these limited perspectives.

For Hayek, capitalist relations were a kind of Burkean institution that had emerged immanently, without conscious planning, from the distributed and diverse activities within populations, and that had withstood the test of time. Hayek stressed that competitive market relations had, like language, "evolved without design" (527) and were in this sense "one of those formations which man has learned to use (though he is still very far from having learned to make the best use of it) after he had stumbled upon it without understanding it" (528).[41]

At the same time, Hayek distinguished his version of liberalism from conservatism, for he insisted that the traditional institution of the market could—and should—be refined through conscious effort.[42] For Hayek, "conservatism, though a necessary element in any stable society, is not a social program," for a conservative movement, by its very nature, is bound to be a defender of established privilege and to lean on the power of government for the protection of privilege, if privilege is understood in its proper and original meaning of the state granting and protecting rights to some which are not available on equal terms to others.[43]

Liberalism, by contrast, *was* a social program, one that aimed to optimize the traditional institution of the market by expanding and securing the realms in which individual differences were expressed. For Hayek, law was a key means by which economic relations were optimized, and it did so by ensuring that rules were explicit and neutrally applied to all individuals. This ensured that individuals focused on their own market initiatives, rather than on activities such as bribing or otherwise seeking to influence government officials.[44]

Hayek's understanding of the market as the systematic linkage of knowledge "dispersed among many people" is, like Mill's liberalism, premised on the importance of differences among people.[45] Though more elliptically than Mill, Hayek also stressed the importance of Humboldt's *The Limits of State Action* for his account of human diversity, and his description of markets as "wholes" that connect the limited perspectives of individuals with one another is reminiscent of Humboldt's stress on the need for individuals to overcome their "one-sidedness" (*Einseitigkeit*) by means of connections with others.[46] Hayek's account of liberalism as the necessary medium for the growth of reason is also reminiscent of both Humboldt and Mill. Hayek contended that "the interaction of individuals, possessing different knowledge and different views, is what constitutes the life of thought. The growth of reason is a social process based on the existence of such differences. It is of its essence that its results cannot be predicted" (179).

Yet Hayek's emphasis on the importance of competition ensures that there are necessarily losers in his vision of human diversity in a way less evident in Mill's or Humboldt's accounts. In both Mill's and Hayek's versions of liberalism, individuals engage in fundamentally speculative behavior, for each individual makes a bet on what way of life will make him or her most happy. Mill acknowledged that, with respect to specifically economic relations, the individual's wager of success in a specific profession necessarily resulted in winners and losers:

> Whoever succeeds in an overcrowded profession, or in a competitive examination; whoever is preferred to another in any contest for an object which both desire, reaps benefit from the loss of others, from their wasted exertion

and their disappointment. But it is, by common admission, better for the general interest of mankind, that persons should pursue their objects undeterred by this sort of consequences. In other words, society admits no rights, either legal or moral, in the disappointed competitors, to immunity from this kind of suffering; and feels called on to interfere, only when means of success have been employed which it is contrary to the general interest to permit—namely, fraud or treachery, and force.[47]

Yet Mill also described "trade" as simply one of many applications of his principle of liberalism, and there were many noneconomic, noncompetitive areas of life in which individuals engaged in experiments in living that did not necessarily produce winners and losers.

While Hayek also stressed that the most important "ends" of our lives are noneconomic, he argued that these ends could only be pursued via economic means. For Hayek, though "the ultimate ends of the activities of reasonable beings are never economic"—that is, though "there are many things which are more important than anything which economic gains or losses are likely to affect, which for us stand high above the amenities and even above many of the necessities of life which are affected by the economic ups and downs"—the *means* to fulfill any noneconomic ultimate end is always and necessarily economic: "What in ordinary language is misleadingly called the 'economic motive' means merely the desire for general opportunity, the desire for power to achieve unspecified ends.'[48] In contemporary society, this results in the desire for money, which "offers us the widest choice in enjoying the fruits of our efforts," but also the hatred of money, for "in modern society it is through the limitation of our money incomes that we are made to feel the restrictions which our relative poverty still imposes upon us" (125). While money could in principle be eliminated within a planned economy, the result would be simply an alternative (and from Hayek's perspective, totalitarian) set of incentives:

> If all rewards, instead of being offered in money, were offered in the form of public distinctions or privileges, positions of power over other men, or better housing or better food, opportunities for travel or education, this would merely mean that the recipient would no longer be allowed to choose and that whoever fixed the reward determined not only its size but also the particular form in which it should be enjoyed. (125)

For Hayek, individuals cannot fully engage in experiments in living unless their economic relations to one another are mediated by the neutral medium of money, for any more specific incentive (for example, privileges or food) limits individual choice. Liberalism is thus, for Hayek, the social system that most fully enables experiments in living, though the necessary cost of such optimization is that experiments in living must take the form of market relations.[49]

Both Mill's and Hayek's accounts of liberalism are premised on the generative potential of individual diversity, and for both, liberalism is the system of social relations that best enables a multitude of experiments in living and makes it possible to gather the results of those experiments. The point upon which they disagree is *how* the results of experiments in living should be gathered. Mill had no real theory of how this might happen and simply gestured toward some sort of government central depository. Hayek, by contrast, contended that only the market could gather together the results of experiments in living and, by synthesizing these, enable collective progress. Hayek reconfigured the concept of collective experimentation and experiments in living such that both were understood as fundamentally economic in orientation. (Or, to put this another way, he suggested that only experiments in living oriented toward the market could contribute to collective progress.) Hayek's liberalism thus aimed to reconfigure actively as many social relations as possible *as* economic relations. The link that Hayek established between individual differences and the price system of competitive economic relations thus set the stage for later, more biologically oriented neoliberal visions such as Lasagna's population-level pharmaceutical collective experiments.

Beck, the Risk Society, and Collective Experimentation

For Mill and for Hayek, individual diversity was understood primarily in terms of conscious choices made by individuals.[50] This stress on conscious choice persisted in Lasagna's neoliberal vision of naturalistic population-wide pharmaceutical drug experiments, for it was central to Lasagna's proposal that individuals be free to choose whether to take potentially beneficial drugs. Yet even as conscious, individual choice is central to Lasagna's proposal, the biological register on which his proposal focused—the effect of pharmaceutical substances on complex human biology—raised the question whether such "naturalistic" collective experimentation might be more effective if the link between individual diversity and conscious choice were severed. If, as Lasagna contended, naturalistic pharmaceutical collective experimentation located useful drugs more efficiently than clinical trials and did so by administering them to larger, more representative populations of biologically unique individuals, why should there be any necessary connection between individuals whose biology best enabled such tests and those who *choose* to take unproven drugs?

Precisely this line of thought has encouraged many contemporary researchers and bioethicists to search for what I would describe as "post-liberal ways" to employ tissue samples and medical information, regardless of whether individuals have consciously opted to be part of such research.[51] These initiatives are postliberal in the sense that though they are committed to the market as the arena within which drug discovery occurs, drug discovery itself now breaks partially with the liberal commitment to individual choice. However, severing the connection between collective experimentation and individual choice also points toward another postliberal relationship between the market and collective experimentation, namely, the *inadvertent* collective experiments produced by modern market-oriented institutions.

Such inadvertent collective experiments were a focus of the German sociologist Ulrich Beck's concept of the "risk society," which points beyond liberalism in a different way. At the same time as Lasagna was advocating for naturalistic, population-wide, market-oriented collective experiments in the United States, Beck highlighted the many collective experiments being run on populations as byproducts of modern market-oriented institutions. Beck developed his reflections on inadvertent collective experimentation in the context of his larger account of modernity. In the first phase of modernity, which ran from the seventeenth to the twentieth century, collective effort had been directed toward "making nature useful," with the goal of "releasing mankind from technical constraints," especially food scarcity.[52] Yet institutions and structures capable of addressing these issues, such as industrial production and nation-states connected by a global market, turned out not to be simply neutral providers of goods but also forces that necessarily produced negative global side effects, including ecological devastation, impending nuclear disaster, and diseases associated with overconsumption of food. Beck argued that his concepts of "risk" and the "risk society" enabled the forces of modernity to become reflexive, for they would facilitate new conceptual categories and practices that would allow us to avoid the "boomerang effect" of social structures narrowly focused on wealth production (23). Or, as Beck put it, his concept of risk is "a *systematic way of dealing with hazards and insecurities induced and introduced by modernization itself*" (21).

Beck claimed that, unlike earlier concepts of risk, his concept is bound up with modern knowledge-producing processes. This is in part a consequence of the fact that risks produced by a global market society are distributed globally and so differ in kind from earlier risks. "Unlike the factory-related or occupational hazards of the nineteenth and the first half of the twentieth centuries," for example, new risks such as "ecological disaster and atomic fallout" (23) "can no longer be limited to certain localities or groups, but rather exhibit a tendency to globalization" (13).[53] However, these risks are also specifically modern because they "are based on *causal interpretations*, and thus initially only exist in terms of the (scientific or anti-scientific) *knowledge* about them" (23). That is, contemporary risks can only be determined by means of what Beck calls "the 'sensory organs' of science" (27).

The concept of the risk society allows us to see that modern global society is, in effect, conducting a series of collective experiments on itself, though it fails to recognize that this is the case. As Beck put it:

> Substances are disseminated in the population in all imaginable ways: air, water, food chains, product chains, etc. . . . [Yet] *the experiment on people that takes place does not take place.* More precisely, it takes place by administering the substance [e.g., the unintended introduction of carcinogens to food as a byproduct of production] to people, as with research animals, in small doses. It fails to take place in the sense that the reactions in people are not systematically surveyed and recorded. . . . For the sake of [corporate] caution, the reactions in people themselves are not even noted, unless someone reports and can *prove* that it is actually *this* toxin which is harming him. (69)

Lasagna had proposed to make medicine naturalistic by removing gatekeeping institutions such as clinical trials and government regulation so that the effects of potential drugs could be tracked within populations. Beck's point was that such naturalistic experimentation is *already* happening in multiple realms of life, for the global population is now constantly exposed to byproducts of modern production processes. Lasagna's specific example of pharmaceuticals suggested that drug companies, if allowed to conduct population-wide experiments, would have financial motivation to determine which experiments were efficacious. However, by focusing on market-oriented production more generally, Beck underscored that it was often not in the financial interests of many companies to understand themselves as engaging in collective experimentation even when they were in fact doing so. As a consequence, the data necessary for assessing the effects of these experiments was never gathered (and efforts to gather this data are often actively opposed by corporations).

Though they disagreed about whether the search for profit was the solution to, or cause of, modern problems, Beck and Lasagna agreed that collective experiments demanded what Beck described as the "democrati-zation" of science. Lasagna objected to government regulations and bodies such as the FDA because they implied a scientific consensus that was never the case (hence, his description of "consensus among Experts" as an "Unholy Grail").[54] In keeping with the vision of Chicago School econo-mists, Lasagna's solution to this problem of differences of opinion among scientists was to eliminate the category of the expert and instead encourage anyone with an opinion and enough funding to enter the marketplace, which latter would determine which scientific opinions were correct. While Beck was not an advocate of "market wisdom," he was nevertheless also critical of the concept of the scientific expert. For Beck, though the modern notion of risk is emphatically scientific, it must be understood as fundamen-tally a democratizing concept, since statements about risks necessarily con-tain normative "statements of the type *that is how we want to live*":

> Even in their highly mathematical or technical garb, statements on risks con-tain statements of the type *that is how we want to live*—statements, that is, to which the natural and engineering sciences . . . can provide answers only by overstepping the bounds of their disciplines. But then the tables are turned. The non-acceptance of the scientific definition of risks [by members of the population] is not something to be reproached as "irrationality" in the popu-lation; but quite to the contrary, it indicates that the cultural premises of acceptability contained in scientific and technical statements on risks *are wrong*. The technical risk experts *are mistaken* in the empirical accuracy of their implicit value premises, specifically in their assumptions of what appears acceptable to the population. . . . One can also view it another way: in their concern with risks, the natural sciences have involuntarily and invisibly *dis-empowered themselves somewhat, forced themselves toward democracy*. (58)

For Beck, the "normative horizons" within which risk assessments neces-sarily operate cannot be determined by science, since such horizons con-cern not facts but values and aspirations. Collective values and aspirations are determined not by experts but by "lay" individuals working together in collectives (for example, community advocacy groups) (28). In this sense, "risk determinations are an unrecognized, still undeveloped symbiosis of the natural and the human sciences, of everyday and expert rationality, of interest and fact" (28).

For Beck, the democratization of science implicit in the concept of the risk society also enables the formation of new forms of community that are oriented toward future transformation. "In the risk society," Beck con-tends, "the past loses the power to determine the present. Its place is taken by the future" (34), by which he means that risk statements orient collectives toward future forms of life that they wish to lead, rather than collectives being oriented by the desire to escape that condition of scarcity that dominated global human existence up until the twentieth century.[55] The risk society enables the formation of new forms of communities—that is, new forms of "we"—which may be based in a shared injury, such as exposure to industrial toxins in a specific geographical region, but which are oriented toward a future condition of transformation and flourishing.

Beck's appropriation of the concept of collective experimentation brings out more fully than the other authors I have considered here the biopolitical dimension of liberalism. While Arbuthnot's smallpox inoculation advocacy was clearly biopolitical, since it bore on biological regularities that occurred both in the absence and presence of inoculation, his proposal was only ambivalently liberal. Burke's, Mill's, and Hayek's accounts are all more clearly liberal, but the biopolitical dimension of each comes out in the margins rather than body of their proposals. The biopolitical dimension of Burke's and Hayek's proposals peeks forth in their shared stress on the *survival* of civilization. Burke justifies the importance of unconscious, gradual political experimentation through the claim that more conscious political experimentation threatens the survival of the polity. Hayek updates this approach with his claim that the market has emerged from the process of human evolution, which demonstrates the evolutionary fitness of the market (and, as a consequence, ensures that one cannot go "beyond" the market). Mill's concept of experiments in living reveals its biopolitical underpinning primarily in his suggestion that a large population of experimenters is necessary for the emergence of a few geniuses, which continues the biopolitical logic of genius that I described in Chapter 1.

Where Burke, Mill, and Hayek imply that these biopolitical concerns are distinct from the key liberal value of individual choice—or, at any rate, that these biopolitical concerns enable individual choice without fundamentally affecting the latter—Beck suggested that, within a market-oriented society, individual choice was one of the means by which unconscious collective experimentation was taking place. Yet for Beck, unconscious collective experimentation does not result, as was the case for Burke, in stable political institutions but instead threatens the polity by producing environmental disasters.

While Beck's account of collective experimentation brings to the surface a biopolitical dimension of liberalism that had remained submerged in Burke's, Mill's, and Hayek's accounts, it is not clear whether this leads Beck to his own version of neoliberalism or to something beyond liberalism.

Like Burke, Mill, and Hayek, Beck was interested in *how*, precisely, the results of collective experimentation were gathered and synthesized. Burke had presented traditions and traditional institutions as the vehicles of such synthesis, Mill hoped that a central governmental repository could take over this function, and Hayek pointed to markets as the synthesizing institution. Because Beck saw modern risks as combinations of scientific statements and value judgments, he stressed that neither government experts nor the market can synthesize the results of collective experimentation. The former cannot do so because they fail to recognize the nonscientific element of values within risk statements, while the market's commitment to one ultimate value—profit—ensures that it is a producer of unconscious collective experiments, rather than the site where the results of these experiments are gathered. In place of Burke's traditions and institutions, Mill's central governmental repository, and Hayek's markets, Beck suggested that scientifically oriented community advocacy groups were the sites at which the results of collective experimentation can be gathered and assessed.

Beck's proposal initially appears antiliberal or at least aliberal. It does so because Beck posits neither individual rights, markets, nor the state as the final arbiter of truth and collective values; rather, interest groups and communities are the agents that determine such values. Moreover, the telos of collective experimentation for Beck is neither the survival of the state (à la Arbuthnot) nor survival of "civilization" (à la Burke, Mill, and Hayek) but rather collectively and democratically determined goals.

At the same time, though, there is an uncanny convergence between Beck's account of the democratizing of the sciences and the attack on scientific expertise that has become a key strategy for contemporary neoliberalism. While Beck stressed that the implicit value component of modern risk statements opens these up to collective debate about final values, he acknowledged that what he described as the "immunity" from scientific expertise that made this possible was a double-edged sword (169). As Beck noted, immunity from science characterizes many forms of contemporary belief:

> Quite generally, ideologies and prejudices, now scientifically armed, are able to defend themselves anew against science. They take recourse to science itself in order to reject its claims. One just has to read *more*, including the alternative investigations. The objections are consumed *before* the results, with advance notice as it were. Keeping a couple of basic (methodological) objections on hand for all cases is enough to make this or that obstinate

scientific news collapse in itself. Until the sixties, science could count on an uncontroversial public that believed in science, but today its efforts and progress are followed with mistrust. People suspect the unsaid, add in the side effects and expect the worst. (169)

When Beck wrote that passage, he was likely not aware of the deliberate use of scientific uncertainty by tobacco companies—and later, companies opposed to climate-change policies—to head off or stall the government regulation that might otherwise seem to be demanded by scientific research on the side effects of their products.[56] Paradoxically, such developments are also examples of Beck's thesis, for these companies employ the results of the sciences selectively *because* they are committed to a nonscientific ultimate value: profit. Just as a community can, in the name of alternative ultimate values, refuse to accept risk assessments from scientific experts about "acceptable" levels of industrial contamination, so too can corporations, in the name of profit, reject the risk assessments of scientific experts concerning the effects of their products on, say, global warming.[57] Though Beck optimistically forecast that the partial dissolution of the distinction between scientific experts and lay people would lead to greater local democracy and to cosmopolitan linkages among local groups, the recent history of the strategic uses of scientific uncertainty by corporations suggests that such a dynamic can also simply further the *bellum omnium contra omnes* logic of liberal market relations.

Immunity, Survival, and Collective Experimentation

Though it is unclear whether Beck is best understood as a critic of or unwitting advocate for neoliberalism, his approach nevertheless opens up new possible futures for the concept of collective experimentation. It does so by breaking partially free of the schema of *survival* that is fundamental to Arbuthnot's, Burke's, Mill's, and Hayek's versions of liberalism and that Roberto Esposito has pinpointed as the means by which liberalism finds itself beholden to the immunitary schema that I described in this book's Introduction. Because modern risk statements align results of the sciences with value statements about desired modes of life, they implicitly allow us to escape the presumption that triage-based notions of survival can constitute ultimate values.

The schema of survival is a red line running through the liberal theories of collective experimentation that I have outlined here. If we take Arbuthnot's advocacy for smallpox inoculation as the original matrix within which the concept of collective experimentation was linked to liberalism, the importance of survival is explicit, since inoculation bore directly upon the life and death of individuals (and, of somewhat less interest to Arbuthnot, the quality of life of survivors). Arbuthnot's suggestion that the state's survival and vitality was the ultimate telos of the collective experiment of smallpox inoculation simply applied the schema of individual survival to the collective body politic. His passing references to Newgate and slave test subjects also underscored the sacrificial, or at least triage-like, dimension of the liberal approach to survival, in the sense that the latter never meant the actual survival of all members of a population but only the survival of its most valued members.

Burke, Mill, and Hayek developed more complicated accounts of the connections among collective experimentation, liberalism, and survival. For Burke, the collective experiment does not bear on the biological survival of individuals—his image of summer flies suggested that humans would survive even if they destroyed their traditional institutions—but on the survival of what he understood as the properly *human* form of life, namely, civilization. The brilliance of Burke's text, though, was to make the survival of civilization feel as though it were equivalent to biological survival. Mill, for his part, broke the link established by Burke between conscious collective experimentation and unconscious adherence to traditional institutions, since for Mill, experiments in living *could* proceed in the absence of traditional institutions, provided that the legal order was oriented around his "one simple principle" of individual autonomy. Yet Mill nevertheless retained the link between collective experimentation and civilization, for he argued that the progress of civilization was threatened by forces, such as strong public opinion, which threatened experiments in living. Hayek's neoliberalism tied both the progress of civilization and the concept of collective experimentation back to a biological basis, and it did so in two ways. He argued, first, that the only effective form, or container, of collective experimentation—the market—was a consequence of human biological evolution and was, for this reason, an institution that could not be surmounted. Second, Hayek argued that failure to protect the market order of civilization would lead to National Socialist– or Stalinist-style totalitarianism, that is, to what Foucault and Esposito described as "thanatological" forms of biopolitical governance, which explicitly justified the elimination of large populations in order to guarantee the survival of the valued population.[58]

This intimate connection between concepts of collective experimentation and survival in liberal theory explains the link, noted by Esposito, between liberalism and the concept of immunity. Concepts of immunity have played two different roles within liberal theory. On the one hand, liberal theorists have consistently opposed an older sense of political immunity, understood as an individual's exemptions from laws that hold for everyone else. Mill, for example, argued that liberalism was the extension to everyone of political immunities originally granted only to a few (which means that liberalism eliminates the very notion of exclusive political immunities).[59] On the other hand, and as I noted in the Introduction to this book, Esposito argues that a different, quasi-medical sense of immunity constitutes the hidden core of liberal theory. Esposito argues that the key liberal value of individual liberty has been, from Locke to the present, a concept of security, rather than an older sense of liberty as creative production. For liberal theorists, liberty names that security that ensures that no "obstacle" will come between an individual "and his will."[60] Yet, Esposito contends, liberal theorists consistently argue that individuals can only be protected against external threats to their wills by the "immunological" procedure of introducing a purportedly small and controllable amount of that same threat of externality in the form of laws and police forces, which are supposed to ensure that individuals are secure enough to enact the actions that they will. The problem, Esposito stresses, is that once the threatening outside is allowed in, even in a small immunitary dose— for example, even just the minimal form of Mill's "one simple principle"— it becomes difficult in practice and arguably impossible in principle to set a firm limit on the extent to which outside forces of law and policing can control the subject. The result, Esposito concludes, is that liberalism is not able "to determine or define liberty except by contradicting it" (75).[61]

While Esposito's account is pitched at a high level of abstraction, the history of liberal approaches to the concept of collective experimentation that I have sketched here clarifies one concrete way that the process of specifying the liberal principle of individual liberty leads to practices that de facto contradict that principle. In Arbuthnot's original formulation, the concept of collective experimentation was literally about (what we would now call) immunization, and the importance of security was equally explicit. Arbuthnot contended that "it would be a most Tyrannical Encroachment" upon "the natural Rights of Mankind" if in the normal course of events individuals were prevented "from the lawful Means of securing themselves from the Fear and Danger of so terrible a Plague" as smallpox. Yet in the same paragraph, Arbuthnot referenced conditions in which the state *could* legitimately negate individual self-determination, namely, states of exception in which the population was threatened by

"Pestilence," for "the Magistrate is forc'd often upon more arbitrary Proceedings in any Pestilence." Arbuthnot's formulation leaves unclear how one could specify in advance which threats constitute states of exception—that is, which threats rise to the level of Pestilence—and the criteria by means of which an authority could make that determination (hence, his claim that any such measures will always seem "arbitrary"). This problem is simply underscored by Arbuthnot's use of the term "Plague" for smallpox and his use of the more or less identical term "Pestilence" for threats that exceed the threat posed by smallpox.[62]

This problem of definition and criteria emerges because Arbuthnot asserted that *both* individuals and the state have a right to security and survival. An individual ought to be able to experiment with smallpox inoculation because of his or her right to security and survival. Yet these are also rights possessed by the state, and as a consequence, the state can, in the name of its own security, deprive individuals of the right to self-determination. This is simply the logical extension of the triage logic of survival. That is, it is an extension of the paradox that, for Arbuthnot, the liberty of individuals in general to choose inoculation was not threatened or disturbed by the fact that some particular individuals—Newgate prisoners and slaves—were forced to undergo inoculation. This same antinomy reappears in Burke's, Mill's, and Hayek's reflections on the relationship between collective experimentation and the health and survival of civilization, for in each author, the survival of the state or system that "protects" individual self-determination always trumps individual self-determination. In other words, for these liberal theorists, survival of the state is the ultimate telos of collective experimentation.

Beck's concept of immunity, though, begins to cut the link between survival and collective experimentation. This is not immediately obvious, since many of the examples by means of which Beck illustrates the modern concept of risk—for example, exposure to industrial byproducts, nuclear waste, and global warming—constitute threats to life and health. From this perspective, it may seem that the risk society is even more focused on survival than the nineteenth- and early-twentieth-century industrial society that preceded it. Yet for Beck, it is *because* individuals in the risk society are "immunized" against automatic deference to science that they are able to formulate "statements of the type *that is how we want to live*" that no longer presume the liberal schema of survival as an ultimate value.[63] The liberal schema of survival operates through the triage logic of sacrificing (or at least not saving) less valued populations so that more valued populations can flourish. This liberal schema of survival is instantiated in risk statements that concern "acceptable" levels of contamination, since such statements implicitly assert that the near-certain death or suffering of a small number of people is acceptable.[64] This liberal schema of survival is contested, though, whenever a community refuses to accept expert risk thresholds, or contends that the experts have failed to take into account other important risks, or asserts that expert decisions unjustly allow risks to fall on some groups and not others. In such cases, individuals or communities reject either the specific assertion that the suffering or death of *this* particular group is an acceptable price to pay or reject the more general liberal schema of survival, in which some must die so that others may live.

Beck's concept of immunity also helps us understand the uncanny way that the affective appeal of contemporary neoliberalism depends upon its capacity to position itself as pursuing an ultimate value *other* than survival even as it employs the triage logic of survival in its daily operations. Though neoliberalism is sometimes presented as a form of naturalism— that is, an attempt to justify markets as "natural" and hence the only mechanisms that ensure collective survival—neoliberalism in fact privileges the ultimate value of economic growth *over* survival.[65] This demotion of survival as ultimate value is especially striking in the case of corporate climate-change denials, since in this case, economic growth is valued over the survival of the entire social system upon which economic growth itself depends (with "survival" understood here in the sense of the continued existence of an entire system and not simply the continued existence of a valued population). As Beck notes, nineteenth-century liberals could sincerely and unreflexively believe that processes of industrialization, capitalism, and economic growth would solve social problems such as poverty without these solutions leading, self-reflexively, to their own problems. Yet Beck stresses that such an unreflexive position has been impossible since the 1960s, as the link between economic growth and collective survival has become contested. Even those who claim that economic growth does *not* endanger collective survival cannot simply assert, but must argue, this point, and must do so, moreover, on the basis of the results of those same sciences they wish to render questionable.

A peculiar consequence is that economic growth can now take on the patina of a nonsurvivalist ultimate value. That is, in the era of climate change, economic growth takes on the status of a transcendent ultimate value that ought to be pursued *even if* it leads to collective destruction (and the transcendence of this ultimate value is proven precisely by its uncertain confrontation with survival). Within neoliberalism, economic growth is, moreover, a transcendent ultimate value that appropriates to itself that quasi-theological affirmation of individual difference and pursuit of perfection that I described in the section on Mill and Humboldt. From this perspective, neoliberalism is able to fulfill a contemporary affective desire for an ultimate value *other* than survival even as it employs the triage logic of survival in its everyday operations.

Conclusion

If, as Beck suggests, we now live in a world of perpetual collective experimentation—experiments with human exposure to industrial waste products, gene-line alterations in foods, and transformations of the environment, to name just a few—how does the history of collective experimentation that I have outlined help us understand better our contemporary condition? I suggest four considerations.

First, as Beck noted, we must ensure that the experiments on people that virtually occur actually occur. Beck proposed that often *"the experiment on people that takes place does not take place,"* by which he meant that people were exposed to all kinds of substances (and so these experiments *do* take place), but it was not in the interests of corporations or even government regulatory bodies to determine the effects of these substances (and thus, the experiment does *not* take place). To turn these virtual experiments into actual experiments means gathering the information that enables them to be assessed, and it likely also means altering the criteria by which cause and effect are established and legal responsibility assigned.

However, and second, Beck's ambivalent critique of liberalism reminds us that this greater commitment to data gathering ought to be coordinated with a conscious articulation of the values that determine *what* data to gather and how such experiments can be judged. We might, for example, combine Beck's critique with Esposito's affirmative understanding of biopolitics, which takes the protection and flourishing of *every* human life as an ultimate value. This in turn suggests that there is *no* level of "acceptable risk"—that is, no level of acceptable suffering and death—from the byproducts of a specific industrial process or from economic growth more generally. This would then encourage data gathering that focuses on how the sacrificial logic of survival employs racial, gender, class, and disability markers to ensure that some groups of people suffer disproportionately. Adopting this hybrid perspective of Beck and Esposito would then also orient this data gathering toward the discovery of measures that allow us to reduce continually—with the constant aim of reducing to zero—the loss of life and health from these processes (which may mean in practice abandoning some of these processes).

However, and third, if my analysis of the affective appeal of neoliberalism is correct, this suggests that loss of life and health must be shifted away from the framework of survival. In other words, life and health cannot be seen as ends in themselves but as the necessary conditions for some other value. From this perspective, "green" attacks on neoliberalism err tactically when they return to survival as the ultimate value. There is no doubt that if we want to ensure the survival of something like contemporary society, we must respect the results of climate science, which reveal that pursuit of economic growth threatens this ultimate goal of survival. Yet if neoliberalism is appealing because it pursues a goal that transcends mere survival, then such an appeal is likely not effectively countered by returning to survival as an ultimate value. Beck's approach suggests, instead, that critics of neoliberalism should emphasize alternative ultimate values that they hope to advance through the *means* of life and health and that those alternative values will be most appealing if they also affirm individual difference and the pursuit of perfection, at least in some form.

One way to begin to do this—and this my fourth and final point—is to use the shifting role of the concept of collective experiments within the history of liberalism as a resource for renewing our understanding of the link between "experimentation" and collectives. Since at least Arbuthnot, liberal theorists have gravitated toward survival as the way to link experimentation and populations, for collective experimentation has been presented as the form into which the dispersed activities of populations must be channeled in order to assure the survival of valued populations of the polity. From this perspective, the basic operation of the collective experiment has been that of triage, that is, distinguishing between more and less valued lives. Yet other ways of understanding the nature and aim of collective experimentation are also evident in the liberal lineage. Burke's defense of tradition, for example, draws upon an earlier linguistic identity between "experiment" and "experience." When Burke defended traditions as a mode of experimental science, he did not mean that traditions employed scientific protocols but rather that they consolidated, preserved, and deployed collective experiences of the world. Burke thus points us toward a notion of collective experimentation as collective experience.[66]

How might we orient the notion of collective experimentation as collective experience away from Burke's use of it to defend the status quo of traditional institutions? Burke described traditional institutions as the results of collective experimentation to stress the necessarily unconscious nature of collective experience and thus to encourage deference toward these institutions. However, as Hayek recognized, the fact that an institution emerged without conscious planning does not mean that it must be respected just as it is, and the point of his *neo*liberalism was to optimize aggressively those market mechanisms that earlier liberals had sought to leave untouched. Hayek hoped that his understanding of the market as a distributed information processor would enable us to better link individual perspectives—ultimately, the individual perspectives of the global population—and thus expand the nature and formatting of collective experience. Yet Hayek's vision of the market as a collective computer ultimately narrows massively the nature of collective experience, since this computer can "store" only prices.

We might, though, reimagine collective experimentation as a matter of framing and storing past common experience so that we end up with something more like Arendt's notion of a "common world" rather than Hayek's common market. Arendt's common world also relies on "the simultaneous presence of innumerable perspectives and aspects," but, *pace* Hayek, there can be "no common measurement or denominator," such as prices, for these innumerable perspectives.[67] Rather, the common world emerges for Arendt, as for Burke, when these innumerable perspectives in the present are connected to the experience of people in the past, and with some guarantee that this connection will be extended to the future. For Burke, such connections between past, present, and future could only be ensured through traditional political institutions. Yet in our age of worldwide computer-mediated storage and communication, we can certainly imagine other ways of maintaining collective experience. We might see this, in fact, as one of the goals of the humanities at present: to engage in a new mode of collective experimentation, which aims to *create* the common world by ensuring that innumerable perspectives, both past and present, are synthesized into common experience. While the "we" engaged by this new mode of collective experimentation would have strong links to its multiple pasts, it would not seek to defend an inviolable core of that past but would instead be oriented more primarily toward a future "that opens the way to richer, more numerous, more diverse, and more flexible relationships with ourselves and others, all the while assuring each of us real autonomy."[68]

6. Life, Self-Regulation, and the Liberal Imagination

> The loss of standards, which does indeed define the modern
> world in its facticity and cannot be reversed by any sort of return
> to the good old days or by some arbitrary promulgation of new
> standards and values, is ... a catastrophe in the moral world only
> if one assumes that people are actually incapable of judging
> things per se, that their faculty of judgment is inadequate for
> making original judgments, and that the most we can demand of
> it is the correct application of familiar rules derived from already
> established standards.
>
> —Hannah Arendt, "Introduction *into* Politics"

In his lecture courses from the mid-1970s, Michel Foucault suggested that the simultaneous emergence of biopolitics and liberalism in the mid-to-late eighteenth century depended upon, among other things, the discovery of *self-regulation* as a principle of social relations.[1] He noted that for eighteenth-century authors, the self-regulation of social relations was something that happened naturally—that is, was a "natural" dynamic that governed human relationships—and yet, at the same time, something that had to be actively enabled by human institutions. One of Foucault's primary examples was the French physiocrats, who contended that humans naturally hoard grain when they think lean times are ahead and naturally sell grain at the highest price they can find.[2] The job of government was thus, for the physiocrats, to develop policies that respected these natural dynamics but also channeled these into a self-regulating form that ensured social stability and prosperity. The aspiration of self-regulation revealed for Foucault the intrinsic connection between biopolitics and liberalism, for both depend upon using legislation to allow some social dynamics, understood as natural, to follow their "own" path, with the overall goal of creating a self-regulating system.

The link that the concept of self-regulation establishes between biopolitics and liberalism has implications for our understanding of Romanticism. Foucault's account helps explain, for example, the centrality of the term "regulation" in a wide variety of late-eighteenth- and early-nineteenth-century authors. The French chemists Antoine Lavoisier and Armand Seguin contended that there were three "principal regulators" (*régulateurs principaux*) of the "animal machine" (*la machine animale*)—namely, respiration, perspiration, and digestion—and argued that the moral order (*l'ordre moral*) must also have regulators, or else human society would long ago have ceased to exist. In British North America, a first step toward the American Revolution was taken by members of the "Regulator" movement in the Carolina colonies, who sought greater self-governance. In Germany, Immanuel Kant contended that the "Ideas" of reason, such as the concept of an uncaused cause (God), must play a *regulative*, rather than constitutive, role in human experience. In Britain, Mary Wollstonecraft hoped to "regulate the passions"; the political economist Thomas Malthus argued against the Poor Laws by outlining a natural logic of population regulation; and Mary Shelley's novel *Frankenstein* begins with Robert Walton's dream that his arctic voyage will produce a discovery capable of "regulat[ing] a thousand celestial observations, that require only this voyage to render their seeming eccentricities consistent for ever."[3] And, with less positive valence, Jane Austen's Lady Catherine in *Pride and Prejudice* is an object of critique in part because of her desire to determine "how every thing ought to be regulated."[4]

Foucault's account of self-regulation also gives us new insight into the role of concepts of nature in works by Romantic-era authors. Within both biopolitics and liberalism, "nature" names both an autonomous dynamic within human relations but also a malleable force that can be shaped by human institutions of self-regulation. Liberalism is a particular way of seeking to produce social order by adjusting and channeling these natural dynamics, though I will argue that we can also locate and theorize a non-liberal version of that same aspiration. As a consequence, the concept of self-regulation allows us to understand the biopolitical, liberal nature of Romanticism not solely through the lens of critique—that is, not simply as another way of exposing the ideologies of Romantic texts—but also as a means of illuminating the potential for positive and more just Romantic forms of biopolitics.[5]

I make this argument in five parts. I begin by distinguishing between two different eighteenth-century models of regulation. One model presumed the existence of what I describe as invariable standards and was linked to a traditional concept of political sovereignty. The other model, by contrast, sought to explain self-regulation, which operated by means of variable standards and a distributed population. While the first section focuses primarily on the importance of this debate for economic questions, such as coinage standards, in the second section I stress that the tension between these two models of regulation was also central to the development of the modern concept of "taste" and especially the debate about whether an invariable "standard of taste" could regulate aesthetic judgments within large populations. In the third and fourth sections, I argue that some of the ambiguities of the model of distributed self-regulation were a focus of Romantic-era interest. The third section explores Malthus's understanding of the population as the subject of self-regulation, and the fourth focuses on Kant's account of the ambiguous relationship between intellect and collective self-regulation and the importance of questions of standards of aesthetic judgment for his account. In the fifth and sixth sections, I outline several legacies of Romantic reflections on regulation in the twentieth century, focusing first on the role of this term within systems ecology and neoliberal economics and then on the uncanny convergence of these two fields within contemporary revivals of the concept of the commons. I conclude by considering how Romantic reflections on the relationship of self-regulation to populations, intellect, and nature help us understand contemporary attempts to produce self-regulation in our era of the Anthropocene.

Though my account of regulation in Romantic literature is primarily intended to help us better understand relationships among Romanticism, liberalism, and biopolitics, I also aim to illuminate the ambiguous nature of this concept in Foucault's work on liberalism and biopolitics. Foucault used the term "regulation" frequently, but in two different senses. In connection with what Foucault called "discipline," regulation meant a process in which every detail of individual behavior is planned and only what is so planned is allowed. For example, in *Security, Territory, Population*, Foucault contended that "by definition, discipline regulates everything [*réglemente tout*]. Discipline allows nothing to escape. Not only does it not allow things to run their course, its principle is that things, the smallest things, must not be abandoned to themselves" (67–68). This sense of regulation underwrites Lady Catherine's desire in Austen's *Pride and Prejudice* and also underwrites important works of Romantic and Victorian literary criticism such as John Mee's *Romanticism, Enthusiasm, and Regulation* and D. A. Miller's *The Novel and the Police*.[6] However, when Foucault used the term "regulation" in connection with his concept of biopolitics, he referred to processes of governance in which individual subjects are largely left to their own devices so long as population-level events can be steered in certain ways. Thus, in *Society Must Be Defended*, Foucault contended that biopolitics requires the development of "regulatory mechanisms" (*mécanismes régulateurs*) that "establish an equilibrium, maintain an average, establish a sort of homeostasis, and compensate for variations within this general population and its aleatory field" (246). This latter sense of biopolitical regulation is opposed to disciplinary regulation, as Foucault stressed in the first volume of his *History of Sexuality*: "The disciplines of the body and the regulations [*les régulations*] of the population constituted the two poles around which the organization of power over life was deployed" (139). When Foucault connected the term "regulation" to discipline, it functioned as a synonym for the sovereign-like control of every aspect of individual behavior; when he used the term in connection with biopolitics, it denoted the use of individual freedom to create regularity at the level of a large aggregate (population). The question raised by these accounts is whether Foucault believed that there were two different modes of regulation—and if so, what was the common operation or term that connected them?—or whether these were two completely different kinds of operations that should have been given two different names.

These ambiguities of Foucault's use of the term "regulation" are not new; they express a long-standing difficulty that has plagued efforts to understand the nature of regulation. Since at least Gottfried Leibniz and Samuel Clarke's debate over the nature of God's regulation of the universe, efforts to distinguish between two models of regulation—regulation by means of invariable standards versus regulation by means of variable standards and a distributed population—have been continually frustrated by what we might think of as the gravitational pull of the schema of regulation as an irresistible sovereign command. The schema receives its emblematic form in the image of the machine that functions necessarily, automatically, and without the intervention of consciousness or choice. The gravitational pull of this schema is such that attempts to think the second model of regulation—which include Foucault's account as well as the efforts of twentieth-century cyberneticians, ecologists, and antiregulation economists—tend to find themselves pulled back toward the first model, with the location of the automatic mechanism simply shifted from one place to another. From this perspective, the virtue of the Romantic authors I consider here is that they make significant progress in thinking through more fully the second model of regulation and, hence, provide us with resources for furthering this process of thought in our own moment.

Regulation and Standards in the Eighteenth Century

There is surprisingly little work on eighteenth- and nineteenth-century concepts of regulation. Or rather, there is fabulous work on specific concepts of regulation, such as the historian of medicine Georges Canguilhem's pre-history of the concept of biological regulation, the legal theorist Randy Barnett's account of the term "regulation" in American Revolution–era periodical debates, and Michael Friedman's account of the role of the concept of regulation in Kant's critical philosophy.[7] But no account brings these separate histories together so that we can understand why this term appeared in so many eighteenth-century discourses, including medicine, political theory, law, philosophy, and political economy. What follows is a necessarily provisional and schematic attempt at such a synthetic account.

The English term "regulation," like its French and German counter-parts, comes from the Latin noun "regula," which meant, among other things, "a rod for drawing straight lines or measuring" or a "basic principle" (*OED*). And whenever the term "regulation" was used, from the seventeenth century to the Romantic era, it meant a technique that makes something else regular or consistent. A political regulation, for example, was intended to produce conformity among political subjects; Descartes's *Regulae*—usually translated into English as "rules"—*for the Direction of the Mind* were intended to bring consistency to philosophy; and the "regulator" of a late-seventeenth-century watch was a mechanical part that employed spring tension to make the time-keeping device run more precisely.[8] In all these cases, as well as the examples with which I opened this chapter, "regulation" means a process or technique that makes something regular. Yet beginning in the late seventeenth century, we find two different models for how a specific realm of experience can be made regular. Each model bound together theological, political, and economic concepts, but in different ways.

The first model proposed that regulation is possible only if all individuals or objects in the relevant area of experience conformed to an invariable standard determined by a sovereign imperative. Legal regulations concerning commodities provide a concrete example of this first model. The 1225 Magna Carta required the creation of standard, invariable measures.[9] In the case of the yard, this demand could be answered via a physical standard, with which individual pieces of commerce were compared.[10] To regulate by means of an invariable standard made a specific kind of social activity regular through activities of comparison and alignment with the standard, which latter was determined by a sovereign political entity.[11]

Though the act of comparing an invariable metal rod with a piece of cloth may seem straightforward, it is worth considering the metaphysical and theological premises of such an activity. In the realm of human affairs,

the invariable standard is understood as a source of illumination, for it—
and it alone—brings into visibility the measure that leads to justice, peace,
and stable commerce. The illumination of the invariable standard reaches
each individual equally, in that both official and layperson can engage in
the activities of comparison, acceptance, and rejection, and these acts of
comparison disseminate illumination outward to the polity as a whole. The
function of law is to ensure that all individuals remain true to the invariable
standard.

Yet even as the invariable standard provides regulating illumination in
the realm of human affairs, the basic schema that gives this model its intel-
ligibility—namely, that of the sovereign command that is obeyed universally and automatically—has a more ambiguous relationship with
"illumination" (and this ambiguity will be especially important when I
consider Malthus). The early eighteenth-century dispute between Gottfried
Leibniz and Isaac Newton's disciple Samuel Clarke concerning the nature
of God's regulation of the universe exemplifies this ambiguity. Leibniz
claimed that God "regulated everything in advance" ("réglant par avance
toutes les choses à la fois") and then left his creation to unfold itself automatically, while Clarke contended (in Georges Canguilhem's gloss) that
"God, after creating the world, continued to watch over it and interfere
providentially."[12] Yet both understood the nature of regulation in the same
manner, namely, as the enactment of a sovereign command.[13] Moreover,
because both were concerned in this debate with law-like processes of
nature, each proposed that divine commands were "obeyed" automatically
by matter, without any need for consciousness or illumination, at least on
the part of matter itself.[14] Hence, even if regulation in human affairs seems
to require a passage of the sovereign command into illumination and
through human consciousness, the aspiration of this first model of regulation is an automatic, unilluminated, and consciousness-free enactment of
divine imperatives.

Beginning at the end of the seventeenth century, a second model of
regulation emerged, which differed significantly in its metaphysical
assumptions. A different set of commercial standards—legal standards for
gold and silver coinage and rates of interest on loans—illuminate this second model. As late-seventeenth-century commentators such as John Locke
noted, monetary standards were vexing because they were often ignored
by many of the large population of individuals (only some of whom resided
in England) who used English coins and made or took loans.[15] For example,
individuals shaved bits of gold or silver from coins, such that they no longer

corresponded to the legal standard, or made under-the-table agreements that exceeded legal rates of interest.

For Locke, these practical problems could be resolved if one understood regulation as a form of order that occurred partially *beyond* the reach of law. In *Some Considerations of the Consequences of the Lowering of Interest, and Raising the Value of Money* (1696), he questioned "whether the Price of the Hire of Money [i.e., the interest rate] can be regulated by *Law*" and responded that "generally speaking, one may say, 'tis manifest it cannot."[16] Instead, Locke claimed, "that alone which regulates" the price of money is "the want of Money" (214). The "rate of Money does not follow the Standard of the Law, but the price of the Market; and Men not observing the legal and forced, but the Natural and Current *Interest* of Money, regulate their Affairs by that" (253). There *was* thus a kind of standard operating in separate economic interactions among individuals and by which each regulated his or her behavior. However, this standard was not determined by legislators but by "the Market," which produced what Locke called a "natural" standard, which was regulated by want (that is, the desire for loans). And since want changes over time, the market's standard, though natural, was not invariable but changed frequently.

How could individuals collectively "regulate their Affairs" by means of this fluctuating natural standard? In a superficial sense, regulation still meant comparing something with an external standard. Yet unlike a physical yard standard, the market standard was not located in a specific place, could not be specified by an authority, and did not remain stable over time. Rather, each individual had to try to locate the standard on the basis of his or her knowledge and best guesses. Where the invariable standard was a positive source of illumination for the polity as a whole, the variable standard could only be intuited negatively, as a sort of limit within the activities of other individuals.[17] Locke invoked the term "regulation," though, in order to stress that the variability and inscrutability of market standards for individuals not only did not prevent but in fact enabled order and regularity at the level of the market as a whole.

Though both of these models of regulation employed the elements of standards, knowledge, and automaticity, they distributed these elements differently. In the first model, an omniscient sovereign freely and knowingly determined a standard, which resulted in an imperative, or law, that then ought to be obeyed universally and automatically. In the second model, by contrast, knowledge is not located in an omniscient sovereign but in large collections of fallible individuals, with the result that knowledge of the standard by any one individual is always partial. However, provided that each individual does *not* act automatically but rather employs his or her limited knowledge to make judgments and act, a natural standard—of interest rates, for example—will then emerge automatically within what Locke calls the market. For Locke, automatic collective self-regulation is possible only when each individual focuses narrowly on his or her own limited sphere of needs, desires, and knowledge.[18]

Locke's concept of extrajudicial collective self-regulation valorized differences among individuals with respect to knowledge, ability, and temperament that were, within the first model of regulation, either of no interest or were understood as limitations to regularity. The first model of regulation presumed that the knowledge necessary to regulate a given realm, whether this is God's knowledge of how best to construct the universe or the human legislator's knowledge of the proper standard for coinage, can be gathered at one time and place in the consciousness of the sovereign, who then translates this knowledge into a corresponding imperative. Locke's model of self-regulation, by contrast, presumed that the relevant knowledge cannot be gathered together at one point but is instead distributed across the limited perspectives of many individuals, who must be linked together through the market for something like knowledge to emerge. Locke's point was not simply that there is no *practical* way for any human to determine, for example, what he calls the "natural interest rate." To understand this limitation as merely practical would imply that the natural interest rate is a relation such as gravitational attraction between two bodies, which is both independent of human needs and wants and presumably could be known by God, at least. For Locke, though, realities such as the natural interest rate for loans do not preexist but only come into being through the mechanism of the market. In this sense, "the market" names certain kinds of linkages among individuals that validate the limited perspective of each market participant, in the sense that each individual makes decisions about needs and desires based on his or her limited perspective and whatever knowledge he or she can obtain about the limited perspectives of others.[19] The schema of the market establishes a conceptual link between limited, individual perspectives; the automatic emergence of a standard; and what would later become the key liberal aspiration of "freedom." Freedom here names the fact that a variable standard and collective order emerge automatically precisely because each individual connected by the market *limits* himself to his own perspective, inclinations, and interests.[20] (Table 1 summarizes the differences between the first and second models of regulation.)

Table 1. Two models of regulation

	First model of regulation	Second model of regulation
figure of **nature**	sovereign imperative	variable forces that humans can partially shift/alter
nature of the **standard**	invariable, visible standard	variable, obscure standard
premise about the relationship of the standard to **knowledge**	standard can be determined only after relevant knowledge is assembled at central, univocal point	standard emerges and persists only through dynamic and ongoing linkage of individual, limited perspectives
role of **the individual**	automatic submission	exercise of (fallible) individual judgment
role of **government**	determining and enforcing invariable standards	empowering individuals to exercise (fallible) individual judgment
key **example**	machine	market

Locke's account of self-regulation exemplifies the beginnings of a form of reasoning that seeks out decentralized processes in large "apolitical" bodies—what will eventually be called populations—and often eventuates in the liberal position that political laws should be framed to allow the standards of these autonomous collective movements to express themselves freely. Locke's specific focus on economic self-regulation was picked up, for example, in mid-eighteenth-century British political economy by authors such as David Hume, Adam Smith, and James Steuart and by French physiocrats such as François Quesnay and Anne-Robert-Jacques Turgot. In *An Inquiry into the Principles of Political Oeconomy* (1767), for example, Steuart reflected at length on what regulated prices and connected this to reflections on "the principles which regulate the distribution of inhabitants into farms, hamlets, villages, towns, and cities" as well as natural and social factors, such as the fertility of soil, which "regulate the multiplication of man, and determine his employment."[21] Smith continued these reflections in his even more influential *The Wealth of Nations* (1776), considering, for example, how the price of corn is regulated by factors relating to silver production and how that form of natural regulation relates to political regulations concerning coinage standards.[22] Yet even as subsequent political economists consistently understood political economy as the search for the laws or principles that "regulated" the variability of the market standards, they also consistently contended that knowledge of such laws could never eliminate the need for the individuals to make their own estimations of variable market standards.[23]

Self-Regulation and the Standard of Taste

While the second model of regulation—that is, a form of immanent regulation enabled by individual perceptions of a variable standard—initially developed around topics over which the state had some power to create and enforce standards, such as interest rates and coinage, the logic and paradoxes of self-regulation were further developed in debates about standards that seemed intrinsically free from government control. For example, mid-eighteenth-century discussions about the existence and nature of a "standard of taste" in texts such as David Hume's essay "Of the Standard of Taste" (1757) and Edmund Burke's "Introductory Discourse Concerning Taste" in *A Philosophical Inquiry into the Origin of Our Ideas of the Sublime and the Beautiful* (1757) further consolidated interest in determining the role of invariable and variable standards in realms of experience in which diversity of opinion was the rule rather than the exception.[24] In the context of this chapter, Burke's and Hume's texts are important because they exemplify two different strategies for dealing with the relationship of invariable and variable standards, and variants of these strategies will appear again in my discussions of Malthus and Kant. Where Burke sought to contain the diversity of judgments of taste by appealing to an invariable standard that was, paradoxically, hidden within the body, Hume suggested that the invariable standard of taste was more like what Kant would later call a regulative ideal, that is, a point of orientation for individuals that, though it could not be instantiated in reality, enabled social unity precisely by providing individuals with a collective point of orientation.

Both Burke and Hume began their texts by noting that if (in Burke's words) taste was understood as "that faculty, or those faculties of the mind which are affected with, or which form a judgment of the works of imagination and the elegant arts," there was a seemingly irreconcilable diversity of judgments of taste within large groups of people.[25] Burke noted that in other realms of experience, differences of opinion are resolved by appeal to an invariable standard: "We find people in their disputes continually appealing to certain tests and standards which are allowed on all sides, and are supposed to be established in our common nature." Yet there "is not the same obvious concurrence in any uniform or settled principles which relate to Taste. It is then commonly supposed that this delicate and aerial faculty . . . cannot be properly tried by any test, nor regulated by any standard" (2). Hume, for his part, contended that "the great variety of Tastes, as well as of opinion, which prevail in the world, is too obvious not to have fallen under every one's observation" and then argued that, when we consider past periods and distant places, this variety of opinion is "still greater in reality than in appearance."[26] However, from the fact of the diversity of taste, Burke and Hume drew different consequences concerning a standard of taste.

In accordance with the first model of regulation, Burke sought to lead the diversity of taste back to an understanding of regulation as conformity to an inflexible standard. Burke argued that, despite the diversity of taste, the latter is in fact grounded in "natural causes of pleasure" that "enable [all humans] to bring all things offered to their senses to that standard, and to regulate their feelings and opinions by it."[27] Burke claimed, for example, that everyone would prefer butter or honey to a "bolus of squills" (15). However, Burke continued, taste—in the sense of aesthetic judgment—is only partly grounded in these "primary pleasures of sense" that are regulated by an invariable standard. In addition to being grounded in these primary pleasures of sense, taste is also grounded in "the secondary pleasures of the imagination, and of the conclusions of the reasoning faculty" (30). These latter differ among people and are partly dependent upon experience. Thus, concluded Burke, "whilst we consider Taste, merely according to its nature and species, we shall find its principles entirely uniform; but the degree in which these principles prevail in the several individuals of mankind, is altogether as different as the principles themselves are similar" (31).

Burke thus ended up with a curious hybrid of the first and second models of regulation. On the one hand, he argued for invariable, universal standards at the base of taste, which seemed to hold out the possibility of achieving social consensus in judgments about "the works of imagination and the elegant arts." On the other hand, he claimed that these invariable standards were so diffracted and dispersed by individual capacities and experience that they effectively become variable standards, in the sense that each individual possessed his or her own idiosyncratic standard (which, moreover, changes with an increase in both experience and "proper and well-directed exercise" of the capacities of judgment [33]). Yet Burke gave no indication that these variable standards of individuals could lead to any sort of immanent self-regulation in the mode described by Locke for monetary phenomena. He simply asserted a distinction between "good

Taste" and "wrong Taste" and claimed that the latter was attributable to a "defect of judgment" arising from either "a natural weakness of understanding" or "a want of proper and well-directed exercise" of this faculty (33). This in turn implied that the only process by means of which one could achieve consensus about objects of taste would be for those possessed of a "wrong Taste" to submit to the judgment of those possessed of "good Taste." What is unfortunately lacking in Burke's account is any standard for deciding among the many claimants to the latter category.

Though Hume also developed a hybrid of the invariable and variable models of regulation, his approach was to employ a consensus about the inaccessibility of an invariable standard of taste as itself the impetus for self-regulation. Burke and Hume each implied that their texts would answer the question of whether there was an invariable standard of taste, yet only Burke in fact answered this question. Hume's essay, by contrast, oscillates between the possibilities that there is or is not an invariable standard of taste and never conclusively commits itself to one of these positions. Rather, Hume's essay is designed to produce a consensus among its readers that there is *likely* such an invariable standard but also that there is no pressing need to determine what that standard might be. This approach to the question is signaled by the essay's first sentence: "The great variety of Tastes, as well as of opinions, which prevail in the world, is too obvious not to have fallen under every one's observation."[28] Though this sentence is apparently a statement of fact, it cannily asserts a solution to the problem of difference announced in the sentence itself. In stressing that the great variety in taste is something we have *all* noticed, the sentence makes explicit a previously only implicit collective agreement (namely, our collective agreement that judgments of taste differ). The rest of the essay is designed not to disturb this consensus that has just been brought into consciousness but rather to present sufficient evidence both for and against the possibility that there is an invariable standard of taste so that the reader can conclude that the question is likely undecidable. This is not intended to be a discouraging conclusion but rather a productive form of skepticism that convinces us that we have achieved a sufficient form of social unity when we acknowledge that our differences of taste occur within a general frame of agreement concerning the desirability, even if not attainability, of an invariable standard of taste. A collective, invariable standard of taste is in this sense an imaginary orientation point for each individual and the collective as a whole, that is, a point that, when it is understood as virtual rather than actual, enables collective unity.

Malthus and the Subject of Self-Regulation

Whereas the model of regulation by invariable standard posed relatively few conceptual problems—disagreements concerned simply what ought to play the role of standard or which individual had the right to determine that standard—the idea of self-regulation by means of variable standards and large collectives was less straightforward and posed at least three key questions. First: What, specifically, is the nature of the collective within which the activity of self-regulation purportedly occurs? Second: If self-regulation occurs when members of that collective orient themselves toward a standard both variable and impossible to perceive directly, how should we understand this form of mental activity that never grasps its object but that, through the effort to grasp it, produces regularity at a collective level? And, finally: If the variable standard that enables self-regulation is, as Locke claimed, "natural"—and if self-regulation is hence a sort of channeling of a natural movement—what concept of *nature* does self-regulation imply?

We can understand the work of many Romantic authors as attempts to answer at least one of these questions. I suggest in this section that Malthus isolated and named the *collective subject* of self-regulation by reformulating the concept of population, and I argue in the following section that Kant delimited the peculiar *kind of thinking* required of self-regulation through his concept of a "regulative" use of reason.

First, then, Malthus, and his determination of "population" as the entity within which self-regulation occurs.[29] Locke had proposed "the market" as the arena in which self-regulation occurred but had not clarified the agent of that self-regulation. However, as I have noted in earlier chapters, by the mid-eighteenth century, the physiocrats in France had connected market activities to the term "population" and did so in part by discussing economic issues, such as food hoarding and famine, which were more clearly biopolitical in nature than Locke's examples of coinage and interest rates.[30] Inoculation debates in France and Britain had also suggested links between the concept of population and diseases such as smallpox, which could not be commanded away by legal decree.[31] Building implicitly on these earlier discussions, Malthus successfully pinned the thought of self-regulation to the term "population," convincing many of his peers (and subsequent commentators and critics) that self-regulation occurs primarily, or perhaps exclusively, within populations, rather than within alternative terms such as "the people" or "the multitude." Each of these terms—"population," "people," "multitude"—had been used since at least the seventeenth century, but Malthus sought to grasp the logic of self-regulation by relocating regulation to the site of populations.

Malthus argued that populations were characterized by two linked modes of self-regulation. He first claimed that the tendency of populations to expand infinitely was naturally regulated by the death of that portion of the population for which no food was available (1992: 21). However, this first form of natural regulation could itself be regulated—that is, made more regular and less acute in its effects—through rigorous commitment to a second form of self-regulation, namely, economic self-regulation of supply and demand.[32] Malthus hoped to convince his readers that these two dimensions of human existence—the biological and the economic— described completely the self-regulatory potential of populations.

Malthus's difficulty, though, was that the natural regulation of population growth by starvation is different in kind than the natural regulation of economic supply and demand, since the latter requires specific human institutions for its operation, such as private property. Malthus was aware of this issue, and it is not necessarily a problem for his account, since his book was intended to intervene in—that is, to help regulate—the natural regulation of population growth. But the difference in kind between these two modes of population regulation required that Malthus clarify how human *awareness of* regulatory processes—an awareness his book was intended to encourage—related to self-regulation.

As Malthus grappled with this question in each subsequent edition of his book, his account became increasingly self-reflexive—that is, increasingly "Romantic" in its poetics (though sadly not in its prose)—as he found his account of the natural self-regulation of human populations more and more bound up with the question of the effect of his text itself on legislators and the more general population of readers.[33] Not surprisingly, this self-reflexivity revolves around the term "regulation." In the first edition of 1798, "regulation" serves Malthus as both a synonym for human legislation—that is, political directives that seek to enforce behavior among legal subjects according to an invariable standard—and for the natural and inflexible relationship between population and food growth that is the central point of his text.[34] However, in subsequent editions, "regulation" increasingly comes to refer to the effects of his text on various kinds of readers. For example, Malthus suggested in the Appendix to the 1806 edition that if British legislators did not think it "advisable" at this point

> to abolish the poor laws, it cannot be doubted that a knowledge of those
> general principles, which render them [the poor laws] inefficient in their

humane interventions, might be so applied so far to modify them and regulate their execution, as to remove many of the evils with which they are accompanied, and make them less objectionable. (1992: 360)

That is, Malthus's text itself should regulate the relation between political regulation and natural regulation, with the goal of making the natural regulatory movements between populations and food operate in a more regular—less unpredictably catastrophic—fashion.

Yet not all of Malthus's readers were legislators, and he also used the concept of regulation to consider the effects of his account on other kinds of readers. Malthus implied in the 1803 edition, for example, that his text might encourage in readers what he called "moral restraint" (1992: 43–44, 71–42). By this Malthus meant that a reader's knowledge of the natural regulatory relation between population and food should encourage the reader to delay sexual reproduction. Yet as William Hazlitt noted, the possibility of widespread moral restraint threatens Malthus's account of the "natural" law regulating population growth, since that account presumes that population *automatically* tends to expand beyond the food base. Hazlitt noted that the point of Malthus's 1798 text was to convince his readers that "excessive population" was an

evil . . . infinitely greater and more to be dreaded than all others [i.e., evils] put together; and that its approach could only be checked by vice and misery . . . and that in proportion as we attempted to improve the condition of mankind, and lessened the restraints of vice and misery, we threw down the only barriers that could protect us from this most formidable scourge of the species, population.[35]

Yet Malthus then "comes forward again with a large quarto, in which he is at great pains both to say and unsay all that he had said in his former volume, that population is in itself a good thing . . . and that the most effectual as well as desirable check to excessive population is *moral restraint*" (45–46).[36]

Malthus could have responded that whereas Hazlitt's critique assumed that Malthus meant that every member of society could regulate him- or herself according to an invariable standard of moral restraint, the concept of a self-regulating population presumed variable and imperceptible standards, which will be differently grasped by members of a population. Malthus may not have fully grasped that implication of his concept of population, but he did stress differences of interpretation—or rather, differences of interpretative paradigms—among his readers. He considered, for example, those readers who found it difficult to coordinate knowledge and affect (that is, readers unable to "regulate their belief or disbelief by their likes or dislikes"). Malthus suggested that many of these readers were convinced "of the truth of the general principles contained in the Essay" but "lamen[ted] this conviction, as throwing a darker shade over our views of human nature, and tending particularly to narrow our prospects of future improvement" (1992: 360). Because such readers felt sincerely that the *only* impediment to social progress was "the perverseness and wickedness of those who influence human institutions," they found themselves "in a constant state of irritation and disappointment," and so were unable to perceive the "regular progress" in fact made by society (361). Malthus encouraged these readers to understand human society as more akin to a state of probation, in which collective human improvement over time is uncertain or even unlikely. Embracing this lack of certainty of future improvement would allow this class of readers to understand our condition as a challenge that calls forth human powers of creativity and perhaps then leads to human improvement after all:

> If . . . [it is] impossible to feel such a confidence [in future improvement], I confess, that I had much rather believe that some real and deeply-seated difficulty existed, the constant struggle with which was calculated to rouse the natural inactivity of man, to call forth his faculties, and invigorate and improve his mind; a species of difficulty which it must be allowed is most eminently and peculiarly suited to a state of probation. (361)

This is a strange moment in the text, for it opens up the possibility that Malthus has painted such a gloomy portrait of the natural regulatory relation between population and food not because he believes it is true but instead for what he hopes the effects of such a representation on his readers will be. By implying that future improvement requires, paradoxically, the absence of certainty that such improvement will occur, Malthus opened up the possibility that the real goal of his text was the destruction of overweening optimism, rather than the transmission of scientific truths.

This is presumably *not* what Malthus wanted to say, for he otherwise seems committed to the truth value of his propositions about natural population regulation. Yet Malthus's efforts to understand the natural movements of populations in terms of regulation encouraged, seemingly despite his intentions, an increasingly self-reflexive consideration of which *representation* of natural regulation would best regulate human thinking about regulation. Malthus's examples of different kinds of readers using the *Essay* to regulate their relations to natural population regulation highlight that the more he considered the role of conscious thought in the operation of self-regulation, the more complicated and problematic his account of natural population regulation became. Or, to put this another way, though Malthus isolated the population as the subject of self-regulation, he found it difficult to grasp what roles consciousness, self-reflection, and thinking could play in a human population's self-regulation. The result, Hazlitt suggested, was a "vibrating backwards and forwards with a dexterity of self-contradiction which it is wonderful to behold."[37]

This vibration within Malthus's efforts to understand the operation of self-regulation resulted from the disjunction between the site to which Malthus had relocated regulation—the population—and the model of regulation to which he was committed. Malthus's stress on the aggregate entity of population seemed to commit him to the second model of regulation. That is, Malthus's population seemed analogous to Locke's aggregate of investors, each of whom actively, consciously, and through the application of rational capacities sought to intuit the variable standard of "want" and through the collective effect of these individual actions produced a regular and coherent market. Yet Malthus employed the concept of population precisely because its biological reference countered the claims of reformers such as Godwin that the regulation of human affairs ought—or even could—pass through human capacities for reason and self-reflection. Despite relocating the site of regulation from the realm of political subjects to a decentralized population, Malthus still understood regulation as a sovereign imperative of nature that required submission to an invariable standard. What submitted to this standard was not the self-consciousness of legal, political, or economic subjects but biological drives that operated like machines, in that they functioned automatically and apart from self-consciousness, thought, and foresight. Malthus and Godwin thus mirrored each other: Where Godwin claimed we must submit to the sovereign dictates of reason, Malthus asserted that our bodies necessarily obey the natural sovereign dictates of biological generation and political economy. Except, of course, that Malthus recognized that these sovereign imperatives of nature were often ignored by humans, and so his text was intended to supplement the sovereign imperatives of nature. This supplemental status of the text kept Malthus oscillating between the two models of regulation, producing that vibration discerned by Hazlitt.

If Malthus's text exemplifies the difficulty of thinking regulation beyond the schema of sovereign command, even when the site of regulation no longer corresponded to this schema, his text also underscored the dangerous conceptual means by which one could attempt to eliminate this vibration between the two models of regulation, namely, what the political theorist Roberto Esposito has called the modern immunitary paradigm.[38] Esposito introduced his concept of immunity in order to clarify both Foucault's account of the emergence of biopolitics in the eighteenth century and its dark trajectory in the nineteenth and twentieth centuries. He contends that the centrality of the immunitary paradigm for biopolitics allows us to distinguish its programs, such as smallpox inoculation and political economy, from, for example, ancient "Egyptian agrarian politics or the politics of hygiene and health of [ancient] Rome."[39] The immunitary paradigm also allows us to understand why biopolitical management of populations increasingly led, in the nineteenth and twentieth centuries, to what he calls "thanatopolitics," that is, the safeguarding of the life of a "good" part of the population by means of the death of "bad" parts of the population. Esposito asserts that the immunitary paradigm "functions precisely through the use of what it opposes. It reproduces in a controlled form exactly what it is meant to protect us from." As a consequence, the immunitary paradigm aims not for "the good" but for security, understood as the limitation of damage: "Instead of something good being acquired, something bad has been taken away."[40] One of the inherent problems with this approach is that "this self-protective syndrome ends up relegating all other interests to the background, including 'interest' itself as a form of life-in-common; the effect it creates is actually the opposite of what is desired. Instead of adapting the protection to the actual level of risk, it tends to adapt the perception of risk to the growing need for protection— making protection itself one of the major risks" (15–16).

While Esposito does not consider Malthus in his accounts of the late eighteenth and early nineteenth centuries' development of the immunitary paradigm, his description illuminates both the source of Hazlitt's perplexity about Malthus's text and the mechanism by means of which Malthus sought to eliminate the conceptual vibration that Hazlitt noted. Hazlitt contended that where previous writers on the topic of population had assumed that "life is a blessing" and that "the object both of the moralist and the politician was to diminish as much as possible the quantity of vice and misery existing in the world," Malthus distanced himself from this framework by linking population to the question of survival.[41] Malthus claimed that the tendency of populations to increase exponentially led to the potential destruction of the population itself. Malthus also asserted, in a rather clear example of Esposito's immunitary paradigm, that this threat could only be countered through the active introduction of a small and controlled amount of death, such as the elimination of the safety net of the

Poor Laws. By focusing attention on the survival of the populace and by drawing on what was becoming an increasingly common immunitary logic that bound survival to the necessity of moderated small doses of death, Malthus could shift attention away from the conceptual dissonance between his claim that regulation occurs in the differential dynamics of population, on the one hand, and his commitment to the model of regulation as sovereign command, on the other.

Kant, Self-Regulation, and Thinking

If Malthus's slide into the immunitary paradigm resulted from his attempt to eliminate the dissonance between his site of regulation (population) and his model of regulation (sovereign imperative), what would it mean to reconceptualize regulation itself from the perspective of populations? While Kant did not address this question directly, he contributed to an answer by considering more rigorously than Malthus what it might mean to regulate thinking and behavior through the thought of regulation itself. The concept of regulation is central to Kant's critical philosophy, which is premised on a distinction between the "constitutive" principles and concepts of the faculty of the understanding and the "regulative" principles and concepts of the faculty of reason.[42] "Constitutive" means, for Kant, subjective principles, concepts, and categories to which experience must conform. For example, a category of the understanding such as "substance" makes experience possible, and we can investigate specific kinds of substances. "Regulative," by contrast, means for Kant a rule or principle that cannot be given in experience but that can guide the investigation of experience. Heuristic rules such as "always seek to create unity" or "see every cause as itself an effect until you find an uncaused cause" exemplify regulative principles.[43]

Kant stressed that humans tend, in a sense unavoidably, to mistake reason's regulative principles for possible objects of experience. The concept of God as an actual entity who created everything else, for example, is the mistaken projection onto the field of experience of the regulative principle of seeking an uncaused cause behind every proximate cause. The concept of God emerges from a capacity of reason essential for the investigation of experience, for without reason's regulative urgings, we would remain benumbed by an unconnected manifold of sensations. Yet for Kant, the progress of science is hindered when we mistake a regulative principle for an object of possible experience. The recognition of regulative principles as such—that is, as principles that guide our thinking rather than as possible objects of experience—is thus essential to making the progress of knowledge more regular. As Gilles Deleuze stressed, Kant linked the concept of regulation to the human ability to see the world in terms of "problems" to be solved. If human reason sometimes "pose[s] false problems"—for example, asking what attributes God has—this is only because reason "is a faculty of posing problems in general."[44] Though the concepts and principles of the faculty of the understanding allow us to find ourselves embedded in a world of specific objects, the regulative principles of reason enable humans to see the world in terms of problems and their possible solutions.

Kant's understanding of regulation in terms of posing problems isolates a key reason for the ambiguities of Malthus's account of regulation. Malthus recognized the relationship between problem solving and regulation but oscillated between attributing this problem-solving capacity to nature and to humans. In the 1798 *Essay*, Malthus suggested that whenever a human population outstrips its food supply, nature approaches this as a problem, which it solves through a variety of techniques, including human vices, disease, and, finally, famine:

> The power of population is so superior to the power in the earth to produce subsistence for man that, unless arrested by the preventative check, premature death must in some shape or other visit the human race. The vices of mankind are active and able ministers of depopulation. They are the precursors in the great army of destruction, and often finish the dreadful work. But should they fail in this war of extermination, sickly seasons, epidemics, pestilence, and plague, advance in terrific array, and sweep off their thousands and ten thousands. Should success still be incomplete, gigantic famine stalks in the rear and, with one mighty blow, levels the population with the food of the world. (1992: 42–43)

Yet Malthus published his *Essay* precisely because this natural mode of problem solving itself poses a problem for humans, which the latter seek to solve through measures such as the Poor Laws. This implies that problems and problem solving are specifically human activities, rather than processes that occur within nature.

Kant's approach to regulation also clarifies the extent to which Malthus's account of population as a threat requires that readers confuse the mental process of projecting population expansion to its infinite limit with the possibility of such expansion actually occurring in experience. Hazlitt intuited this confusion in Malthus's account and sought to expose it through the following thought experiment:

Let us also suppose that these checks [to population growth] are for a time removed, and that mankind become perfectly virtuous and happy. Well, then, according to the former supposition, this would necessarily lead to an excessive increase of population. Now the question is, to what degree of excess it would lead, and where it would naturally stop. Mr. Malthus, to make good his reasoning, must suppose a miracle to take place; that after population has begun to increase excessively, no inconvenience is felt from it, that in the midst of the "immanent and immediate" evils which follow from it, people continue virtuous and happy and unconscious of the dangers with which they are surrounded; till of a sudden Mr. Malthus opens the flood-gates of vice and misery and they are overwhelmed by them, all at once. In short he must suppose either that this extraordinary race of men, in proportion as population increases, are gradually reduced in size, "and less than smallest dwarfs, in narrow room, throng numberless, like that pygmean race beyond the Indian mount, or fairy elves;" or that they have some new world assigned them as a breeding-place, from which attempting to return they are immediately squeezed to death, like people rushing into a crowded theatre.[45]

Hazlitt's spatial images of rapid transformation—a population of people who slowly grow smaller as their numbers increase, until suddenly, after a certain number, they swiftly grow large, or a population swiftly relocated from one site to another—expose the ways that Malthus encouraged readers to confuse the logical terminus of a regulative rule (a population growing toward an infinite number of members) with that of the actual time- and space-based process of population expansion (which, according to Malthus's own account, could never reach that logical limit). Kant's account of the regulative ideals and principles of reason outlines the more general possibilities for this kind of confusion, of which Malthus's account is one example.

Kant's understanding of regulation as "problematic" keeps the concept of regulation helpfully balanced at the intersection of human creative potential and those natural movements that the sciences reveal. The sciences may provide elements of provisional solutions to human problems by developing figures through which to understand the natural world: for example, the figure of pressure that underwrites much of Malthus's account of the relationship between population growth and food.[46] But Kant's understanding of regulation demands careful attention to possible confusions of regulative principles with objects of experience and stresses that any particular solution to a problem, such as Malthus's claim that the Poor Laws must be eliminated, can never be warranted by nature, since problems and their solutions by *their* very nature emerge from the faculty of reason and not directly from nature in its phenomenally given sense. Or, to put this another way, there can be for Kant no purely natural self-regulation; rather, self-regulation always names a collective human effort to solve a problem.

However, beyond providing grounds for criticizing Malthus's account of regulation, Kant's account of the regulative principles of reason also helps us think more clearly about what it could mean to think population by means of the second model of regulation. This may seem like a strange claim, since "population" was not a concept that Kant employed explicitly in either his critical works or essays. Moreover, Kant's tendency to understand reason as a univocal, invariable standard that speaks in the same voice through every individual seems to preclude any positive appraisal of the diversity of opinions and capacities with which concepts of population had been connected in the seventeenth and eighteenth centuries. Yet by drawing on Hannah Arendt's reading of Kant, we can see that his basic assumptions about the nature of philosophy, his reflections on the principle of purposiveness that he understood as the foundation of judgment, and his approach to the question of a standard of taste oriented him—albeit tentatively—toward a model of self-regulation as a process that took place among the members of populations and did so by means of variable, and hence only partially illuminated, standards.

Arendt stressed that Kant was unusual among philosophers in understanding philosophy not as a discipline for "the few" who would, through its practice, achieve that highest form of life traditionally called the *vita contemplativa* (life of the mind) but rather as a basic human need. Arendt noted that "philosophizing, or the thinking of reason, which transcends the limitations of what can be known, the boundaries of human cognition, is for Kant a human 'need,' the need of reason as a human faculty. It does not oppose the few to the many."[47] For Kant, this basic human need for philosophy resulted from the fact that humans were characterized by an "unsocial sociability" but, as inhabitants of a globe, could not escape from one another. The "unsocial sociability" of humans means that they have an innate "tendency to come together in society," "coupled . . . with a continual resistance which constantly threatens to break this society up."[48] For Kant, as for Malthus, the difficulties presented by unsocial sociability could not be solved by emigration, for humans have "common possession of the earth," and since "the earth is a globe, they cannot disperse over an infinite area, but must necessarily tolerate one another's company."[49] Though Kant's critical philosophy was oriented toward an understanding of "man" as (in Arendt's gloss) a "reasonable being, subject to the laws of practical reason which he gives to himself, autonomous, an end in himself, [and] belonging to a *Geisterreich*, [i.e., a] realm of intelligible beings," Kant *also* intended his texts for human beings understood as "earthbound creatures, living in communities, endowed with common sense, *sensus communis*, [i.e.,] a community sense," and who were "not autonomous" but rather required "each other's company."[50]

The importance of implicit concepts of population, variable standards, and self-regulation for Kant's interpretation of humans as "earthbound creatures, living in communities, [and] endowed with common sense" emerged in several places, including his reflections in the *Critique of Judgment* on the origins of the standards that individuals employ to make judgments of beauty. Kant asked how an individual comes to have a "standard idea" (*Normalidee*) of the various entities that he or she frequently encounters in the world, and answered that the individual produces this standard by abstracting from a population of perceptions. He exemplified this process through the example of a population of human beings:

> When the mind wants to make comparisons, [it] can actually proceed as follows, though this process does not reach consciousness: the imagination projects, as it were, one image onto another, and from the congruence of most images of the same kind it arrives at an average that serves as the common standard for all of them. For instance: Someone has seen a thousand adult men. If now he wishes to make a judgment about their standard size, to be estimated by way of a comparison, then (in my opinion) the imagination projects a large number of the images (perhaps the entire thousand) onto one another. . . . Now if in a similar way we try to find for this average man the average head, for it the average nose, etc., then it is this shape which underlies the standard idea of a beautiful man in the country where this comparison is made. That is why, given these empirical conditions, a Negro's standard idea of the beauty of the [human] figure necessarily differs from that of a white man, that of a Chinese from that of a European. (*CJ* 82 [234])

Against the background of Burke's and Hume's earlier reflections on aesthetics and standards, two aspects of Kant's claims about standards are important. First, though Kant, like Burke, posited an unconscious dimension for the standard of beauty—for Kant, the process of imaginative projection and abstraction "does not reach consciousness"—unconscious activity plays a role opposite that which it played in Burke's account. For Burke, the foundation of the standard of taste is unconscious because it is located in the register of the body, which renders it both universal (everyone has the same standard) and possessed of a sovereign-like imperative (one cannot help but to prefer honey to squills). For Kant, by contrast, a standard idea is *generated* by means of an unconscious *activity*. The unconscious nature of this activity underscores the dependence of a particular individual's standard of beauty upon his or her experience, which in turn implies that this standard can—and should—be revised by means of additional experience. Though Kant distinguished among the standards of beauty of different groups (Negro, white man, Chinese person, European), his point is that each human individual has a unique standard of beauty generated from, and hence limited to, his or her individual experience with that subpopulation of the human species with whom he or she has come into contact. Given Kant's assertions about the inescapable unsocial sociability of humans who are bound to a globe, his account of the genesis of the standard idea of beauty suggests that an individual ought to understand his or her judgments of beauty as provisional and always open to revision upon the basis of a wider experience of the globe and its inhabitants.[51]

The individual and dynamic nature of the standard in Kant's account also underscores—again, contra Burke—that the standard idea operates not as a sovereign imperative that compels individuals to judge in a certain way but rather as an inherently open movement between instance and class. Thus, Kant wrote that

> this *standard idea* is not derived from proportions that are taken from experience *as determinate rules*. Rather, it is in accordance with this idea that rules for judging become possible in the first place. It is the image for the entire kind, hovering between all the singular and multiply varied intuitions of the individuals, the image that nature used as the archetype on which it based its productions within any one species, but which it does not seem to have attained completely in any individual. (*CJ* 82–83 [234–35])

For an individual, the standard idea emerges within experience, and thus though it may "dictate" rules for judging (for example, "individuals who look like this are beautiful, while individuals who look like that are not"), these rules will change on the basis of greater experience with the multitudes of humans who populate the globe.

Even more significant than Kant's hypotheses about the genesis of individual aesthetic standards are his reflections on the regulative "principle of purposiveness" at the heart of both the faculty of judgment and the diversity of aesthetic judgments one finds in a population.[52] Kant takes more seriously than either Burke or Hume the fact of the diversity of judgments of beauty. However, Kant did not aspire, as did Burke, to discover a hidden natural standard that might in principle allow individuals to make identical judgments of taste, nor did he follow Hume in sidestepping the diversity of judgments of taste in order to stress a shared agreement concerning that diversity. *Pace* Burke, Kant believed that the diversity of judgments of taste was unavoidable. *Pace* Hume, Kant contended that it is vital to recognize the *claim* for universality that every judgment of beauty implies. Kant contended that if the universal aspiration of every judgment of beauty was properly understood, it would, as Hume suggested, reveal to us why a *specific* universal standard of taste was beside the point but would do so because it revealed that our thoughts and even perceptions were collective in nature.[53]

Kant moved away from Burke's and Hume's emphasis on an invariable standard of taste by means of his "principle of purposiveness." Kant asserted that whenever we are presented with a novel object or process and we must creatively discover the rule, principle, or law of which that particular is an instance, we are guided by judgment's principle of "the *purposiveness of nature* in its diversity" (*CJ* 20 [Ak. 180]). This is the principle that the individual elements of nature form a unity of the kind that can be understood by our mental faculties (*CJ* 19 [Ak. 180]).[54] Kant stressed that we have no warrant for concluding that nature actually has a coherent and meaningful unity that we can cognize; it might be the case, for example, that nature's order is so complex and alien that it becomes incomprehensible after a certain point. The principle of purposiveness is simply the immanent (a priori) principle of our faculty of judging, rather than a principle of nature. However, this principle is also the only means by which we are able to move from a novel individual to a rule, principle, or law of which that individual would be an instance.

Kant asserted that judgments of taste result when, in the absence of any attempt to make use of or even gain specific knowledge about an object, the *form* of that object brings the faculties of imagination and understanding into harmony with each other. This produces pleasure because the object seems to confirm that harmony between nature's order and our mental powers upon which the principle of purposiveness is premised.[55] Because this harmony is discovered outside of any attempt to obtain knowledge about the object, the faculties relate to one another harmoniously, rather than in what Kant calls a "law-governed" (*gesetzlich*) relationship (*CJ* 162 [Ak. 295]).

> Only where the imagination is free when it arouses the understanding,
> and the understanding, without using concepts, puts the imagination
> into a play that is regular [*ein regelmäßiges Spiel versetz*], does the presentation

communicate itself not as a thought but as the inner feeling of a purposive state of mind. (*CJ* 162 [Ak. 296])

Judgments of beauty result, in other words, when an object facilitates "freedom" and "regular play"—which we might translate as "self-regulated play"—(*regelmäßiges Spiel*) between faculties.[56] Kant's stress on the regularity of this play turns his quasi-equation between freedom and play into an image of regulation. Yet it is an image of regulation divorced from the sovereign schema, for here each faculty can take on the rule of the other.

Because judgments of beauty depend for Kant upon this self-regulated play, the search for a standard of taste in Burke's and Hume's senses is beside the point, for such a search misunderstands the relationship of taste to universality. Burke understood the standard of taste as a set of determinate, universally valid rules that dictated the conditions under which an object would automatically cause a judgment of taste. For Kant, by contrast, a judgment of taste results when the schema of the sovereign imperative is displaced in favor of the schema of regulative play (for example, the specific form of a particular flower enables a self-regulating play between my faculties of imagination and understanding).

From this perspective, Burke's hope to locate universality in a determinate standard of taste mistook the conditions under which judgments of taste are made. For Kant, universality characterizes the mode, rather than the content, of judgment. To judge that a flower is beautiful is, Kant claimed, implicitly to judge that the grounds for that judgment—the feeling of pleasure that results when this object allows my faculties to engage in free play—are available to every individual, though one knows from experience that many individuals will not in fact judge this object in the same way. For Kant, every time an individual makes a judgment of taste, she implicitly proposes that judgment to every human. This is true even when a judgment of taste is made in private and is not shared with anyone else, for the universality inherent in such a judgment necessarily orients it toward "everyone else."[57] For Kant, the "universality" of judgments of taste is their capacity *to orient us toward one another.* Kant captured this sense of orientation toward everyone else in his idea of the *sensus communis,* which is

> the idea of a sense *shared* [by all of us], i.e., a power to judge that in reflecting takes account (a priori), in our thought, of everyone else's way of presenting [something], in order *as it were* to compare our own judgment with human reason in general. . . . Now we do this as follows: we compare our judgment not so much with the actual as rather with the merely possible judgments of

others, and [thus] put ourselves in the position of everyone else, merely by abstracting from the limitations that [may] happen to attach to our own judging. (*CJ* 160 [AK 293–94])

Kant's point was not that, in a judgment of taste, we in fact "compare our judgment" of an object "with the merely possible judgments of others" but rather that a judgment of taste requires our sense that we are judging on the basis of the regulative play of the faculties.

As Arendt—and following Arendt, Linda Zerilli—have noted, Kant's account of the role of *sensus communis* in judgments of taste underscores his more general understanding of thinking as inherently social and hence also as capable of enabling collective human relations determined by something more like regulative play than determinate legislation.[58] This social dimension of thinking is even more evident in what Kant called "common human understanding," in which one explicitly seeks to "think from the standpoint of everyone else" (*CJ* 160 [Ak. 294]), which means "transferring himself to the standpoint of others" (*CJ* 161 [Ak. 295]). As Zerilli notes, this does *not* mean thinking from an abstract, "universal" position but instead denotes attempts to think from the position of concrete human beings, especially those who differ most significantly from me.[59] This process of thinking from the position of other people is, in effect, the conscious, deliberative version of the unconscious, automatic process by means of which each individual's standard idea of the human form emerges. Where the latter process automatically creates a standard idea by running through the specific corporeal particularities of that population of individuals that I have encountered, the former is the process by which a uniquely situated individual consciously attempts to think from the standpoints of many other unique individuals.

In ways that will become useful for my conclusion to this chapter, Arendt developed Kant's reflections on the human plurality implicit in the *sensus communis* into a more general theory of plurality and the collective composition of the common world. As I noted at the end of Chapter 5, Arendt described "the common world" as what comes into being when the individual perspectives of concrete human beings, which cannot be subordinated to any common standard, are brought into connection with one another by means of common objects and institutions:

The reality of the public realm relies on the simultaneous presence of innumerable perspectives and aspects in which the common world presents itself and for which no common measurement or denominator can ever be devised. For though the common world is the common meeting ground of all, those who are present have different locations in it, and the location of one can no more coincide with the location of another than the location of two objects. Being seen and being heard by others derive their significance from the fact that everybody sees and hears from a different position.[60]

For Arendt, the common world must be actively created, in the sense that only by means of common objects and concrete common political practices and institutions can a "common meeting ground" be created, enabling those "innumerable perspectives and aspects in which the common world presents itself" to be brought together. This active and continued creation of the common world enables something like a quasi-immortality of human works and actions, connecting the works and actions of the past, present, and future generations. The common world is "what we enter when we are born and what we leave behind when we die," and as such, it

> transcends our life-span into past and future alike; it was there before we came and will outlast our brief sojourn in it. It is what we have in common not only with those who live with us, but also with those who were here before and with those who will come after us. But such a common world can survive the coming and going of the generations only to the extent that it appears in public. It is the publicity of the public realm which can absorb and make shine through the centuries whatever men may want to save from the natural ruin of time. (55)[61]

Arendt's theory of the common world takes even more seriously than either Malthus or Kant the fact of our common inhabitation of a globe. For Malthus, the fact that we share a world from which emigration is not possible led him to the conclusion that nature and political economy must regulate our relationships to one another; for example, we must "consider chiefly the mass of mankind and not individual instances," which meant, in practical terms, that those who have access to food must steel themselves against the emotional pleas of those without.[62] For Kant, the primary significance of our status as globe dwellers is that we are crowded together, which forces both sides of our contradictory nature—our unsociable sociability—into conflict. For Arendt, by contrast, the common world does not denote simply the fact of a crowded globe. Rather, the common world—or rather, a common world—emerges when concrete, embodied individuals who share a geographic location are connected to one another through specific objects and things, such as agriculture, buildings, and works of art, and employ this connection for the sake of individual and collective judgments. For Arendt, the common world is not identical with the earth (that is, "the limited space for the movement of men") or nature ("the general condition of organic life" and the realm of all processes that appear to us to be automatic).[63] Rather, the common world depends upon "human artifacts"—objects that have been created by human hands, such as buildings—and those "affairs which go on among those who inhabit the man-made world together. To live together in the world means essentially that a world of things is between those who have it in common, as a table is located between those who sit around it" (52). A group's common world must take into account both what Arendt calls earth and nature: For example, buildings will be designed in order to endure the effects of weather and use. But because a common world is the place in which individuals show *who* (and not simply what) they are, the common world explicitly distinguishes itself from all natural and automatic processes. The common world is the site of collective regulation, which results in part from "transferring [oneself] to the standpoint of others" (*CJ* 161 [Ak. 295]).

Post-Romantic Self-Regulation in the Twentieth Century I: Systems Ecology and Neoliberalism

Though concepts of regulation were of widespread importance in the Romantic era, they have been even more central to twentieth- and twenty-first-century understandings of the interrelationships among nature, individuals, and collectives. Concepts of regulation have attained centrality in our own moment along at least three paths. First, since the early twentieth century, the physiological body has been understood in terms of self-regulative processes.[64] Second, this image of physiological regulation was inspirational for population and ecosystem ecologists, who, beginning in the 1940s, used the concept of regulation both to denote circular natural processes, such as the carbon cycle, that linked living beings and the natural environment and to update versions of the population dynamics outlined by Malthus. Third, regulation, this time in the sense of explicitly framed political laws and government agencies, has been a persistent point of critique for an influential wing of Chicago School economists who have argued that the self-regulatory dynamics of economic processes render political regulation both unnecessary and counterproductive.

Though there are significant differences among these more recent concepts of self-regulation, they are nevertheless all characterized by that same oscillation that Hazlitt first isolated in Malthus. This oscillation results from the persisting difficulty of squaring the desire to locate a natural mechanism that makes it possible to remove self-regulation from its passage through human self-consciousness with the fact that such *accounts* of self-regulation necessarily pass through that human consciousness and capacity for action. To underscore the extent to which Romantic approaches to regulation illuminate both the aporias and potentials of several of these twentieth-century reflections on self-regulation, I will briefly discuss two fields—namely, ecosystems ecology and Chicago School economics—in which this oscillation has been especially evident; both bear directly on the questions explored by Malthus and Kant.

Self-Regulation in Ecosystems Ecology

As the historian of ecology Sharon E. Kingsland has noted, up until the 1950s, "ecology had developed largely as a biological subject, in which plants and animals were studied, but humans were ignored."[65] Though nineteenth- and early-twentieth-century ecologists were interested in the impact of human activity on natural environments, they generally understood human activity as an external influence on the internal dynamics of natural processes. Frederic E. Clements's early-twentieth-century concept of "ecological succession," for example, proposed that, in a given region—and in the absence of human engagement—different species succeeded one another until they reached a stable "climax" community.[66] The development of the ecosystem concept in the 1940s and 1950s was, in principle, a point at which this might have changed, for the ecosystem concept itself emerged in attempts to understand how radioactive isotopes such as strontium cycled through different living beings and their environments, and this question was bound up with concern over the potential impact on humans of nuclear weapons.[67] Yet even ecosystem ecologists initially treated humans not as integral parts but primarily as potential disturbers of ecosystems. Given the history of regulation I have sketched here, this was, I suggest, a predictable consequence of the fact that the concept of ecosystem drew heavily both on a paradigm of automatic bodily regulation drawn from physiology and on machine-oriented concepts of systems and regulation drawn from cybernetics.

The link between cybernetics and what would become systems ecology was especially clear in the contribution of the Yale ecologist G. Evelyn Hutchinson to a conference on "Teleological Mechanisms." The purpose of the conference was to reveal the ways that mechanistic processes produced the *appearance* of goal-directed behavior.[68] Hutchinson's paper, "Circular Causal Systems in Ecology," used the terms "regulation" and "self-regulation" to describe systems that "corrected" themselves by returning to a specific state when the system was disturbed by outside influences. Hutchinson outlined many ecological processes that could be described and quantified through this understanding of self-regulation, ranging

> from cases in which at least part of the self-regulatory mechanism depends on purely physical aspects of the structure of the earth, such as the disposition of oceans and continental masses, to cases where the self-regulatory mechanism depends on very elaborate behavior on the part of organisms or groups of organisms.[69]

He considered the cycling of carbon and of methane through the biosphere, the cycling of phosphorus in lakes, and the ways that populations of living beings regulate themselves (with an explicit reference to Malthus: 236). In an intriguing final section, Hutchinson also described both the regulation of what he metaphorically called the "birth" and "death" rates of commodities in "nearly saturated capitalist communities" (243) and the growth of scientific knowledge (243–44). Though these final sentences of Hutchinson's paper pointed toward the possibility that human knowledge is *not* bound by the same kinds of deterministic mechanisms he had sketched out in the rest of the paper, the vast majority of Hutchinson's examples are intended to align the concept of self-regulation with automatic, nonconscious processes.

This latter approach guided influential ecosystem ecologists such as Eugene Odum, Howard T. Odum, and Francis Evans in their formulation of an ecosystem as a "self-regulating entity."[70] For these ecosystem ecologists, an ecosystem was a linkage of living beings and external environment that maintained its identity by means of automatic and nonconscious regulatory mechanisms and processes. As Kingsland notes, the ecosystem ecologist's reassertion of this link between regulation and automaticity meant that though it was possible "to include humans as part of the environment . . . the ecosystem ecologist perceived those humans as operating mostly in opposition to nature's strategy" (203). If ecosystem self-regulation was automatic and machine-like, humans could align themselves *with* nature's strategy only by subordinating human activities to some natural standard or goal. In Tom Odum's case, for example, this meant aligning human activity with what he described as the natural tendency of ecosystems to "maximize[e] 'power,' or the rate of flow of useful energy."[71]

What remained difficult to think within ecosystem ecology, in other words—and what has remained difficult to think even in much more recent schools of ecological thought—was the role of human thought and deliberation in the self-regulation of ecosystems. Kingsland puts this point clearly:

> Since humans affect the operation of ecosystems, understanding ecosystems must involve understanding humans, including how humans relate to nature and how societies function. The study of how societies function must include how science is perceived and used by societies. Therefore ideas, which influence behavior, are also part of ecosystems. (199)

As I will note in the penultimate section of this chapter, the recent concept of the "Anthropocene" is intended precisely to understand human activities as part of ecological processes. However, it is not always clear that even this concept rises to the challenge, suggested by Kingsland, of rethinking the very concept of the self-regulation of the global ecosystem in such a way as to include thought.

Self-Regulation, Markets, and Chicago School Economics

Chicago School neoclassical economics is another important twentieth-century field in which self-regulation emerged as a key reference, though in this case as an object of critique rather than embrace. Ecosystem ecologists relied on images of automatic, machine-like regulation, which had the effect of limiting the role of conscious human thought within ecosystems to that of disturbance. Chicago School economists, by contrast, argued that conscious reflection *was* necessarily part of social self-regulation but also argued that such reflection must be limited to those modes of thought that characterized the economic field. Or rather, the modes of cognition traditionally associated with the economic field—for example, viewing the world in terms of investments and profit—must be understood as so fundamental to human cognition that every kind of human judgment was fundamentally economic in nature. Hence, for these economists, human self-regulation would become possible when every human institution was reenvisioned through and reconfigured by an economic lens.

As historians of economics such as Philip Mirowski, Edward Nik-Khah, and Robert Van Horn document, the economic theory developed within the Chicago School of economics beginning in the 1950s was consciously linked to a broader neoliberal program for social transformation that found its focus in the Mont Pèlerin Society.[72] For the founders of this society, such as Friedrich Hayek and Milton Friedman, Western market-oriented society was under attack both from without, such as the threats posed by Communist Russia and China, but equally from within by reformers who hoped to introduce more government planning and regulation to Western Europe and the United States. Hayek and his compatriots blamed this latter tendency in part on the failure of nineteenth-century liberals to understand that strong capitalist economies would not thrive simply because governments stepped back from intervention in markets; rather, they could thrive only if governments actively formulated law and policy in order to encourage market relations.[73] Those associated with the Mont Pèlerin Society thus sought to formulate the principles of a new liberalism—a neoliberalism—that would guide the public and governments in creating this new liberal order.

Government regulation over industries such as electrical power utilities or interstate trucking was a particular point of dissatisfaction for many neoliberal economists, for such regulation represented for them an effort to "plan" market relations. The Chicago economist George Stigler, for example, sought to prove that government regulation had, at best, no positive effect on prices or supply within a given industry and often had the unintended effect of "artificially" increasing prices and decreasing supply.[74] Stigler argued that regulation was in fact a means by which government favored one company over another, since regulation over an industry inevitably limited competition and determined prices. Businesses, recognizing this fact, worked to "capture" government regulation and regulatory bodies for their own ends.[75] For Stigler, economic regulations and regulatory bodies such as the Food and Drug Administration never worked in the interest of the public but instead to the advantage of some companies—and hence against economic competition.[76]

Though one might object that government regulation expressed the democratic will of voters, Stigler contended that democratic politics was simply not capable of directing economic activity in any virtuous way. Stigler presumed that individuals approached every aspect of life through an economic lens, in the sense of seeking out information in order to make choices that maximize individual interests. This worked well in explicitly economic life, for an individual could then "vote" by purchasing one item rather than another:

> A consumer chooses between rail and air travel, for example, by voting with his pocketbook: he patronizes on a given day that mode of transportation he prefers. A similar form of economic voting occurs with decisions on where to work or where to invest one's capital. The market accumulates these economic votes, predicts their future course, and invests accordingly.[77]

Stigler argued that when it came to political life, though, an individual had no incentive to gather the relevant information concerning whether a particular proposed law regulating a given industry was in his or her interest. Moreover, regulatory measures and agencies were generally created by elected representatives, which further diluted the relationship between voters-cum-consumers and industry regulations (and made it possible for companies to "capture" regulatory bodies). Hence, as Nik-Khah notes,

> Stigler denied that democratic results such as the public's willingness to countenance an expansion of government regulation were an outcome of reasoned reflection, holding instead that they were the inevitable outcome of the poor instincts possessed by the vast majority of people. . . . Yet rather than call for the public to rethink its views and eliminate regulation (a prospect Stigler believed to be unrealistic in most cases), Stigler sought to immunize government policy from the public, for example, by developing for regulators a set of "intelligent guides" [namely, Chicago School–trained economists] and subjecting regulators to performance audits.[78]

Though Stigler would have preferred to eliminate government regulation of industry entirely, he believed that a more practical (albeit second-best) solution was to ensure that regulators were guided by those who understood the ultimate inefficacy of regulation itself and who could therefore hollow out regulatory bodies from within.

Seen in the context of the history of the concept of regulation that I have sketched in this chapter, the animus of Chicago School economists toward government regulation of industry is significant because these economists in fact opposed the view of regulation as sovereign imperative. Their opposition to this understanding of regulation was quite literal, for they presented contemporary government regulation of industry as essentially no different than the decisions of early modern sovereigns to grant royal licenses to specific guilds. In place of the view of regulation as sovereign imperative, they instead promoted a view of population-based collective decision making, which—like Locke—they called "the market." However, for Chicago School economists, the market was the *only* possible site of self-regulation, and attempts to develop other sites—for example, within or by means of democratic politics—were bound to fail. Their solution was to transform all social relations into market relations, so that every aspect of human existence could partake of those virtues of self-regulation enabled by market relations.

Post-Romantic Self-Regulation in the Twentieth Century II: The Commons

While both ecosystem ecology and neoliberal economics largely recapitulate Romantic aporias of self-regulation—the former by cleaving to the schema of sovereign imperative and the latter by limiting the role of human consciousness in self-regulation to market relations—other commentators have sought to move beyond these conceptual problems by drawing on concepts of "the commons." The commons is a concept familiar to scholars of Romantic literature, for multiple studies of Romanticism have underscored the relationship of many British eighteenth-century and Romantic aesthetic concepts (including the concept of the aesthetic itself) to the great process of enclosure of the commons in the name of "improvement" and "progress" that took off in the eighteenth century.[79] More recently, left-leaning commentators, pointing to the ways that global capitalism has exacerbated class, racial, and gender inequities and encouraged environmental devastation, have advocated for recognition of the self-regulatory capacities of people committed to common resources.[80] At the same time, and perhaps more surprising, the commons has also appealed to neoliberal authors, who see in studies of the commons empirical confirmation that government regulation cannot achieve its purported ends.[81] Finally—and for my purposes here, most productive—Bruno Latour has recently oriented his science studies "actor-network" theory toward what he describes as "the progressive composition of the common world," which has much in common with Arendt's claims about the collective creation of the common world.[82] For a wide variety of critics, ranging from neoliberals to those on the far left, the commons has emerged as the solution to the theoretical problems of the concept of self-regulation.

This unanimity may seem more superficial than real, for there are significant and relatively obvious differences in how each commentator understands "the commons." The differences are often announced by the kinds of communities each employs to exemplify a virtuous commons. Left-leaning commentators tend to exemplify the commons with traditional societies (for example, the wild rice–harvesting practices of "the peoples of the Wabigoon Lake Ojibway Nation of Ontario") or point to newer community groups that have organized in opposition to environmental threats such as "toxic dump sites and proposed nuclear plants."[83] Neoliberals, by contrast, illustrate the commons through examples such as "condominium associations and private (sometimes gated) communities that have spread rapidly over recent years across the USA and East Asia."[84]

Despite these differences, both left-leaning and neoliberal commentators use the concept of the commons to denote the same basic aspirations of local embeddedness within a specific geographic site, autonomy (understood as freedom from centralized control), and collective decision making, which is connected to the capacity for learning from experience over time. That is, commentators on both the right and left valorize the commons precisely because they see in this mode of social organization a form of self-regulation that is *not* "automatic"—that is, does not execute a sovereign imperative determined elsewhere—but rather requires the active deliberation of members of a human population. Where these commentators differ fundamentally is on the question of whether this kind of self-regulation can coexist with capitalism. Those on the left argue that the concept of the commons eliminates the distinction between "private" and "public" upon which capitalism relies, while neoliberals argue that commons are fundamentally compatible with the distinction between private and public uses of land and property (see Table 2).

From the perspective of the history of Romantic concepts of self-regulation I have developed here, perhaps the most intriguing and promising of these recent proposals for a renewed understanding of the commons is Bruno Latour's project of "political ecology." Latour's political ecology is, I suggest, a fundamentally neo–Romantic rethinking of self-regulation, for he takes up and recombines all of the threads I have noted, including the progress of the sciences (Kant), the role of aesthetics (Burke, Hume,

Table 2. Relationships of markets and commons in contemporary models of self-regulation

	Second model of regulation
figure of **nature**	variable forces that humans can partially shift/alter
nature of the **standard**	variable, obscure standard
premise about the relationship of the standard to **knowledge**	standard emerges and persists only through dynamic and ongoing linkage of individual, limited perspectives
role of **the individual**	exercise of (fallible) individual judgment
role of **government**	empowering individuals to exercise (fallible) individual judgment
key **example**	market + commons (neoliberals)
	OR
	~~market~~ commons (left-leaning commentators)

Kant), the importance of an expanded notion of population (Malthus), the relationship between economics and the state (Locke, Hume, Smith, Steuart), the avoidance of the immunitary paradigm (Kant), and the relationship of all of these elements to human consciousness and intentional activity (Kant). However, though the commons also names for Latour the site of the reconfiguration of self-regulation, he provocatively suggests that the commons can be assembled only by *rejecting* the very concept of self-regulation.

Latour's political ecology continues his long-standing attempt to develop an antimodern—or perhaps more accurately, amodern—understanding of the relationship between science and politics.[85] For Latour, as for many other commentators, "modernity" begins in Europe in the seventeenth and eighteenth centuries, but Latour argues that modernity should be understood as a fundamental contradiction between theoretical orientation and actual achievement. Modernizers past and present, Latour claims, are committed to the Enlightenment premise that human subjects face a natural world of objects, about which they can establish objective knowledge by means of the sciences, and that the telos of such knowledge is the emancipation of humans from any unchosen dependencies on the natural world. Yet Latour claims that this aspiration is belied by what modernizers in fact produce. Since the seventeenth century, the sciences have not emancipated humans from but rather multiplied attachments to the natural world. As Latour notes,

> science, technology, markets, etc. have *amplified*, for at least the last two centuries, not only the *scale* at which humans and nonhumans are connecting with one another in larger and larger assemblies, but also the *intimacy* with which such connections are made. Whereas at the time of ploughs we could only scratch the surface of the soil, we can now begin to fold ourselves into the molecular machinery of soil bacteria. While three centuries back we could only dream, like Cyrano de Bergerac, of traveling to the Moon, we now run robots on Mars and entertain vast arrays of satellites to picture our own Earth. While in the past, my Gallic ancestors were afraid of nothing except that the "sky will fall on their heads," metaphorically speaking, we are now afraid quite literally that the climate could destroy us.[86]

This increasing entwinement of humans and nature is especially evident and unavoidable in the late twentieth and early twenty-first century, as "miracle" technologies such as asbestos or air conditioning, which developers hoped would emancipate humans from undesired changes in the external natural environment, turn out to have long-lasting detrimental effects on both human health and the ecological cycles within which humans are embedded.[87]

For Latour, examples such as asbestos and global warming reveal that we would be better off recognizing that "we have never been modern" and instead work to create a new, amodern form of political ecology. Giving up on the idea of modernity means abandoning the model of human subjects confronting an objective, external, stable nature (a model, Latour argues, shared by both resolute modernizers *and* their environmentalist opponents).[88] In place of this model, Latour argues, we must begin with the premise that there are many kinds of agents, or what he calls "actants"— individual human beings, human institutions, animals, microbes, gravity, and many others—which can be encouraged into alliances with one another, and that the sciences are a key method by which such alliances are created and maintained.[89] Such a premise, Latour argues, allows us to shift our interest from the modernist obsession with "matters of facts"—that is, purportedly neutral, objective scientific statements about the natural world upon which politicians would then base their moral and policy considerations—to "matters of concern," which enable groups of humans and their nonhuman allies to trace out patiently their many and changing forms of connection.[90]

Latour suggests that this shift in perspective makes possible a new political ecology, within which scientists, politicians, economists, and populations each have a role but that collectively enables a very specific kind of commons-creation, what Latour calls "the progressive composition of the common world."[91] This is not precisely a return to past forms of commons but rather a collective effort of "associations of humans and nonhumans" to decide, by means of "an explicit procedure . . . what collects them and what unifies them in one future common world" (41). The "explicit procedure" that Latour proposes is a modification of democratic representative parliamentary procedure, with scientific work "speaking for" the interests and effects of nonhuman actants. Our task, Latour contends, is "to find out what equipment has to be available to populations in order for them to assemble into a viable collective," with "population" understood as an aggregate that includes both humans and nonhumans. Latour contends that the progressive composition of the common world is the only productive way for humans to engage our current era of the Anthropocene, for it is only through an intensification and amplification of our interconnections with the collectives of the world that groups of humans (and their nonhuman allies) will be able to address realities such as global warming.

Latour suggests, though, that the progressive composition of the common world requires that we abandon the concept of self-regulation. Though modernists sought to emancipate humans from natural constraints, Latour claims that they also always located a core "natural law"—what I have described as the schema of sovereign imperative—that purportedly commanded certain forms of behavior from human beings. Latour implies that the concept of self-regulation is completely dominated by this schema of sovereign command and hence cannot be salvaged. The "notion of self-regulating markets," for example, encourages contemporary economists to believe that it "will be possible to do without the question of government altogether, since the relations that are internal to the collective are going to be similar to those which connect predators and their prey within ecosystems."[92] Latour is equally critical of the concept of self-regulation employed within ecological theory, arguing that while we should embrace James Lovelock's ecological "Gaia" theory, we must pry it loose from the concept of self-regulation that seemed to be so important to Lovelock himself.[93] While the aim for political ecology—namely, the "art of governing without mastery"—sounds something like the Kantian sense of self-regulation that I outlined earlier, Latour suggests that a commanding God and his imperatives are always hidden in the "self" of self-regulation.[94]

As if to underscore that he is reconfiguring the conceptual matrix that I mapped in the first half of this chapter, Latour contends that pursuing the progressive composition of the common world by abandoning the concept and aspiration of self-regulation is equivalent to developing a more "radical" form of liberalism. "If it is true," Latour writes,

> that the word "economy" and the word "liberty" have been linked throughout history, then this liberty should be expanded—yes, radically expanded—to all the devices, experiments, instruments, voting mechanisms, shares and stocks that constitute the makeshift, artificial and constantly reengineered armamentarium of the economy. Liberalism means "not letting anything go, not letting anything pass."[95]

By rejecting the concept of self-regulation but retaining that of liberalism, we can link an expanded understanding of population with both the reflexivity of human thought and care for the ways that humans are attached to the nonhuman agencies of the earth and cosmos.

Latour's political ecology–cum–liberalism seeks to prevent the elements previously gathered under the second model of regulation from being drawn back into the first model. Thus, in place of "the individual," Latour employs the concept of "actants," which expands the concept of population beyond humans to include nonhuman agents and puts the stress on alliances between actants (rather than the simple exercise of individual human judgment). Convinced that the first concept of regulation will always embed itself within the elements of the second model, he seeks to detach his newly configured liberalism from the concept of regulation entirely: a liberalism without self-regulation, in effect. Where more traditional versions of the second model of regulation presumed that a virtuous form of collective automaticity resulted when individuals exercised "freedom" within a specific kind of institution (for example, the market or the commons), Latour extends this sense of virtuous automaticity to *all* semistable collectives of devices, humans, nonhumans, and knowledge. Yet because virtuous automaticity is not limited to one institution, such as the market, but takes place wherever new, stable alliances among actants emerge, there is no longer any need for the dimension of divine imperative and its implication of perpetual stability, which Latour implies underwrites the figure of self-regulation (see Table 3).

Despite the virtues of Latour's proposal, there are nevertheless several difficulties inherent in his liberalism without self-regulation. The first is its extraordinary abstraction, which—despite the thousands of pages that he devotes to this topic across many books and articles—makes it very difficult to determine what, exactly, a "parliament of things" might mean in practice and what its "explicit procedures" would concretely entail.

Table 3. Bruno Latour's liberalism without self-regulation

	All models of regulation/ self-regulation	**Parliament of things**
figure of **nature**	sovereign imperative	variable forces that humans can partially shift/alter
nature of the **standard**	invariable, visible standard	variable, obscure standard
premise about the relationship of the standard to **knowledge**	standard can be determined only after relevant knowledge is assembled at central, univocal point	standard emerges and persists only through dynamic and ongoing linkage of individual, limited perspectives
role of **actants**	automatic submission	**pursuit of interests and development of automatized alliances**
role of **government**	determining and enforcing invariable standards	**enabling a "parliament" that makes a common world possible**
key **example**	machine	(market) (commons)

Second, as Philip Mirowski notes, Latour's "liberalism without self-regulation" is strikingly similar to plain old neoliberalism. Latour's attack on the concept of "society," for example, echoes Margaret Thatcher's claim that "There is no such thing as Society." Latour's elimination of the "Nature/Society divide was characteristic" of ur-neoliberal Friedrich Hayek, and Hayek's "doctrines of 'spontaneous order' and 'complexity' are trademark enthusiasms of the Latourist canon." Mirowski makes a compelling case that these resonances "qualif[y] Latour to be considered a fellow traveler of the neoliberals, at minimum."[96] Given the closeness of Latour's political ecology to neoliberal positions, his unwillingness to stake out a position on the question of the compatibility between capitalism and his version of the commons is especially unfortunate, since precisely this issue divides left-leaning and neoliberal advocates of the commons.[97]

Yet Latour's unwillingness to stake out a position on this particular topic is emblematic of his unwillingness to stake out *any* concrete political position, which is arguably the most problematic aspect of his rejection of the concept of self-regulation.[98] Latour objects to concepts of self-regulation because he believes that they always presume a "natural law" that dictates behaviors. Yet my survey of different Romantic theories of self-regulation and, particularly, my analysis of Arendt's reading of Kant suggest that self-regulation can also denote collective processes of determining specific shared goals. From this perspective, self-regulation can function more in the spirit of a collectively posited Kantian regulative ideal, rather than as a natural law promulgated by a divine sovereign. This in turn would mean that the concept of self-regulation not only allows but in fact demands precisely that articulation of concrete political aspirations that Latour avoids.

Conclusion: Composing the Common World

Though Latour's political ecology is troubled by both its proximity to neoliberalism and (perhaps as a consequence) his unwillingness to articulate concrete political goals, it remains a powerful analytical tool for understanding the possibilities for the concept of self-regulation in our era of neoliberalism and global warming. Our understanding of the threats that result from large-scale human transformation of the natural environment has emerged from ecosystems ecology and its successors, while many contemporary attempts to solve these problems through economic means—for example, cap-and-trade carbon emissions trading—are underwritten by the assumptions of Chicago School economics. Not surprisingly, then, questions of self-regulation tend to dominate accounts of the relationship of neoliberalism and the Anthropocene, whether in the form of the critique of capitalism as committed to unregulatable growth or in the hope that capitalism can be transformed into a harmonious, self-regulating, and ecologically neutral system through carbon-offsetting mechanisms or through new forms of artificial intelligence, which seek to channel population-level differences algorithmically in "smart electrical grids" or "smart cities."[99] While Latour's attempt to reconfigure the elements associated with concepts of regulation—automaticity, standards, individuals, collectives, and government—remains ambiguous, his work underscores those aspects that must be engaged in any renewal, or elimination, of the concept of self-regulation.

We can build upon the strengths of Latour's approach while avoiding its problems by linking his project with Arendt's more Kantian-inspired emphasis on plurality, the common world, and what she calls the human condition. There are significant differences between Latour's and Arendt's projects, not least Latour's emphasis on nonhuman actants, which seems in principle opposed to Arendt's commitment to human exceptionality, and Latour's embrace of the modern sciences, which contrasts with Arendt's claim that the modern sciences are grounded in a suspicion about "givenness" that effectively undermines belief in a common world.[100] Yet several important commonalities trump those differences. Arendt and Latour each stress plurality, and for both, the common world is not something that emerges automatically but instead requires collective composition, care, and the materiality of bodies and things. Is it possible to mix Latour and Arendt in ways that point toward a new understanding of self-regulation?

We can begin by focusing on Arendt's and Latour's shared goal of isolating and delimiting the relationship of automaticity to the common world. For both Arendt and Latour, the common world enables what is in effect the "other" of automaticity. For Arendt, a common world enables what she calls *action* rather than the automaticity characteristic of *behavior*; for Latour, the common world enables parliamentary discussion rather than submission to natural law. Both thus agree that nature—whether understood as laws of physics, population dynamics, political economy, or any other automatic process—cannot "regulate" the activities enabled by the common world.

With that said, though, the common world can itself draw on forms of automaticity in order to ensure its persistence and stability. For Arendt, for example, the common world is composed primarily of human-made objects and institutions that persist through time, and they can do so only because they exploit some "automatic" processes of nature at the expense of others (for example, the ability of stone to persist over long periods of time despite significant seasonal changes of temperature and humidity). Moreover, the objects of the common world—and hence, the common world itself—cannot persist if one ignores natural processes, including those instantiated in new contemporary realities such as global warming, and ensuring the persistence of common objects and institutions in the present requires coordination with these natural processes and realities. Such coordination is not "self-regulation" but rather a condition for enabling a common world within which the exchange of individual perspectives can occur. This exchange of perspectives is itself self-regulating, though only in the sense that it remains distinct from behavior and automaticity.

Latour, for his part, approaches the relationship of the common world and automaticity from the other side, arguing that what has traditionally been called nature is *not* the site of automatic processes but is rather composed of actants who collectively struggle and negotiate with one another. Hence, for Latour too, the common world cannot be understood in terms of automatism or self-regulation. At the same time, though, Latour stresses that these processes of struggle and negotiation can take place only on the basis of "habits." Habits are not precisely automatisms, since habits require a mode of attention and can hence always be changed. Yet habits also provide sufficient "veiling" for projects to be undertaken and for things to occur.[101] In short, for both Latour and Arendt, the common world (Latour) or a common world (Arendt) enables something other than "automatic" self-regulation, yet the (or a) common world itself also requires, in its composition, forms of habit or automaticity.

A key question is that of the telos or teloi that determine how the common world is composed. The implicit telos that underwrites Latour's distinction between better and worse common worlds is the quasi-quantitative regulative ideal of "greatest possible composition"; that is, for Latour, one *ought* to create ever more inclusive common worlds. Latour stresses that political ecology does not aim, as does traditional ecological thinking, at "'total connectivity,' the global system, the catholicity that wants to embrace everything," since political ecology is willing to exclude entities that threaten common worlds.[102] Nevertheless, Latour locates what he calls "virtue" in the movement "from state n to state $n + 1$ that takes into account a greater number of beings or that at least does not lose too many beings along the way" (199). He does not clarify, though, *why* an increase of connections among actants is virtuous. Is it because such an increase better respects the independent rights of each actant to be acknowledged by and connected to the collective? Or is it because, as Arendt suggests, the common world itself necessarily embodies an aspiration of something like "immortality," in the sense that every common world connects multiple generations?

Arendt's emphasis on immortality is compatible with Latour's emphasis on ever-greater inclusion of actants within the common world, if only because contemporary life and earth sciences suggest that without such inclusion, we risk the kind of ecological destruction that makes the collective composition of common worlds very difficult. However, Arendt's perspective clarifies that the goal of the common world is not simply to include more actants but to do so for the sake of enabling "works" that can persist through generational time. Such an approach would in turn allow us to distinguish between neoliberalism and that common project at which Arendt and Latour aim but that Latour misrecognizes as a new form of liberalism.

An important issue on which Arendt and Latour diverge is the question of whether the common world is one or many. Modern science seems to provide for Latour a thread that is able to knit together an increasingly expansive common world and in this way enable linear progress in which a collective, presumably global "we" takes into account an ever-greater number of actants. For Arendt, by contrast, common worlds are always first and foremost local, since they depend on embodied presence in the same architecture and institutions and are best exemplified by the classical Greek city-states (for example, Athens), the ward system of the early US republic, and the briefly lived council and *Räte* systems of early-twentieth-century socialist revolutions.[103] A key question is whether the fact that all locally based common worlds now face a common threat of global warming can itself be enough commonality to produce linkages—and perhaps even a meta–common world—among these common worlds. Drawing on Arendt, Latour, and Esposito, we can say that perhaps this is the case—but only if what unites these common worlds is not a threat but rather a promise (to each) of greater flourishing in the future.

Acknowledgments

In keeping with the emphasis of this book on populations, I would like to thank the large collective of people who helped bring this project to fruition. These include Nancy Armstrong, Alan Bewell, Mav Block, Timothy C. Campbell, James Castell, Chris Catanese, Russell Coldicutt, Kimberley Dimiatridis, Stefani Engelstein, Grace Francese, Lauren Gillingham, Amanda Jo Goldstein, Kevis Goodman, Sara Guyer, Orit Halpern, Kate Hayles, Nathan Hensley, Noah Heringman, Jerry Hogle, Anton Kirchhofer, Deanna Koretsky, Catherine Lee, Anne K. Mellor, John Mulligan, Benjamin Murphy, Julie Murray, Thomas Pfau, Dahlia Porter, Rosalind Powell, Tilottama Rajan, Alan Richardson, Tim Rickert, Hannah Rogers, Alexander Schlutz, Viv Soni, Phil Stillman, Charlotte Sussman, Len Tennenhouse, Gabe Trop, Keir Waddington, Stefan de la Peña Waldschmidt, Haley Walton, Leif Weatherby, and Martin Willis. My thanks as well to the wonderful staffs at the National Humanities Center and the Hanse-Wissenschaftskolleg Institute for Advanced Study (Delmenhorst, Germany), and to both institutions for fellowships that supported parts of this book. I am also very grateful to Sara Guyer, Brian McGrath, Thomas Lay, Robert Fellman, John Garza, Eric Newman, and Fordham University Press for their support of this book, and to the two anonymous reviewers of the manuscript, who provided extremely detailed and helpful responses.

Finally, a most special thanks to Inga, Kaia, and Nankea, for their patience and love.

Notes

Preface

1. I was tempted during the copy-editing process to introduce various COVID-19 references into the chapters themselves: for example, comparing eighteenth-century efforts to alter the "normal" curve of smallpox mortality by means of the then-new practice of inoculation (see Chapters 1 and 2) with our own efforts "to flatten the curve" of COVID-19 infections; illustrating my point about the need of biopolitical campaigns to create "surfaces" that can both track and alter population-level dynamics (see Chapters 1, 2, 3) with our own often politically contentious efforts to develop contact-tracing systems that can accurately track who has been infected with COVID-19 and social-distancing protocols that can alter those infection patterns; and comparing concerns in the eighteenth century about the effects of smallpox infection on commerce and trade (see Chapters 1 and 5) with our own efforts to halt COVID-19 infections without simultaneously destroying businesses and national economies. However, I ultimately decided that these parallels would likely be fairly obvious to readers as well, and so they remain implicit in the chapters that follow.

2. The intuition that liberatory politics could not be dissociated from collective healthcare issues—that is, from biopolitics—was shared in the 1960s and 1970s by feminist groups, the Black Panthers, and gay rights activists, among others. On the importance of healthcare for 1970s feminist activists, see Sandra Morgen, *Into Our Own Hands: The Women's Health Movement in the United States, 1969–1990* (New Brunswick, NJ: Rutgers University Press, 2002); on the Black Panthers, see Alondra Nelson, *Body and Soul: The Black Panther Party and the Fight against Medical Discrimination* (Minneapolis: University of Minnesota Press, 2011); on gay rights activists, see Melinda Cooper, *Family Values: Between Neoliberalism and the New Social Conservatism* (New York: Zone, 2017), 167–214. Cooper provides an illuminating account of how neoliberals (often working in collaboration with neoconservatives) appropriated many elements of these movements by endorsing their suspicion of government and medical and pharmaceutical experts while at the same time dulling or eliminating their progressive and collectivist dimensions.

Introduction

1. William Wordsworth, *Lyrical Ballads, with Other Poems. In Two Volumes*, 1st ed. (London: Printed for T. N. Longman and O. Rees, Paternoster-Row, 1800), 1:xx, v.

2. See Harold Bloom, "The Internalization of Quest-Romance," in *Romanticism and Consciousness: Essays in Criticism*, ed. Harold Bloom (New York: Norton, 1970), 3–24.

3. E. M. Forster's classic account of round and flat characters appears in *Aspects of the Novel* (New York: Harvest, 1956), but Deidre Lynch's *The Economy of Character: Novels, Market Culture, and the Business of Inner Meaning* (Chicago: University of Chicago Press, 1998) provided a much-needed historical contextualization of the emergence of this distinction. On the increasingly numerous and individualized populations of nineteenth-century literature, see Alex Woloch, *The One vs. the Many: Minor Characters and the Space of the Protagonist in the Novel* (Princeton, NJ: Princeton University Press, 2003).

4. On the emergence of our modern concept of "literature," see Douglas Lane Patey, "The Eighteenth Century Invents the Canon," *Modern Language Studies* 18, no. 1 (1988): 17–37; M. H. Abrams, *Doing Things with Texts: Essays in Criticism and Critical Theory*, ed. Michael Fischer (New York: Norton, 1989), 144–46; John Guillory, *Cultural Capital: The Problem of Literary Canon Formation* (Chicago: University of Chicago Press, 1993), esp. 124–33; Terry Eagleton, *Literary Theory: An Introduction*, anniversary ed. (Malden, MA: Blackwell, 2008), 1–18; and Raymond Williams, *Keywords: A Vocabulary of Culture and Society*, rev. ed. (New York: Oxford University Press, 2015), 134–38. On the rise of the concept of experimental literature, see Robert Mitchell, *Experimental Life: Vitalism in Romantic Science and Literature* (Baltimore, MD: Johns Hopkins University Press, 2014).

5. See Mark Rose, *Authors and Owners: The Invention of Copyright* (Cambridge, MA: Harvard University Press, 1993); and James Boyle, *Shamans, Software, and Spleens: Law and the Construction of the Information Society* (Cambridge, MA: Harvard University Press, 1996).

6. See Richard D. Altick, *The English Common Reader: A Social History of the Mass Reading Public, 1800–1900*, 2nd ed. (Columbus: Ohio State University Press, 1998); and Jon Klancher, *The Making of English Reading Audiences, 1790–1832* (Madison: University of Wisconsin Press, 1987).

7. See E. P. Thompson, *The Making of the English Working Class* (New York: Pantheon, 1964); Raymond Williams, *The Country and the City* (London: Chatto and Windus, 1973); Ian Watt, *The Rise of the Novel: Studies in Defoe, Richardson, and Fielding* (Berkeley: University of California Press, 1957); and Franco Moretti, *The Way of the World: The Bildungsroman in European Culture* (London: Verso, 1987).

8. See John B. Bender, *Imagining the Penitentiary: Fiction and the Architecture of Mind in Eighteenth-Century England* (Chicago: University of Chicago Press, 1987); D. A. Miller, *The Novel and the Police* (Berkeley: University of California Press, 1988); and Nancy Armstrong, *Desire and Domestic Fiction: A Political History of the Novel* (New York: Oxford University Press, 1987).

9. Watt, *The Rise of the Novel*, 92; Fredric Jameson, *The Political Unconscious: Narrative as a Socially Symbolic Act* (Ithaca, NY: Cornell University Press, 1981), 17–102, esp. 34; Armstrong, *Desire and Domestic Fiction*, 23–24; Miller, *The Novel and the Police*, x.

10. Michel Foucault, *Society Must Be Defended: Lectures at the Collège de France, 1975–76*, trans. David Macey (New York: Picador, 2003), 242–43. Foucault distinguished between *disciplinary* and *biopolitical* power: "We have . . . two technologies of power which were established at different times and which were superimposed. One technique is disciplinary; it centers on the body, produces individualizing effects,

and manipulates the body as a source of forces that have to be rendered both useful and docile. And we also have a second technology which is centered not upon the body but upon life: a technology which brings together the mass effects of a population, which tries to control the series of random events that can occur in a living mass, a technology which tries to predict the probability of those events (by modifying it, if necessary), or at least tries to compensate for their effects. . . . Both technologies are obviously technologies of the body, but one is a technology in which the body is individualized as an organism endowed with capacities, while the other is a technology in which bodies are replaced by general biological processes" (249).

11. Russell Hardin, *Liberalism, Constitutionalism, and Democracy* (Oxford: Oxford University Press, 1999), 43. Hardin claims that political liberalism "began in the seventeenth century with the effort to establish a secular state in which some religious differences would be tolerated," while economic liberalism emerged in the eighteenth century and "came into being without a party or an intellectual agenda" (42). The "core concern of political liberalism," Hardin claims, "is the individual, while the dominant concern in the main, long tradition of economic liberalism that passes through Smith is focused on the general prosperity of the society, not on individual advantage" (43).

12. *Free to Choose: A Personal Statement* is the title of Milton and Rose D. Friedman's extraordinarily successful neoliberal manifesto, first published in 1980 and subsequently turned into a ten-part US Public Broadcasting Station (PBS) special.

13. Foucault's approach thus also provides another way of approaching the question of the "communal" dimension of liberal approaches to social relations. For non-Foucauldian discussions of these issues, see L. T. Hobhouse, *Liberalism and Other Writings* (Cambridge: Cambridge University Press, 1994); Stefan Collini, *Liberalism and Sociology: L. T. Hobhouse and Political Argument in England, 1880–1914* (Cambridge: Cambridge University Press, 1979); and Alan Ryan, *The Making of Modern Liberalism* (Princeton, NJ: Princeton University Press, 2012), 91–106.

14. Thomas Malthus, *An Essay on the Principle of Population, as It Affects the Future Improvement of Society. With Remarks on the Speculations of Mr. Godwin, M. Condorcet, and Other Writers* (London: J. Johnson, 1798). I return to Malthus's account of population, as well as competing models of population, in Chapters 1, 2, 3, and 6.

15. This point is arguably implicit in a text such as Armstrong's, for she contends that the eighteenth- and nineteenth-century British novel did not mirror an existing set of norms but rather *created* new norms, to which readers then conformed. My focus in this book is upon what we might think of as the conditions of possibility for such norm creation, and my argument is that these conditions of possibility exceed the bourgeois, liberal frame within which literary critics have tended to restrict them.

16. Because my goal is to reclaim and rehabilitate biopolitics, rather than liberalism, and to do so by positioning liberalism as simply one mode of biopolitics, my goals diverge from the Victorian literary critic David Russell's efforts to "reclaim or rehabilitate" aspects of Victorian liberalisms eclipsed either in that period or by subsequent versions of liberalism (most prominently, neoliberalism); David Russell, "Aesthetic Liberalism: John Stuart Mill as Essayist," *Victorian Studies* 56, no. 1 (2013): 7–307. Russell points to David Wayne Thomas's attempts to revive a liberal

understanding of agency that keeps its distance from both "liberalism as imperialism or as atomistic individualism" and to Amanda Anderson's efforts to "recuperat[e] a liberal ethos of a rigorous critical reason that has been foreclosed by the successes of laissez-faire neoliberalism and poststructuralist theory alike" (7). Russell himself describes and promotes a version of "aesthetic liberalism" that, he argues, appears in John Stuart Mill's essays. Though I do not seek to reclaim or rehabilitate earlier liberalisms, the fact that liberalism has grasped, arguably more fully than any other political tradition, the importance of individual difference for biopolitics means that my recuperative project also attempts to take seriously the various schools and histories of liberalism.

17. Samuel Taylor Coleridge and William Wordsworth, *Lyrical Ballads, with a Few Other Poems* (London: Printed for J. & A. Arch, Gracechurch-Street, 1798), i; Edmund Burke, *Reflections on the Revolution in France, and on the Proceedings in Certain Societies in London Relative to That Event. In a Letter Intended to Have Been Sent to a Gentleman in Paris* (London: Printed for J. Dodsley, in Pall-Mall, 1790), 188.

18. An important exception is Jerome Christensen, *Romanticism at the End of History* (Baltimore, MD: Johns Hopkins University Press, 2000), which takes up explicitly the question of whether "Romantic liberalism" is an anachronism. Christensen concludes that, even if it is, we should nevertheless actively engage in such anachronism (see, e.g., 145–47). Christensen also stresses the different modes of Romantic liberalism (for example, the "corporate liberalism of Coleridge" versus the "bureaucratic liberalism of Scott" [8]), and his interest in "formulat[ing] a policy of cryptoliberalism that will motivate . . . a reconsideration of the kind of work that humanists should be doing" (8) resonates with my own interest in rethinking liberalism and biopolitics. Other recent discussions of relationships between Romanticism and liberalism include Peter L. Thorslev Jr., "Post-Waterloo Liberalism: The Second Generation," *Studies in Romanticism* 28, no. 3 (1989): 437–61; Jonathan David Gross, *Byron: The Erotic Liberal* (Lanham, MD: Rowman & Littlefield, 2001); Julie Murray, "Company Rules: Burke, Hastings, and the Specter of the Modern Liberal State," *Eighteenth-Century Studies* 41, no. 1 (2007): 55–69; Anne Frey, *British State Romanticism: Authorship, Agency, and Bureaucratic Nationalism* (Stanford, CA: Stanford University Press, 2010); Brent Lewis Russo, "Romantic Liberalism," PhD diss., University of California–Irvine, 2014; Daniel Stout, *Corporate Romanticism: Liberalism, Justice, and the Novel* (New York: Fordham University Press, 2017); and Jamison Kantor, "Immortality, Romanticism, and the Limit of the Liberal Imagination," *PMLA* 133, no. 3 (2018): 508–25. Scholars have more often linked Geman Romanticism to liberalism, though often by distinguishing Romanticism from liberalism and conservatism; see, for example, Frederick C. Beiser, *Enlightenment, Revolution, and Romanticism: The Genesis of Modern German Political Thought, 1790–1800* (Cambridge, MA: Harvard University Press, 1992). The relevant philosophical, historical, and literary-critical scholarship on eighteenth-, nineteenth-, and twentieth-century liberalism is vast, but I have found the following particularly helpful. Pierre Manent's *An Intellectual History of Liberalism* (Princeton, NJ: Princeton University Press, 1994) and Ryan's *The Making of Modern Liberalism* provide good philosophical-historical accounts of liberalism. Collini's *Liberalism and Sociology*, Michael Freeden's *The New Liberalism: An Ideology of Social Reform* (Oxford: Oxford

University Press, 1986), and Mitchell Dean's *The Constitution of Poverty: Toward a Genealogy of Liberal Governance* (New York: Routledge, 1991) are helpful intellectual-historical accounts of eighteenth- to early-twentieth-century liberalisms. On the historical consolidation of the trinity of the "ideologies" of liberalism, conservatism, and Marxism in the nineteenth century, see Immanuel Maurice Wallerstein's *Unthinking Social Science: The Limits of Nineteenth-Century Paradigms* (Philadelphia: Temple University Press, 2001) and *The Modern World-System IV: Centrist Liberalism Triumphant, 1789–1914* (Berkeley: University of California Press, 2011). On the importance of built environments and technologies for liberalism, see Chris Otter, "Making Liberal Objects: British Techno-Social Relations, 1800–1900," *Cultural Studies* 21, no. 4–5 (2007): 570–90; and Chris Otter, *The Victorian Eye: A Political History of Light and Vision in Britain, 1800–1910* (Chicago: University of Chicago Press, 2008). On whether there is a specifically "liberal" aesthetics, see Lionel Trilling, *The Liberal Imagination* (New York: NYRB, 2008); Russell, "Aesthetic Liberalism"; and Amanda Anderson, *Bleak Liberalism* (Chicago: University of Chicago Press, 2016). For an account of the ways that the inherent paradoxes of the classical liberal, tort-based legal order led directly to population-based (and hence biopolitical) social insurance systems in the United States, see John Fabian Witt, *The Accidental Republic: Crippled Workingmen, Destitute Widows, and the Remaking of American Law* (Cambridge, MA: Harvard University Press, 2004). I have also profited immensely from Domenico Losurdo, *Liberalism: A Counter-History*, trans. Gregory Elliott (New York: Verso, 2011), which develops a well-researched and compelling account of liberalism always having been committed to a division between the small number of those "fit" for individual liberty and the much greater number of human beings who must be ruled in illiberal ways in order to secure freedom for the privileged few. I note at several points in this book the convergence between Losurdo's history of liberalism and Esposito's account of the "immunitary logic" of modernity, to which I return in this introduction.

19. See, e.g., Manent, *An Intellectual History of Liberalism*; and Ryan, *The Making of Modern Liberalism*.

20. John Stuart Mill, "Autobiography," in *Collected Works of John Stuart Mill*, 33 vols., ed. John M. Robson (Toronto: University of Toronto Press, 1963–1991), 1:149–53; and John Stuart Mill, *On Liberty*, in *Collected Works*, 18:261–62, 300. I discuss Mill at length in Chapter 5.

21. Recent examples of Victorian literary critical interest in liberalism include Lauren M. E. Goodlad, *Victorian Literature and the Victorian State: Character and Governance in a Liberal Society* (Baltimore, MD: Johns Hopkins University Press, 2003); Elaine Hadley, *Living Liberalism: Practical Citizenship in Mid-Victorian Britain* (Chicago: University of Chicago Press, 2010); and Kathleen Frederickson, *The Ploy of Instinct: Victorian Sciences of Nature and Sexuality in Liberal Governance* (New York: Fordham University Press, 2014). The centrality of liberalism for Victorian literary scholars is partly explained by the fact that it was during this literary period that members of a political party began referring to themselves as "Liberals" and created an explicitly titled "Liberal Party" in 1859. However, to restrict liberalism to the emergence of a party bearing that name risks discounting the long history of earlier uses of the term "liberal" as well as the importance of "the emancipatory movements of the *liberals* in

Spain and the insurgents in Italy and Greece, which erupted in the second and third decades of the nineteenth century and which were supported spiritually and materially by Shelley, Byron, Hazlitt, and Hunt." Christensen, *Romanticism at the End of History*, 146. Focusing on the emergence of the Liberal political party also risks missing the earlier development of those concepts and techniques that were adopted by the Liberal Party. For a brief, but helpful, account of uses of the term "liberal" in the late eighteenth century as both an adjective (e.g., "a liberal system of policy") and a nominative (e.g., a 1780 letter in the *Pennsylvania Packet* advocated for the abolition of slavery and was signed "A Liberal"), see Losurdo, *Liberalism*, 58–59, 241–46.

22. In 1970, Carl Woodring described his book *Politics in English Romantic Poetry* (Cambridge, MA: Harvard University Press, 1970) as in part a response to "the oddity that almost all students of English literature equated romanticism with revolt and that almost all social scientists equated romanticism with conservative reaction" (vii). Woodring's work, as well as subsequent accounts of the politics of English Romantic poetry by other critics, shifted the disciplinary split that Woodring noted into Romantic literary criticism itself, in the sense that Romantic literary critics often subsequently sought to determine when a given poet shifted his political allegiances from radical to conservative (e.g., Nicholas Roe, *Wordsworth and Coleridge: The Radical Years* [New York: Oxford University Press, 1990]), or whether a poet earlier taken as radical was in fact always a conservative (e.g., James K. Chandler, *Wordsworth's Second Nature: A Study of the Poetry and Politics* [Chicago: University of Chicago Press, 1984]), or what kinds of text ought properly to be called radical (e.g., Kevin Gilmartin, *Print Politics: The Press and Radical Opposition in Early-Nineteenth-Century England* [Cambridge: Cambridge University Press, 1996]).

23. Ryan, *The Making of Modern Liberalism*, 21.

24. Ryan contrasts classical liberalism—which includes Locke and Smith but also the twentieth-century economist Friedrich Hayek—with "modern liberalism," exemplified by authors such as John Stuart Mill and L. T. Hobhouse. Ryan suggests that modern liberals were committed not simply to the premise that each individual ought to determine his or her own best interests but also believed that the individual—and society more generally—should "progress," which required that the individual be free from "the fear of hunger, unemployment, ill health, and a miserable old age," and that the individual perpetually seek to improve herself. As Ryan notes, these positions tended to encourage support for a powerful state focused on the welfare of its citizens and hence undercut the position that property is "sacrosanct." Ryan, *The Making of Modern Liberalism*, 25.

25. My Foucault-inspired approach to liberalism leaves me unconvinced by attempts to distinguish between "political" and "economic" liberalism. As I noted above, Hardin distinguishes in *Liberalism, Constitutionalism, and Democracy* between "political liberalism" (the "core concern" of which is the individual) and "economic liberalism" (which focuses "on the general prosperity of the society, not on individual advantage" [43]). Such an attempt to distinguish between these two modes of liberalism obscures the fact that key "political" liberal theorists such as John Locke and John Stuart Mill were also economic theorists/liberals. More important, this

distinction also overlooks the political implications of the epistemological shift stressed by Foucault, which is common to both Hardin's political and economic liberalisms, and suggests that, rather than being fully distinct modes of liberalism, political and economic liberalisms instead are simply two different tactics for achieving the same end of limiting sovereign power.

26. See, e.g., Sara Emilie Guyer, *Reading with John Clare: Biopoetics, Sovereignty, Romanticism* (New York: Fordham University Press, 2015); Ron Broglio, *Beasts of Burden: Biopolitics, Labor, and Animal Life in British Romanticism* (Albany: State University of New York Press, 2017); Amy Mallory-Kani, "'Contagious Air(s)': Wordsworth's Poetics and Politics of Immunity," *European Romantic Review* 26, no. 6 (2015): 699–717; and the essays collected in the "Romanticism and Biopolitics" Praxis Series special issue of *Romantic Circles* (https://www.rc.umd.edu/praxis/biopolitics).

27. In her astute "Response" in the "Romanticism and Biopolitics" Praxis Series special issue of *Romantic Circles*, Eva Geulen makes a similar point, noting that though "many analyses tend to use the perspective of biopolitics to indict or, at least, challenge the cherished icons of the Romantic tradition," the "authors collected in this volume are determined to relieve Romanticism from any biopolitical charges and suspicions" (para. 7).

28. See Giorgio Agamben, *Homo Sacer: Sovereign Power and Bare Life*, trans. Daniel Heller-Roazen (Stanford, CA: Stanford University Press, 1998); and Giorgio Agamben, *Remnants of Auschwitz: The Witness and the Archive* (New York: Zone, 1999).

29. See especially Roberto Esposito, *Bíos: Biopolitics and Philosophy*, trans. Timothy Cambell (Minneapolis: University of Minnesota Press, 2008); Roberto Esposito, *Immunitas: The Protection and Negation of Life*, trans. Zakiya Hanafi (Malden: Polity, 2011); and Roberto Esposito, *Third Person: Politics of Life and Philosophy of the Impersonal*, trans. Zakiya Hanafi (Malden: Polity, 2012).

30. Esposito, *Bíos*, 56.

31. I also find helpful Alexander G. Weheliye's critique of the color-blindness of several important existing accounts of biopolitics, especially that developed by Agamben; see his *Habeas Viscus: Racializing Assemblages, Biopolitics, and Black Feminist Theories of the Human* (Durham, NC: Duke University Press, 2014). However, the notable absence of Esposito from Weheliye's critique encourages my sense that Esposito resolves some of those difficulties in Agamben's approach to which Weheliye points.

32. Benjamin Constant's 1819 "The Liberty of the Ancients Compared with That of the Moderns," in *Political Writings* (Cambridge: Cambridge University Press, 1988), 308–28, provides the classic liberal distinction between ancient and modern senses of liberty.

33. Esposito, *Bíos*, 70.

34. As Esposito stresses in *Bíos* (70–71), even the supposedly "positive" liberty described by Isaiah Berlin in his famous "Two Concepts of Liberty" essay is a fundamentally negative concept, defined primarily by what one wishes to avoid (in Berlin's words, "external forces of any kind," 178).

35. Esposito's own understanding of the relationship of liberalism and biopolitics is not entirely clear, in part because *Bíos* and *Third Person* present significantly different accounts of this relation. In *Bíos*, Esposito claims that liberalism's innate "tendency to

intervene legislatively" and its commitment to self-preservation ensured that "liberal individualism" transformed into nationalism in the nineteenth century and into totalitarianism in the twentieth century (76). This minimal (and, to my eyes, not convincing) account of the transformation of liberalism into totalitarianism is rejected in *Third Person*, in which Esposito contends that "liberalism . . . came out as the real winner in the epochal double battle against Nazism and communism," and as a consequence, we must not "blur the clear boundary that separates the bio-thanatopolitics of the Nazi State from the individual biopolitics of the liberal type, which represents its clear reversal. While the first is based on an increasingly totalized restriction of freedom, the second is devoted to the progressive expansion of freedom. But it does remain bound to the same imperative, which is to manage life productively: in the first case, to benefit the racial body of the chosen people; and in the second, to benefit the body of the individual subject who becomes its master" (91). This latter understanding of the biopolitical constellation of liberalism is much more convincing and guides my reflections in subsequent chapters.

36. Esposito, *Bíos*, 12. With that said, Esposito does approach something like a positive claim about immunity and biopolitics in the final chapter of *Immunitas*, 145–77.

37. My approach also aligns with an apparent shift in Michel Foucault's thinking, for though in *The Birth of Biopolitics: Lectures at the Collège de France, 1978–79*, trans. Graham Burchell, ed. Michel Senellart (New York: Palgrave Macmillan, 2008) he positioned "liberalism as the general framework of biopolitics" (22n), he sought in his later lecture series, devoted to practices of "the care of the self," to theorize something like an affirmative biopolitics. On this point, see especially Thomas Lemke, "Beyond Foucault: From Biopolitics to the Government of Life," in *Governmentality: Current Issues and Future Challenges*, ed. Ulrich Bröckling, Susanne Krasmann, and Thomas Lemke (New York: Routledge, 2011); and Miguel Vatter, "Foucault and Hayek: Republican Law and Civil Society," in *The Government of Life: Foucault, Biopolitics, and Neoliberalism*, ed. Vanessa Lemm and Miguel Vatter (New York: Fordham University Press, 2014), 163–84.

38. Useful accounts of the origins and nature of neoliberalism include Foucault, *The Birth of Biopolitics*, 51ff.; Philip Mirowski and Dieter Plehwe, eds., *The Road from Mont Pèlerin: The Making of the Neoliberal Thought Collective* (Cambridge, MA: Harvard University Press, 2009); and Philip Mirowski, *Never Let a Serious Crisis Go to Waste: How Neoliberalism Survived the Financial Meltdown* (New York: Verso, 2014). David Harvey's *A Brief History of Neoliberalism* (New York: Oxford University Press, 2005) provides a helpful historical account of neoliberalism, though Harvey is not convinced of the utility of the term itself.

39. Mirowski discusses some reasons for this in *Never Let a Serious Crisis Go to Waste*; see also Naomi Klein, *The Shock Doctrine: The Rise of Disaster Capitalism*, 1st ed. (New York: Metropolitan, 2007).

40. For an overview of the logic of smartness, see Orit Halpern, Robert Mitchell, and Bernard Dionysius Geoghagen, "The Smartness Mandate: Notes toward a Critique," *Grey Room* 68 (2017): 106–29.

41. The influence of Newton was often explicit, as in David Hume's approving citation in the *Enquiry Concerning the Principles of Morals* (London: Printed for

A. Millar, 1751) of "Newton's second rule of philosophizing": that is, "where any principle has been found to have great Force and Energy in one Instance, to ascribe to it a like Energy in all similar Instances" (61). For a helpful discussion of the ways that political economists such as Hume and Adam Smith both followed but also departed from Newton's method, see Leonidas Montes, "Newton's Real Influence on Adam Smith and Its Context," *Cambridge Journal of Economics* 32 (2008): 555–76.

42. John Arbuthnot, *Mr. Maitland's Account of Inoculating the Smallpox Vindicated, from Dr. Wagstaffe's Misrepresentations of That Practice, with Some Remarks on Mr. Massey's Sermon* (London: Printed and sold by J. Peele, at Lock's Head in Paternoster-Row, 1722), 39. I return to these debates about smallpox inoculation in Chapters 1 and 5.

43. Philip Mirowski has documented in *More Heat Than Light: Economics as Social Physics; Physics as Nature's Economics* (Cambridge: Cambridge University Press, 1989) the ways that late-nineteenth-century economists sought to secure economics as a science by drawing on the theoretical frameworks and equations of physics. These economists argued that though authors such as Hume, Smith, and Malthus had *aspired* to create a real science, they had unfortunately lacked the mathematical skills and tools to do so. In *Machine Dreams: Economics Becomes a Cyborg Science* (Cambridge: Cambridge University Press, 2002) and *Never Let a Serious Crisis Go to Waste*, Mirowski tracks the subsequent (and never resolved) battles of twentieth-century and twenty-first-century economists to present their project as really a "hard" natural science.

44. Michel Foucault, "Sexuality and Solitude," in *The Essential Works of Foucault, 1954–1984*, vol. 1: *Ethics*, ed. Paul Rabinow (New York: The New Press, 1997), 177. Foucault explicitly distinguished technologies of the self from "techniques that permit one to determine the conduct of individuals"—that is, from "techniques of domination," which have traditionally been the focus of Marxist analyses (but also, as Foucault notes, of his own earlier accounts of "discipline" in a work such as *Discipline and Punish*). Foucault developed brief discussions of governmentality in his lectures "Governmentality," in *The Essential Works of Foucault, 1954–1984*, vol. 3: *Power*, ed. James D. Faubion (New York: The New Press, 2000), 201–22; and "'Omnes et Singulatim,'" in *Essential Works*, 3:298–325; and extensive analyses of different ancient and early Christian practices of self-government in the lecture series *On the Government of the Living: Lectures at the Collège de France, 1979–80*, trans. Graham Burchell, ed. Michel Senellart (New York: Palgrave Macmillan, 2014); *Hermeneutics of the Subject: Lectures at the Collège de France, 1981–82*, trans. Graham Burchell, ed. Frédéric Gros (New York: Palgrave Macmillan, 2005); *The Government of Self and Others: Lectures at the Collège de France, 1982–83*, trans. Graham Burchell, ed. Frédéric Gros (New York: Palgrave Macmillan, 2011); and *The Courage of the Truth: Lectures at the Collège de France, 1983–1984*, trans. Graham Burchell, ed. Frédéric Gros (New York: Palgrave Macmillan, 2011).

45. See especially Lemke, "Beyond Foucault"; and Thomas Lemke, "The Risks of Security: Liberalism, Biopolitics, and Fear," in *The Government of Life: Foucault, Biopolitics, and Neoliberalism*, ed. Vanessa Lemm and Miguel Vatter (New York: Fordham University Press, 2014), 59–74.

1. Biopolitics, Populations, and the Growth of Genius

1. Michel Foucault reflects on biopolitics in *Society Must Be Defended: Lectures at the Collège de France, 1975–76*, trans. David Macey (New York: Picador, 2003); in *Security, Territory, Population: Lectures at the Collège de France, 1977–78*, trans. Graham Burchell, ed. Michel Senellart (New York: Palgrave Macmillan, 2007); and in *The Birth of Biopolitics: Lectures at the Collège de France, 1978–79*, trans. Graham Burchell, ed. Michel Senellart (New York: Palgrave Macmillan, 2008), which latter also contains his most extensive discussion of liberalism.

2. Frances Ferguson developed an early, important discussion of eighteenth-century population theory and literature in "Malthus, Godwin, Wordsworth, and the Spirit of Solitude," in *Literature and the Body: Essays on Populations and Persons*, ed. Elaine Scarry (Baltimore, MD: Johns Hopkins University Press, 1988), 106–24; more recent discussions of literature and Malthusian concepts of population include Maureen N. McLane, *Romanticism and the Human Sciences: Poetry, Population, and the Discourse of the Species* (Cambridge: Cambridge University Press, 2000); Catherine Gallagher, *The Body Economic: Life, Death, and Sensation in Political Economy and the Victorian Novel* (Princeton, NJ: Princeton University Press, 2006); and Scott R. MacKenzie, *Be It Ever So Humble: Poverty, Fiction, and the Invention of the Middle-Class Home* (Charlottesville: University of Virginia Press, 2013). Charlotte Sussman's recent *Peopling the World: Representing Human Mobility from Milton to Malthus* (Philadelphia: University of Pennsylvania Press, 2020) considers the implications of both pre-Malthusian and Malthusian population theory for literary study.

3. Foucault, *The Birth of Biopolitics*, 317.

4. William Petty, *Another Essay in Political Arithmetick, Concerning the Growth of the City of London with the Measures, Periods, Causes, and Consequences Thereof, 1682* (London: Printed by H. H. for Mark Pardoe, 1683), 38.

5. John Guillory, *Cultural Capital: The Problem of Literary Canon Formation* (Chicago: University of Chicago Press, 1993), 85–133.

6. William Petty, *Political Arithmetic, or a Discourse Concerning, the Extent and Value of Lands, People, Buildings [Etc.]* (London: Printed for Robert Clavel at the Peacock and Hen. Mortlock at the Phenix in St. Paul's Church-yard, 1690), preface, unnumbered page. Petty pursued this project in other texts, such as *Another Essay in Political Arithmetick* and *Five Essays in Political Arithmetick* (London: Printed for Henry Mortlock, 1687). Ted McCormick's *William Petty and the Ambitions of Political Arithmetic* (Oxford: Oxford University Press, 2009) provides a helpful introduction to political arithmetic and an account of the differing roles played by Petty in various histories of economic theory. See also Keith Tribe's brief but helpful contextualization of political arithmetic in *Land, Labour, and Economic Discourse* (London: Routledge & Kegan Paul, 1978), 86–88. Important literary-critical approaches to political arithmetic include Mary Poovey, *A History of the Modern Fact: Problems of Knowledge in the Sciences of Wealth and Society* (Chicago: University of Chicago Press, 1998), 120–37; David Glimp, *Increase and Multiply: Governing Cultural Reproduction in Early Modern England* (Minneapolis: University of Minnesota Press, 2003), 146–80; and Charlotte Sussman, "The Colonial Afterlife of Political Arithmetic: Swift, Demography, and Mobile Populations," *Cultural Critique* 56 (2004): 96–126.

7. Petty, *Political Arithmetic*, 1.

8. See, e.g., Joseph Schumpeter, *History of Economic Analysis* (New York: Oxford University Press, 1954), 209–12.

9. On population growth figures, see the entirety of Petty, *Another Essay*, as well as *Political Arithmetic*, 97; on the costs of keeping laborers alive, see *Political Arithmetic*, 102.

10. See Foucault's discussion of the shifting relationship between "population" and "people" in *Security, Territory, Population*, 42–44.

11. Francis Bacon, *The Essays, or Councils, Civil and Moral, of Sir Francis Bacon, Lord Verulam, Viscount St. Alban with a Table of the Colours of Good and Evil, and a Discourse of the Wisdom of the Ancients* (London: Printed for H. Herringman, R. Scot, R. Chiswell, A. Swalle, and R. Bentley, 1696), 38.

12. Petty, *Another Essay*, 28.

13. Petty, *Another Essay*, 29.

14. As George R. Havens notes in "Rousseau, Melon, and Sir William Petty," *Modern Language Notes* 55, no. 7 (1940): 499–503, Rousseau certainly knew of Petty and likely read at least one of Petty's texts on political arithmetic.

15. Jean-Jacques Rousseau, *Politics and the Arts: Letter to M. D'Alembert on the Theatre*, trans. Allan Bloom (Ithaca, NY: Cornell University Press, 1960), 60.

16. The relationship between political arithmetic and political economy has been contested since the mid-eighteenth century. Adam Smith famously distanced political economy from political arithmetic by claiming in *An Inquiry into the Nature and Causes of the Wealth of Nations*, ed. R. H. Campbell and A. S. Skinner (Indianapolis, IN: Liberty Fund, 1981) that he had "no great faith" in political arithmetic (1:534.30), while Karl Marx equally famously contended in *A Contribution to the Critique of Political Economy*, in Karl Marx and Friedrich Engels, *Collected Works*, 50 vols. (New York: International Publishers, 1859) that Petty was "the father of English political economy" (294n). More recent commentators have been equally divided, with some arguing that Petty laid important groundwork for political economy (e.g., Schumpeter, *History of Economic Analysis*, esp. 209–15; William Letwin, *The Origins of Scientific Economics: English Economic Thought, 1660–1776* [London: Methuen, 1963]), and others stressing significant differences between the two sciences (e.g., McCormick, *William Petty and the Ambitions of Political Arithmetic*).

17. See Foucault, *Security, Territory, Population*, 35–49; and *The Birth of Biopolitics*, 1–32. My reading of political arithmetic as liberal (in the sense given to that latter term by Foucault) is, I believe, consistent with the account of political arithmetic in *William Petty and the Ambitions of Political Arithmetic*, in which McCormick stresses the importance of Petty's Baconian ambitions for political arithmetic and his desire to reconfigure the sense of what it meant to govern (168–208).

18. I consider Hume's, Steuart's, and Smith's contributions to political economy at more length in Chapter 7.

19. See Martha Woodmansee, "The Genius and the Copyright: Economic and Legal Conditions of the Emergence of the 'Author,'" *Eighteenth-Century Studies* 17, no. 4 (1984): 425–48; Martha Woodmansee, *The Author, Art, and the Market: Rereading the History of Aesthetics* (New York: Columbia University Press, 1994); Mark Rose, *Authors and Owners: The Invention of Copyright* (Cambridge, MA: Harvard University

Press, 1993), esp. 6, 113–29; and Zeynep Tenger and Paul Trolander, "Genius versus Capital: Eighteenth-Century Theories of Genius and Adam Smith's *Wealth of Nations*," *MLQ* 55 (1994): 169–89. Woodmansee and Rose argue that insofar as mid-eighteenth-century texts on genius located the source of this capacity in an author's individuality, these accounts buttressed authorial property right claims and thus helped ensure that literary texts would be understood as properly part of the market. Tenger and Trolander stress, by contrast, that where political economists such as Adam Smith located the key to wealth and progress in a market-oriented division of labor, authors such as Sharpe, Gerard, Duff, and Young saw the key to social order and progress in the many forms of genius that nature providentially provided: "What labor and capital were to Adam Smith, genius was to Sharpe, Duff, and Gerard; it made work, wealth, and progress possible" (174). Tenger and Trolander cite Adam Smith's dismissive comments about genius in *The Wealth of Nations* as evidence of the competition between these two ways of understanding the source of wealth and progress and suggest that the discourse of genius petered out in the 1770s precisely because political economy had by then won this discursive battle.

20. William Sharpe, *A Dissertation upon Genius: Or, an Attempt to Shew, That the Several Instances of Distinction, and Degrees of Superiority in the Human Genius Are Not, Fundamentally, the Result of Nature, but the Effect of Acquisition* (London: Printed for C. Bathurst, 1755), 93.

21. Sharpe, *Dissertation upon Genius*, 132.

22. Alexander Gerard, *An Essay on Genius* (London: Printed for W. Strahan; T. Cadell in the Strand, 1774), 3. As Tenger and Trolander stress in "Genius versus Capital," almost all eighteenth-century commentators agreed that genius occurred in both the sciences and arts (169).

23. Edward Young, *Conjectures on Original Composition. In a Letter to the Author of Sir Charles Grandison* (London: Printed for A. Millar, in the Strand, and R. and J. Dodsley, in Pall-Mall, 1759), 47.

24. Sharpe, *Dissertation upon Genius*, 10.

25. Young, *Conjectures on Original Composition*, 42.

26. As Ronnie Young notes in "James Beattie and the Progress of Genius in the Aberdeen Enlightenment," *Journal for Eighteenth-Century Studies* 36, no. 2 (2013): 245–61, for Scottish Enlightenment authors such as Alexander Gerard, William Duff, and James Beattie, debates on genius were part of curricular reforms at Marischal College in Aberdeen, Scotland; see also Paul B. Wood, *The Aberdeen Enlightenment: The Arts Curriculum in the Eighteenth Century* (Aberdeen: Aberdeen University Press, 1993).

27. Joseph Addison and Richard Steele, *The Spectator*, ed. Donald Frederic Bond (Oxford: Clarendon, 1965), 2:129.

28. Young, *Conjectures on Original Composition*, 12. "Imitations," by contrast, "are often a sort of *Manufacture* wrought up by those *Mechanics, Art,* and *Labour,* out of preexistent materials not their own."

29. M. H. Abrams, *The Mirror and the Lamp: Romantic Theory and the Critical Tradition* (New York: Oxford University Press, 1953), 198–200.

30. Young, *Conjectures on Original Composition*, 46–47.

31. Foucault, *The Birth of Biopolitics*, 43–44; see also the classic account in Albert O. Hirschman, *The Passions and the Interests: Political Arguments for Capitalism before Its Triumph*, 20th anniversary ed. (Princeton, NJ: Princeton University Press, 1997).

32. Thomas Gray, *An Elegy Written in a Country Church Yard*, p. 7 (ll. 45–52). Page numbers refer to Gray's 1751 published text; line numbers in brackets refer to the version of the poem in Roger Lonsdale, Thomas Gray, William Collins, and Oliver Goldsmith, *The Poems of Thomas Gray, William Collins, Oliver Goldsmith* (New York: Norton, 1972), 103–41.

33. William Empson, *Some Versions of Pastoral* (New York: New Directions, 1974), 4.

34. While both Empson and Guillory stress the *Elegy*'s extensive canonization, surprisingly neither provides documentation to support this claim. However, Catherine Robson provides an extensive documentation of nineteenth- and twentieth-century school use of Gray's *Elegy* in *Heart Beats: Everyday Life and the Memorized Poem* (Princeton, NJ: Princeton University Press, 2012), 123–90.

35. Guillory, *Cultural Capital*, 120, my emphasis. More specifically, Guillory contends that "the place of the *Elegy* in the world of cultural production is just at the intersection of two opposing forces: the homogenizing forces expressed by the commonplaces and the common language; and the differentiating forces expressed by the nostalgic evocation of the pastoral genre and the valorized withdrawal from the public sphere. . . . This unique place of rest, the place which is the poem, renders no reader illiterate by 'refinements of subtlety and the dogmatism of learning.' To every common reader is given the pleasure of the commonplace and the common language, and at the same time, the pleasure of the withdrawal from the (urban) place—the scene of Ambition, Luxury, and Pride—where this language is formed as the product of a specific kind of struggle, the agon of social mobility. The *Elegy* is thus at once peculiarly accessible to a wide reading public at the same time that its narrative reinscribes this access as innate rather than acquired" (120–21; internal quote from Samuel Johnson).

36. I draw the term "surface" from Foucault's suggestion that the emergence of biopolitics in the eighteenth century depended upon a new understanding of a population as "a set of elements that, on one side, are immersed within the general regime of living beings and that, on another side, offer a surface on which authoritarian, but reflected and calculated transformations can get a hold" (*Security, Territory, Population*, 75). The term "surface" is also indigenous to late-eighteenth- and early-nineteenth-century liberal theory, as is evident in Benjamin Constant's suggestion that "the conquerors of our day, peoples or princes, wish their empire to present a unified surface [*une surface unie*] upon which the proud eye of power may travel without meeting any unevenness that could offend or limit its view. The same code of law, the same measures, the same regulations and if they could contrive it gradually, the same language, this is what is proclaimed to be the perfect form of the social organization." Benjamin Constant, "The Spirit of Conquest and Usurpation and Their Relation to European Civilization [1814]," in *Political Writings* (Cambridge: Cambridge University Press, 1988), 73, trans. modified. Constant, a key early-nineteenth-century French theorist and proponent of liberalism, critiqued the

creation of "unified surfaces" under "conquerors" such as Napoleon because he felt such surfaces destroyed individual differences. However, Foucault's work suggests that the creation of unified surfaces was also a key liberal means for *locating* differences and putting these latter to work.

37. W. E. B. Du Bois, "The Talented Tenth," in Booker T. Washington et al., *The Negro Problem: A Series of Articles by Representative American Negroes of Today* (New York: J. Pott & Company, 1903), 33; Virginia Woolf, *A Room of One's Own* (New York: Harvest, 2005), 48. None of these surfaces are premised on the idea that everyone can become a genius but rather that everyone (or at least many people) must be "tested" for such potential so that those few who have this capacity can be identified. As Gray noted in an August 19, 1748, letter to Thomas Warton, he believed that the proper "alliance" of education and government "must necessarily concur to produce great & useful Men." Lonsdale et al., *The Poems of Thomas Gray, William Collins, Oliver Goldsmith*, 85.

38. Percy Bysshe Shelley, *Prometheus Unbound, a Lyrical Drama in Four Acts, with Other Poems* (London: C. and J. Ollier, 1820), 137. My thanks to Greg Lynall for reminding me of this passage in Shelley's play.

39. Cleanth Brooks, *The Well Wrought Urn: Studies in the Structure of Poetry* (New York: Harcourt, Brace, Jovanovich, 1975), 116. The secular, redemptive potential of this common body also helps resolve one of the *Elegy*'s other major ambivalences, namely, that the poem's rural would-be Hampdens, Miltons, and Cromwells are potentially disruptive figures, as prone to "Luxury and Pride" as their urban counterparts. Gray, *An Elegy*, p. 9 (l. 71). Through these figures of expressed but diverted gifts, the poem encourages not only a desire to identify but also to regulate potential genius.

40. See Genevieve Miller, *The Adoption of Inoculation for Smallpox in England and France* (Philadelphia: University of Pennsylvania Press, 1957), 146–56.

41. On the history of this hospital, see Miller, *The Adoption of Inoculation*, 146–56.

42. Isaac Maddox, *A Sermon Preached before His Grace Charles, Duke of Marlborough, President, the Vice-Presidents and Governors of the Hospital for the Small-Pox, and for Inoculation, at the Parish-Church of St. Andrew Holburn, on Thursday, March 5, 1752* (London: Printed by H. Woodfall, 1753), 11.

43. John Green, *A Sermon Preached before His Grace George, Duke of Marlborough, President, the Vice-Presidents, the Treasurer, &C. Of the Hospitals for the Small-Pox. On Tuesday, April 26, 1763. By the Right Reverend Father-in-God John Lord Bishop of Lincoln* (London: Printed by H. Woodfall, in Paternoster-Row, 1763), 17.

44. Samuel Squire, *A Sermon Preached before His Grace Charles, Duke of Marlborough, President, the Vice-Presidents, the Treasurer, &C. Of the Hospitals for the Small-Pox, on Thursday, March 27, 1760* (London: Printed by H. Woodfall, 1760), 7.

45. Richard Eyre, *A Sermon, Preached, at St. Andrew's, Holborn, April the 25th 1765. On the Anniversary Meeting, of the Governors, of the Small-Pox Hospitals* (London: Printed by H. Woodfall, 1765), 17–18.

46. Brownlow North, *A Sermon, Preached before His Grace Augustus Henry Duke of Grafton, President, the Vice-Presidents, and Treasurer, &C of the Hospitals for the Small-Pox and Inoculation, on Thursday May, the 6th, 1773, by Brownlow, Lord Bishop of*

Lichfield and Coventry, and Published at Their Request (London: Printed by William Woodfall, 1773), 19.

47. Young, *Conjectures on Original Composition*, 14–15.

48. William Wordsworth, *Lyrical Ballads, with Other Poems. In Two Volumes*, 1st ed. (London: Printed for T. N. Longman and O. Rees, Pasternoster-Row, 1800), 1:xviii–xix.

49. Useful accounts of the narrowing of the concept of literature include (in addition to Guillory) Douglas Lane Patey, "The Eighteenth Century Invents the Canon," *Modern Language Studies* 18, no. 1 (1988): 17–37; M. H. Abrams, *Doing Things with Texts: Essays in Criticism and Critical Theory*, ed. Michael Fischer (New York: Norton, 1989), 144–46; Terry Eagleton, *Literary Theory: An Introduction*, anniversary ed. (Malden, MA: Blackwell, 2008), 1–18; and Raymond Williams, *Keywords: A Vocabulary of Culture and Society*, rev. ed. (New York: Oxford University Press, 2015), 134–38.

50. Guillory, *Cultural Capital*, 133.

51. Guillory discusses the topic of genius only in passing and only in connection with literacy; see 364n17.

52. William Godwin, *The Enquirer, Reflections on Education, Manners, and Literature. In a Series of Essays* (London: Printed for G. G. and J. Robinson, Paternoster-Row, 1797), vi. On the differences in method, style, and aims between *Political Justice* and *The Enquirer*, see Jon Mee, "'The Use of Conversation': William Godwin's Conversable World and Romantic Sociability," *Studies in Romanticism* 50 (2001): 572, 580; Jon Klancher, "Godwin and the Republican Romance: Genre, Politics, and Contingency in Cultural History," *Modern Language Quarterly* 56, no. 2 (1995): 153–54; and Victoria Myers, "William Godwin's *Enquirer*: Between Oratory and Conversation," *Nineteenth-Century Prose* 41, no. 1–2 (2014): 335–78. See also Tilottama Rajan, *The Supplement of Reading: Figures of Understanding in Romantic Theory and Practice* (Ithaca, NY: Cornell University Press, 1990), 167–70.

53. Godwin, *The Enquirer*, 4.

54. In a note, Godwin suggested that though the production of genius is not currently the work of the preceptor, it might at some future point be (30n).

55. The likelihood that Godwin's late-eighteenth- and nineteenth-century readers would recall Gray's *Elegy* here was increased by Godwin's explicit citation of lines from both Gray's *Elegy* and "Ode on a Distant Prospect of Eton College" (*The Enquirer*, 207, 70). Godwin's trilogy of politician, philosopher, and poet also recalled, even if it did not map exactly to, Gray's trilogy of Hampden, Cromwell, and Milton.

56. Godwin applied this description specifically to classical authors who wrote in Latin, but it seems to function as the ideal to which he hoped other literature would aspire.

57. In *The Enquirer*, Godwin positioned the institution of law as the antithesis of optimized mental thinking—Godwin claimed that law doesn't "shorte[n] my course" but rather "multiplies my difficulties a thousandfold" (224)—and, hence, a lawyer, as agent of this institution, was an "evil genius" (227).

58. Pointing to Godwin's claim in *The Enquirer* that "Literature, taken in all its bearings, forms the grand line of demarcation between the human and animal

kingdoms" (31), McLane argues in *Romanticism and the Human Sciences* that Godwin is part of the ideologically suspect project of "literary anthropology," which defines the human in terms of the capacity for, and acquisition of, the newly narrowed category of literature (10–42). This critique has the effect of bringing Godwin's account of literature back into the fold of Guillory's argument. Though I share McLane's suspicion of Godwin's human/animal distinction, *The Enquirer* seems to me more divided in its aims than it does to McLane, for her critique does not seem able to account for Godwin's understanding of literature as leading to precisely those kinds of critical accounts exemplified by McLane's own text.

59. On Wordsworth's interest in the apocalyptic implications of the new sciences of geology, see David Collings, "After the Covenant: Romanticism, Secularization, and Disastrous Transcendence," *European Romantic Review* 21, no. 3 (2010): 345–61.

60. William Wordsworth, *The Prelude, 1799, 1805, 1850: Authoritative Texts, Context and Reception, Recent Critical Essays* (New York: Norton, 1979), 154, ll. 38–40, 41, 45.

2. Imagining Population in the Romantic Era: *Frankenstein*, Books, and Readers

1. For the source of this chapter's epigraph, see http://www.zazzle.com/government_an_evil_usurpation_bumper_sticker-128678877475618014.

2. William Hazlitt, *A Reply to the Essay on Population, by the Rev. T. R. Malthus. In a Series of Letters. To Which Is Added, Extracts from the Essay, with Notes* (London: Printed for Longman, Hurst, Rees, and Orme, Paternoster-Row, 1807), 46; Percy Bysshe Shelley, *Prometheus Unbound, a Lyrical Drama in Four Acts, with Other Poems* (London: C. and J. Ollier, 1820), xiv. On Romantic critiques of Malthus, see Philip Connell, *Romanticism, Economics, and the Question of "Culture"* (Oxford: Oxford University Press, 2001), 13–61.

3. Karl Marx, *Capital: A Critique of Political Economy*, trans. Ben Fowkes (New York: Vintage, 1977), 1:767.

4. Donella H. Meadows, Dennis L. Meadows, Jørgen Randers, and William W. Behrens III, *The Limits to Growth: A Report for the Club of Rome's Project on the Predicament of Mankind* (New York: Universe, 1972); Garrett Hardin, "The Tragedy of the Commons," *Science* 162, no. 3859 (1968): 248. For an account of Hardin within the history of ecology, see Sharon E. Kingsland, *The Evolution of American Ecology, 1890–2000* (Baltimore, MD: Johns Hopkins University Press, 2005), 220–21. On reactions to *The Limits of Growth* report, see Elodie Vieille Blanchard, "Modelling the Future: An Overview of the 'Limits to Growth' Debate," *Centaurus* 52 (2010): 91–116; and Melinda Cooper, *Life as Surplus: Biotechnology and Capitalism in the Neoliberal Era* (Seattle: University of Washington Press, 2008), 15–18.

5. See, e.g., David Warsh, *Knowledge and the Wealth of Nations: A Story of Economic Discovery* (New York: Norton, 2006), 50–51, 202.

6. See Philip Mirowski, *Science-Mart: Privatizing American Science* (Cambridge, MA: Harvard University Press, 2011); Luc Boltanski and Eve Chiapello, *The New Spirit of Capitalism*, trans. Gregory Elliott (New York: Verso, 2005); and Bruno Latour, "Why Has Critique Run out of Steam? From Matters of Fact to Matters of Concern," *Critical Inquiry* 30 (2004): 225–48.

7. This dynamic is not restricted to literary criticism; for a compelling account of how feminism has, since the 1970s, established its "smartness" by rejecting biology, see Elizabeth Wilson, "Underbelly," *differences: A Journal of Feminist Cultural Studies* 21, no. 1 (2010): 194–208.

8. For accounts of the continued impact of Shelley's novel, see George Lewis Levine and U. C. Knoepflmacher, *The Endurance of Frankenstein: Essays on Mary Shelley's Novel* (Berkeley: University of California Press, 1979); and Susan Tyler Hitchcock, *Frankenstein: A Cultural History* (New York: Norton, 2007). I consider the enduring relevance of this novel at more length in "*Frankenstein* and the Sciences of Self-Regulation" (forthcoming); see also the other contributions to this special issue of the *Huntington Library Quarterly*.

9. Michel Foucault, *Security, Territory, Population: Lectures at the Collège de France, 1977–78*, trans. Graham Burchell, ed. Michel Senellart (New York: Palgrave Macmillan, 2007), 67.

10. Frances Ferguson, "Malthus, Godwin, Wordsworth, and the Spirit of Solitude," in *Literature and the Body: Essays on Populations and Persons*, ed. Elaine Scarry (Baltimore, MD: Johns Hopkins University Press, 1988), esp. 106–11. Among Malthus's contemporaries, Hazlitt made this same point, noting in *A Reply to the Essay on Population* that the "common notions that prevailed on this subject, till [Malthus's] first population-scheme tended to weaken them, were that life is a blessing, and that the more people could be maintained in any state in a tolerable degree of health, comfort and decency, the better" (44). I return to Hazlitt's critique of Malthus in Chapter 6.

11. See, e.g., Daniel Bernoulli, "Essai d'une nouvelle analyse de la mortalité causée par la petite vérole," in *Histoire de l'Académie Royale des Sciences* (Paris: De l'imprimerie royale, 1766), 2–45. On the importance of the concept of populations for the physiocrats, see Joseph J. Spengler, *French Predecessors of Malthus: A Study in Eighteenth-Century Wage and Population Theory* (Durham, NC: Duke University Press, 1942); and Foucault, *Security, Territory, Population*, 70–79.

12. Foucault, *Security, Territory, Population*, 62–63.

13. William Godwin, *Of Population: An Enquiry Concerning the Power of Increase in the Numbers of Mankind* (London: Printed for Longman, Hurst, Rees, Orme and Brown, 1820).

14. Foucault, *Security, Territory, Population*, 63.

15. For further discussion of the reasons that I find Foucault's term "surface" useful, see Chapter 1.

16. Ernst Mayr, "Darwin and the Evolutionary Theory in Biology," in *Evolution and Anthropology: A Centennial Appraisal*, ed. Betty J. Meggers (Washington, DC: Anthropological Society of Washington, 1959), 2.

17. See, for example, Ernst Mayr, *Animal Species and Evolution* (Cambridge, MA: Belknap Press of Harvard University Press, 1965), 360–99, 481–515.

18. Ernst Mayr, "Speciation and Selection," *Proceedings of the American Philosophical Society* 93, no. 6 (1949): 516.

19. The account generally accepted by historians of science is that Darwin arrived at population thinking by combining Malthus's understanding of a population as a collection of individuals who compete *against* one another for food with

Francis Galton's stress on the variations of individuals within a population. See Elliott Sober, "Evolution, Population Thinking, and Essentialism," *Philosophy of Science* 47, no. 3 (1980): 350–83; and Piers J. Hale, "Finding a Place for the Anti-Malthusian Tradition in the Victorian Evolution Debates," in *New Perspectives on Malthus*, ed. Robert J. Mayhew (Cambridge: Cambridge University Press, 2016), 182–207. As Jody Hey points out in "Regarding the Confusion between the Population Concept and Mayr's 'Population Thinking,'" *Quarterly Review of Biology* 86, no. 4 (2011), Darwin in fact rarely used the term "population" (256–57), and so "population thinking" in Mayr's very specific sense required several additional post-Darwin conceptual innovations (258–62).

20. With that said, the historian of science Jacques Roger, in *Buffon: A Life in Natural History*, trans. L. Pearce Williams (Ithaca, NY: Cornell University Press, 1997), suggested that the influential eighteenth-century naturalist Georges-Louis Leclerc, Comte de Buffon ended up with something like Mayr's model of population thinking. Roger argued that Buffon's concept of species departed from the typological premise that characterized most other eighteenth-century reflections on species. While Buffon proposed, like many of his contemporaries, a "general prototype of each species on which every individual is modeled," he stressed that no individual member of a species "is entirely similar to any other individual, or consequently to the model whose imprint it carries." Buffon, cited in Roger, *Buffon*, 297. Roger contends that this "destroyed all definitions of a species as a collection of absolutely similar beings. It is therefore tempting to see here the origin of the 'idea of populations' in the sense used by Ernst Mayr, that is, a conception of the species as a population composed of individuals all differing among themselves" (297). See also John C. Greene, "Aristotle to Darwin: Reflections on Ernst Mayr's Interpretation in *The Growth of Biological Thought*," *Journal of the History of Biology* 25, no. 2 (1992): 257–84. Alan Bewell, "Jefferson's Thermometer: Colonial Biogeographical Constructions of the Climate of America," in *Romantic Science: The Literary Forms of Natural History*, ed. Noah Heringman (Albany: State University of New York Press, 2008), 111–38, stresses the importance of Buffon's population approach for eighteenth- and early-nineteenth-century natural history.

21. Romantic-era interest in the transformative potential of anomalies was not restricted to the effects of government intervention; as Denise Gigante has noted in "The Monster in the Rainbow: Keats and the Science of Life," *PMLA* 117 (2002): 433–48, Romantic-era authors reconceived the very category of "monstrosity" by seeing in it no longer a falling away from proper form but rather a vital excess that was immanent to life and which brought new species and forms of life into being. See also Robert Mitchell, *Experimental Life: Vitalism in Romantic Science and Literature* (Baltimore, MD: Johns Hopkins University Press, 2014), 144–89.

22. Franco Moretti, *The Way of the World: The Bildungsroman in European Culture* (London: Verso, 1987), 44. Similar comparisons of primitive and modern technologies of socialization play an important, if often understated, role in other Marxist accounts of the institution of literature—see, e.g., Fredric Jameson, *The Political Unconscious: Narrative as a Socially Symbolic Act* (Ithaca, NY: Cornell University Press, 1981), 69–70, 77–79—as well as in Friedrich Kittler's account in "Über die

Sozialisation Wilhelm Meisters," in *Dichtung als Sozialisationsspiel: Studien zu Goethe und Gottfried Keller*, ed. Gerhard Kaiser and Friedrich Kittler (Göttingen: Vandenhoeck & Ruprecht, 1978), 13–124.

23. Moretti, *The Way of the World*, 45.

24. I will take up Moretti's subsequent attempt to consider textual variants from the perspective of evolutionary populations in what follows.

25. Ian Watt, *The Rise of the Novel: Studies in Defoe, Richardson, and Fielding* (Berkeley: University of California Press, 1957), 92; Jameson, *The Political Unconscious*, 34, 17–102; Nancy Armstrong, *Desire and Domestic Fiction: A Political History of the Novel* (New York: Oxford University Press, 1987), 23–24.

26. Maureen N. McLane, *Romanticism and the Human Sciences: Poetry, Population, and the Discourse of the Species* (Cambridge: Cambridge University Press, 2000), 87.

27. Mary Wollstonecraft Shelley, *Frankenstein, or, the Modern Prometheus*, 3rd ed., ed. David Lorne Macdonald and Kathleen Dorothy Scherf (Peterborough: Broadview, 2012), 80–81.

28. This is Moretti's reading of *Frankenstein*, for he contends that the novel seeks to reassure its readers that the events it depicts are simply an anomalous "case," out of keeping with the flow of history; in this way, the novel validates the dominant normative beliefs of early nineteenth-century social relations. Franco Moretti, *Signs Taken for Wonders*, rev. ed. (New York: Verso, 1988), 89.

29. Though Shelley engaged the topic of population in *Frankenstein* indirectly via Victor's reflections on the consequences of reproduction, she engaged the term and concept much more explicitly in her third novel, *The Last Man* (1826; Peterborough: Broadview, 1996). This novel begins with Lionel Verney's quasi-political arithmetical reflection that though England is tiny when compared to the rest of the globe, "yet, when balanced in the scale of mental power, [it] far outweighed countries of larger extent and more numerous population" (7). The novel then tracks the effects of a plague that destroys more and more of the global human population, first to the point that the narrator lives "on an earth whose diminished population a child's arithmetic might number" (306) and then to the point that Verney is literally the last man. This plot provides Shelley with many occasions to discuss epidemic-related population measures and to make more general reflections on populations (in addition to the references cited previously, see esp. 17, 31, 82, 117, 153, 179, 185, 186, 187, 204, 217, 232, 238, 240, 358, 361). Yet *Frankenstein*, precisely because of its more indirect engagement with the topic of population, allows us to recognize more easily than in *The Last Man* that claims about populations are always based on *models* of populations. Or, to put this another way, the plot of *The Last Man* commits itself to a specific model of population, whereas *Frankenstein* emphasizes the modeling activity itself that is bound up with claims about populations.

30. On Richardson's claim to have invented a new "species," see Michael McKeon, *The Origins of the English Novel, 1600–1740* (Baltimore, MD: Johns Hopkins University Press, 1987), 410; and William Park, "What Was New About the 'New Species of Writing'?" *Studies in the Novel* 2, no. 2 (1970): esp. 112–19.

31. Walter Scott, "Remarks on Frankenstein," *Blackwood's Edinburgh Magazine* 2, no. 12 (1818): 613. This review is also available at the website Romantic Circles,

"Mary Wollstonecraft Shelley," https://romantic-circles.org/editions/frankenstein/MShelley/mshelley.

32. A search in the British Periodicals database (http://www.proquest.com/en-US/catalogs/databases/detail/british_periodicals.shtml) for articles that appeared between 1790 and 1822 and contained both the words "novel*" and "species" suggests that it was around 1818 that it became commonplace to refer to (sub)species within the more general species of the "novel."

33. See, e.g., Scott, "Remarks on Frankenstein," 614; the anonymous review in *Literary Panorama and National Register* 8 (1818): 411–41; and (arguably) the anonymous review in *Belle Assemblée; or Bell's Court and Fashionable Magazine* 17 (March 1818): 139–42. All of these reviews are available in Romantic Circles, "Mary Wollstonecraft Shelley."

34. Scott, "Remarks on Frankenstein," 620.

35. Anon., "Review of Frankenstein; or the Modern Prometheus," *Edinburgh Magazine and Literary Miscellany* (1818): 249, 53.

36. An extreme example of this awareness of the effects of reviews on authorial production was P. B. Shelley's claim that the reviewers of the *Quarterly Review* had effectively killed John Keats with bad reviews; see Percy Bysshe Shelley, *Shelley's Adonais: A Critical Edition*, ed. Anthony D. Knerr (New York: Columbia University Press, 1984), 5–6.

37. This paragraph is indebted to an unpublished response that Alan Bewell provided at the Pre-Conference on the Romantic Life Sciences for the 2017 North American Society for the Study of Romanticism (NASSR) annual conference. Though Bewell was responding to an early version of Chapter 3 of this book, his stress on the multiple political valences of concepts of population is relevant to many of my chapters, and I have taken up his points here.

38. Montesquieu discussed populations in chapters CXII–CXVIII of his *Lettres persanes* (1721)—see *Persian Letters; Trans. Mr. Ozell* (London: Printed for J. Tonson, 1722), 2:150–78—and in the twenty-nine chapters of book 23 of *The Spirit of Laws* (1748); see *The Spirit of Laws; Translated from the French of M. De Secondat, Baron De Montesquieu, by Mr. Nugent*, trans. Thomas Nugent (London: Printed for J. Nourse, and P. Vaillant, 1752), 2:125–60. David Hume discussed population in "Of the Populousness of Antient Nations," in *Essays and Treatises on Several Subjects* (Edinburgh: Printed for A. Kincaid, and A. Donaldson, 1753), 155–262. For discussions of Montesquieu's understanding of the links among population growth, governmental type, and economics, see Joseph J. Spengler, *French Predecessors of Malthus: A Study in Eighteenth-Century Wage and Population Theory* (Durham, NC: Duke University Press, 1942), 212–23; and David B. Young, "Libertarian Demography: Montesquieu's Essay on Depopulation in the *Lettres persanes*," *Journal of the History of Ideas* 36, no. 4 (1975): 669–82. On Hume's theory of population, see Rotwein's introduction to "Of the Populousness of Antient Nations" in Hume, *Writings on Economics* (Madison: University of Wisconsin Press, 1955), lxxxviii–xc; Miller's notes to Hume's essay in *Essays, Moral, Political, and Literary*, ed. Eugene F. Miller (Indianapolis, IN: LibertyClassics, 1987), 377–464; and Ernest Campbell Mossner, "Hume and the Ancient-Modern Controversy, 1725–1752: A Study in Creative

Scepticism," *University of Texas Studies in English* 28 (1949): 139–53. For more general discussion of pre-Malthus debates about population, see D. V. Glass, "The Population Controversy in Eighteenth-Century England. Part I. The Background," *Population Studies* 6, no. 1 (1952): 69–91; Spengler, *French Predecessors of Malthus*; and Charles Emil Stangeland, "Pre-Malthusian Doctrines of Population: A Study in the History of Economic Theory," *Studies in History, Economics and Public Law* 21, no. 3 (1904).

39. See, for example, Richard Price, *An Essay on the Population of England, from the Revolution to the Present Time*, 2nd ed. (London: Printed for T. Cadell, 1780), 26–29. For a reading of Price as, in fact, a "bourgeois radical"—i.e., liberal—rather than a republican, see Isaac Kramnick, *Republicanism and Bourgeois Radicalism: Political Ideology in Late Eighteenth-Century England and America* (Ithaca, NY: Cornell University Press, 1990), 176–80.

40. Richard Price was also a central figure in the development of economically viable life insurance models, a project he pursued in texts such as *Observations on Reversionary Payments; on Schemes for Providing Annuities for Widows, and for Persons in Old Age*, 3rd ed. (London: T. Cadell, 1773). On the importance of Price for the development of the mathematics of probability in the eighteenth century, see Lorraine Daston, *Classical Probability in the Enlightenment* (Princeton, NJ: Princeton University Press, 1995), 179–82.

41. On the nineteenth-century explosion of statistical surveys of various kinds of populations, see Theodore M. Porter, *The Rise of Statistical Thinking, 1820–1900* (Princeton, NJ: Princeton University Press, 1986); Ian Hacking, *The Taming of Chance* (New York: Cambridge University Press, 1990); and Alain Desrosières, *The Politics of Large Numbers: A History of Statistical Reasoning*, trans. Camille Naish (Cambridge, MA: Harvard University Press, 1998).

42. See Franco Moretti's *Signs Taken for Wonders*, 262–78; "The Slaughterhouse of Literature," *Modern Language Quarterly* 61, no. 1 (2000): 207–27; *Graphs, Maps, Trees: Abstract Models for a Literary History* (New York: Verso, 2005); and *Distant Reading* (New York: Verso, 2013).

43. Moretti narrates this story in the headnotes to the essays collected in *Distant Reading*, esp. 1–2, 63–65, 121–22, 37–38. See also *Signs Taken for Wonders*, 262–78; "The Slaughterhouse of Literature"; and *Graphs, Maps, Trees*. Moretti explicitly references Mayr's accounts of evolution, populations, and speciation in *Graphs, Maps, Trees*, 76, 90; and *Distant Reading*, 148–49, 179.

44. Moretti, *Signs Taken for Wonders*, 265. More specifically, Moretti contended that the process of harsh selection in the nineteenth century was encouraged by "industrial and political convulsions," which posed for a European readership the problems of "redraw[ing] the territory of individual expectations . . . defin[ing] anew its 'sense of history,' and its attitude toward the values of modernity. For all sorts of reasons, the *Bildungsroman* was the symbolic form most apt to solve these problems—the fittest for surviving in the new, selective context. And the *Bildungsroman* did indeed survive, while the *Erziehungsroman* and the *Entwicklungsroman* and the *Künstlerroman*, the allegorical, the lyric, the epistolary and the satirical novel, all perished in that veritable struggle for literary life" (265).

45. In his *Atlas of the European Novel, 1800–1900* (New York: Verso, 1998), Franco Moretti also read markets through the lens of evolution, though in this case much more implicitly, limiting himself to the use of Stephen Jay Gould's account of the limits on the number of biological species within a habitat (159).

46. Moretti, *Graphs, Maps, Trees*, 3–33.

47. In *Graphs, Maps, Trees*, Moretti also used the model of speciation to explain the movement, during the nineteenth century, of the literary device of free indirect discourse from its origin in British literature to the new "geographies" of, for example, French, Russian, and Latin American literature (81–91).

48. Christopher Prendergast, "Evolution and Literary History: A Response to Franco Moretti," *New Left Review* 34 (2005): 40–62. Moretti responded to many of Prendergast's points in *Distant Reading*, 137–58. In response to Prendergast's charge that his method makes it impossible in principle to explain the nature of consumer preferences, Moretti contended that he was simply assuming there an explanation he had provided in earlier texts, namely, "the idea that literary genres are problem-solving devices, which address a contradiction of their environment, offering an imaginary resolution by means of their formal organization. The pleasure provided by that formal organization . . . is the vehicle through which a larger symbolic statement is shaped and assimilated. When readers of detective fiction 'like' clues, in other words, it is because the structure provided by clues makes them feel that the world is fully understandable, and rationalization can be reconciled with adventure, and individuality is a great but dangerous thing" (141). However, since Moretti also stressed in his response that he himself had already begun to have doubts about his method when Prendergast's critique appeared, his responses do not seem intended to salvage any of his method of thinking populations of texts through the lens of evolutionary theory (139).

49. Moretti, *Graphs, Maps, Trees*, 8.

50. For Moretti's reflections on his desire to make literary history scientific, see Moretti, *Graphs, Maps, Trees*, 1–2.

51. Alternatively, one could take Mayr's model of population and speciation much more seriously than does Moretti. New kinds of readers—for example, urban workers, boys, and girls—would then be analogous to those geographic divisions that Mayr stressed in his account of speciation and that enable new populations of novels to emerge. It would probably also make sense to think of novelistic genres not as analogous to a Mayrian *species*—which would mean that they could not "mix" with members of another novelistic species—but rather as analogous with subpopulations of the general species of the novel. This would in turn allow for the possibility, seemingly amply exemplified by the nineteenth-century history of the novel, of crossings and mixings of different genres (e.g., historical gothic novels). This approach would also likely require relating the emergence of new novelistic genres to that process, which began in the late eighteenth century, through which "Literature" was separated from other species of writing (history, philosophy, etc.), and which I discussed at the end of the last chapter.

52. In Chapter 5, I return to this history from the perspective of concepts of "collective experiments."

53. Friedrich A. von Hayek, "The Use of Knowledge in Society," *American Economic Review* 35, no. 4 (1945): 519.

54. Or, to put this another way, competition is the most "efficient" means of planning, since it makes the "full[est] use of the existing knowledge" that is possible (521).

55. In the 1980s, Hayek explicitly connected his claims about the wisdom of markets to evolutionary biology and to Mayr specifically; see, for example, his 1983 lecture "Evolution and Spontaneous Order," https://www.youtube.com/watch?v=yQhqZ-iWMRM; and *The Fatal Conceit: The Errors of Socialism* (Chicago: University of Chicago Press, 1989), 45.

56. On the neoliberal transformation of scientific research, see Philip Mirowski, *Science-Mart: Privatizing American Science* (Cambridge, MA: Harvard University Press, 2011). I expand on ways that scientific research instantiates population approaches in Robert Mitchell, "Biopolitics and Population Aesthetics," *South Atlantic Quarterly* 115, no. 2 (2016): 367–98.

57. Though employing a different concept of population than that which I outline here, Nancy Armstrong and Len Tennenhouse come to similar conclusions in "The Problem of Population and the Form of the American Novel," *American Literary History* 20, no. 4 (2008): 667–85; and the expansion of this argument in *Novels in the Time of Democratic Writing: The American Example* (Philadelphia: University of Pennsylvania Press, 2018).

3. Freed Indirect Discourse: Biopolitics, Population, and the Nineteenth-Century Novel

1. Hannah Arendt, *The Human Condition*, 2nd ed. (Chicago: University of Chicago Press, 1998), 39.

2. Ian Watt, *The Rise of the Novel: Studies in Defoe, Richardson, and Fielding* (Berkeley: University of California Press, 1957); John B. Bender, *Imagining the Penitentiary: Fiction and the Architecture of Mind in Eighteenth-Century England* (Chicago: University of Chicago Press, 1987); Nancy Armstrong, *Desire and Domestic Fiction: A Political History of the Novel* (New York: Oxford University Press, 1987); D. A. Miller, *The Novel and the Police* (Berkeley: University of California Press, 1988). See Chapters 1 and 2 for further discussion of these literary critics.

3. I draw the term "surface" from Foucault's suggestion that, for eighteenth-century authors, a population was "a set of elements that, on one side, are immersed within the general regime of living beings and that, on another side, offer a surface on which authoritarian, but reflected and calculated transformations can get a hold." Michel Foucault, *Security, Territory, Population: Lectures at the Collège de France, 1977–78*, trans. Graham Burchell, ed. Michel Senellart (New York: Palgrave Macmillan, 2007), 75. For further discussion of the utility of the term "surface," see Chapter 1.

4. Francis Bacon, *The Essays, or Councils, Civil and Moral, of Sir Francis Bacon, Lord Verulam, Viscount St. Alban with a Table of the Colours of Good and Evil, and a Discourse of the Wisdom of the Ancients* (London: Printed for H. Herringman, R. Scot, R. Chiswell, A. Swalle, and R. Bentley, 1696), 77–86.

5. See, e.g., Ted McCormick, "Population: Modes of Seventeenth-Century Demographic Thought," in *Mercantilism Reimagined: Political Economy in Early Modern Britain and Its Empire*, ed. Philip J. Stern and Carl Wennerlind (Oxford: Oxford University Press, 2014), 25–45; Charlotte Sussman, "The Colonial Afterlife of Political Arithmetic: Swift, Demography, and Mobile Populations," *Cultural Critique* 56 (2004): 96–126; and Chapter 1 of this volume.

6. Foucault, *Security, Territory, Population*, 62–63.

7. See Foucault, *Security, Territory, Population*, 30–79.

8. For an account of the actual construction of such data in Britain in the 1720s, largely through the Royal Society and its organ, the journal *Philosophical Transactions of the Royal Society*, see Genevieve Miller, *The Adoption of Inoculation for Smallpox in England and France* (Philadelphia: University of Pennsylvania Press, 1957), 100–133, esp. 111–23.

9. Michel Foucault, "Sexuality and Solitude," in *The Essential Works of Foucault, 1954–1984*, vol. 1: *Ethics*, ed. Paul Rabinow (New York: The New Press, 1997), 177.

10. Alex Woloch, *The One vs. the Many: Minor Characters and the Space of the Protagonist in the Novel* (Princeton, NJ: Princeton University Press, 2003), 32.

11. Woloch's focus on adult humans is highlighted by the parallel he draws between George Eliot's desire in her novels "to preserve a singular protagonist *and* to extend narrative attention to a broad mass of characters" and what he describes as John Stuart Mill's "strange compromise position on universal suffrage" (31), according to which Mill wished to grant the right to vote to "every adult human being" but proposed to weight those votes according to the voter's knowledge (32). "Mill," Woloch writes, "imagines a franchise that is both stratified and universal: all citizens would receive voting power but to unequal degrees, just as *Middlemarch* includes many characters, while configuring them in various ways" (31–32). Yet Woloch does not comment at all on Mill's restriction of voters to *adult* human beings, nor does he even consider nonhuman agents in the novels that he discusses. While Woloch's emphasis is valid for a novel such as *Middlemarch*, since Eliot resolutely restricts her agents to human beings, it does not work for many other nineteenth-century novelists.

12. For Zola's own democratic political leanings, see Susanna Barrows, *Distorting Mirrors: Visions of the Crowd in Late Nineteenth-Century France* (New Haven, CT: Yale University Press, 1981), 93–113.

13. Émile Zola, *Germinal*, ed. and trans. Roger Pearson (New York: Penguin, 2004), 510; French original in *Les Rougon-Macquart, histoire naturelle et sociale d'une famille sous le Second Empire*, 5 vols., ed. Armand Lanoux and Henri Mitterand (Paris: Bibliothèque de la Pléiade), 1571.

14. In *Germinal*, the hereditary crack is described as "la lésion héréditaire" (1571), while in *La bête humaine* (*The Human Beast*), it is described as a "fêlure héréditaire" (*La bête humaine*, in *Les Rougon-Macquart*, 1043; ed. and trans. Roger Pearson [Oxford: Oxford University Press, 2009]). Further references to the French originals of *Germinal* and *La bête humaine* will be noted parenthetically following the English page numbers. On the nature and role of hereditary cracks in Zola's work, see Gilles Deleuze's appendix on "Zola and the Crack-Up" in *The Logic of Sense*, trans. Mark

Lester with Charles Stivale, ed. Constantin V. Boundas (New York: Columbia University Press, 1990), 321–33.

15. Daniel Defoe, *Robinson Crusoe: An Authoritative Text, Contexts, Criticism*, ed. Michael Shinagel (New York: Norton, 1994), 4, 13; Herman Melville, *Moby-Dick: A Norton Critical Edition*, ed. Hershel Parker (New York: Norton, 2018), 158, 148.

16. My approach has been assisted by recent literary-critical work on the role of population models within eighteenth-century fiction and prose, such as Charlotte Sussman's *Peopling the World: Representing Human Mobility from Milton to Malthus* (Philadelphia: University of Pennsylvania Press, 2020), and on the role of population-based statistics within Victorian fiction, such as Jesse Rosenthal, "The Large Novel and the Law of Large Numbers; Or, Why George Eliot Hates Gambling," *ELH* 77 (2010): 777–811; Emily Steinlight, "Dickens's 'Supernumeraries' and the Biopolitical Imagination of Victorian Fiction," *Novel: A Forum on Fiction* 43, no. 2 (2010): 227–50; and Nancy Armstrong and Len Tennenhouse, *Novels in the Time of Democratic Writing: The American Example* (Philadelphia: University of Pennsylvania Press, 2018).

17. Frank Norris, *The Octopus: A Story of California* (New York: Bantam, 1971), 39.

18. Theodore Dreiser, *The Financier* (New York: Penguin, 2008), 3. My interest in animals as minor characters is indebted to earlier historical and literary-critical work on the role of animals in nineteenth-century literature, including Harriet Ritvo, *The Animal Estate: The English and Other Creatures in the Victorian Age* (Cambridge, MA: Harvard University Press, 1987); Deborah Denenholz Morse and Martin A. Danahay, *Victorian Animal Dreams: Representations of Animals in Victorian Literature and Culture* (Aldershot: Ashgate, 2007); James Turner, *Reckoning with the Beast: Animals, Pain, and Humanity in the Victorian Mind* (Baltimore, MD: Johns Hopkins University Press, 1980); and Christine Kenyon-Jones, *Kindred Brutes: Animals in Romantic-Period Writing* (Aldershot: Ashgate, 2001). Yet because I see plants, diseases, and transportation technologies as equally likely minor characters in nineteenth-century novels, the lens through which I consider animals as characters differs from much of the work listed here. In thinking about these literary examples, I have found especially useful Phillip Thurtle's discussion of trotter horses in Wharton's *The Age of Innocence* and the squid and the lobster in Dreiser's *The Financier*. Phillip Thurtle, *The Emergence of Genetic Rationality: Space, Time, and Information in American Biological Science, 1870–1920* (Seattle: University of Washington Press, 2007). Thurtle helpfully demonstrates how the animals in these novels are connected to larger questions of populations, breeding, heredity, and the emerging sciences of genetics. I have also found useful Ivan Kreilkamp, "Dying Like a Dog in *Great Expectations*," in Morse and Danahay, *Victorian Animal Dreams*, 81–94 (in part because Kreilkamp explicitly connects animals to Woloch's concept of minor characters); and Ron Broglio, *Beasts of Burden: Biopolitics, Labor, and Animal Life in British Romanticism* (Albany: State University of New York Press, 2017).

19. Frances Ferguson argues that, for the Russian formalists, "agency became such a capacious and formally empty notion that one no longer needed human actors or characters to achieve it; animals and pots and kettles could carry the narrative action as well as a human could. Action, in other words, displaced character, and any sense of characterological depth looked misplaced in an analysis in which both

animals and inanimate objects might play active roles." Frances Ferguson, "Jane Austen, *Emma*, and the Impact of Form," *Modern Language Quarterly* 61, no. 1 (2000): 158. Ferguson's larger argument is that Foucauldian-inspired literary criticism has followed the same route of "dispatch[ing] character to the shadows" (158), for "discursive regimes . . . become the pots and kettles of Proppian analysis, the actors that make it clear that activity in no way requires actual persons" (158–59). See also Woloch's discussion of the antinomy between structuralist and referential approaches to novels in *The One vs. the Many*, 15–16.

20. Zola, *Germinal*, 60 [1182], 501–2 [1564], my emphasis.

21. Since Woloch takes for granted that all characters are humans, he does not engage the basic question of how one identifies a novelistic character and from what other novelistic elements a character might be distinguished. Kreilkamp, who is interested in treating animals as minor characters, in Woloch's sense of that term, engages this question more fully. However, Kreilkamp arguably also begs this question via his claim that "animals in the Victorian period . . . are often treated as semi-human in the realm of culture and as semi-characters in the realm of literature" (82–86). He suggests that this is a consequence of the fact that some animals in novels are given nicknames and something like speech is attributed to them and of the fact that minor human and animal characters both appear and disappear suddenly and without explanation. I agree with this analysis and see these as good reasons to engage animals as characters. Yet it is not clear from this account why names and speech are the minimum criteria for character, nor whether Kreilkamp believes that *any* novelistic entities that have at least some of these same characteristics (names and attributed speech) should also be understood as characters. For example, the coal mine in Zola's *Germinal* is given a name and attributed something like intentionality, but it is not clear to me whether Kreilkamp would therefore understand that entity as a minor character.

22. Woloch, *The One vs. the Many*, 13.

23. As Kreilkamp astutely notes, a novelist's decision to name and attribute subjective interiority to nonhuman entities, especially animals, could in some cases determine the genre of the text: "When pets and especially dogs feature as characters in Victorian narratives, those narratives tend to fall into the orbit of one of two minor generic categories, either children's literature or the anecdote" (83). I am interested here in uses of nonhuman characters that did not relegate novels to these "minor" genres.

24. Zola, *La Bête Humaine*, 147 [1128].

25. Thomas Mann, *Buddenbrooks*, trans. John E. Woods (New York: Vintage, 1993), 725–26; German original from *Buddenbrooks* (Frankfurt am Main: Fischer Taschenbuch Verlag, 1989), 754.

26. Because I define characters in terms of explicitly attributed agency, I read the coal mine of *Germinal* as failing to rise to the level of character, for seeming attributions of agency are nearly always qualified as subjective illusions, as the italicized words in the following quotations suggest: "the pit looked to [Etienne] *like* some monstrous and voracious beast [lui semblait avoir un air mauvais de bête goulue] crouching there ready to gobble everyone up" (Zola, *Germinal*, 7 [1135], my italics);

the sound of steam hissing is *"as though* [qui était comme] the monster were congested and fighting for breath" (8 [1136], my italics). By contrast, the narrator directly attributes agency to the mob (*la bande/la foule*) of striking mineworkers and their families: "And so, out on the open plain that lay white with frost beneath the pale winter sun, the mob [la bande] departed [s'en allait] along the road, spilling out on both sides into the fields of beets" (330 [1417]); "The crowd, easily led, [La foule entraînée], was already turning, even though Étienne protested and begged them not to stop the drainage" (331 [1418]). For the roles of crowds and mobs in nineteenth-century British and French literature, see John Plotz, *The Crowd: British Literature and Public Politics* (Berkeley: University of California Press, 2007); and Barrows, *Distorting Mirrors*, respectively.

27. See Émile Zola, "The Experimental Novel," in *The Experimental Novel and Other Essays* (New York: Haskell House, 1964), 1–54.

28. As Woloch beautifully demonstrates in *The One vs. the Many*, the "realistic" referential dimension of novelistic characters does not prevent these latter from bearing allegorical and symbolic meanings (18–20). To extend Woloch's analysis, in *Germinal*, Battle can both refer literally to the use of animal labor in mines and serve as an allegory of the "animalization" of human laborers.

29. Bruno Latour, *The Pasteurization of France*, trans. Alan Sheridan and John Law (Cambridge, MA: Harvard University Press, 1988), 159.

30. Significantly, Latour begins *The Pasteurization of France* by drawing explicitly and heavily on the account of a battle that Leo Tolstoy developed in his novel *War and Peace* (3–5).

31. For Lukács's attacks on Zola, see Georg Lukàcs, "Narrate or Describe?," in *Writer & Critic and Other Essays*, ed. A. D. Kahn (New York: Universal Library, 1971), 110–48; and the chapters on Zola in Georg Lukács, *Studies in European Realism*, trans. Edith Bone (New York: Howard Fertig, 2002).

32. Lukàcs, "Narrate or Describe?," 123.

33. Dorrit Cohn, *Transparent Minds: Narrative Modes for Presenting Consciousness in Fiction* (Princeton, NJ: Princeton University Press, 1978). My approach to free indirect discourse here has more in common with Ann Banfield's suggestion in *Unspeakable Sentences: Narration and Representation in the Language of Fiction* (Boston: Routledge & Kegan Paul, 1982) that free indirect discourse often produces "unspeakable sentences," though I do not adopt Banfield's structuralist approach. Both Cohn and Banfield refer to "free indirect style," rather than "free indirect discourse." For reasons that will become clear in what follows, I stress the discursive, rather than stylistic, dimension of this literary device and so use the term free indirect discourse, which keeps the focus on differences among direct discourse, indirect discourse, and free indirect discourse.

34. Jane Austen, *Emma: An Authoritative Text, Contexts, Criticism*, ed. George Justice (New York, Norton, 2012), 15.

35. This developmental telos established by the narrator's use of free indirect discourse also helps us understand why Mr. Knightley is, ultimately, the proper object of Emma's love within the novel, for Mr. Knightley's style and mode of observations come closest to those of the narrator.

36. I find useful Erich Auerbach's classic claim in *Mimesis: The Representation of Reality in Western Literature* (Princeton, NJ: Princeton University Press, 1953), 482–86, about the "bitter" emotional atmosphere that pervades a dinner between Emma Bovary and her husband in Flaubert's *Madame Bovary*. Auerbach stressed that this description is *not* a "representation of the content of Emma's consciousness, of *what* she feels *as* she feels it*." Though Emma "doubtless has such a feeling [of bitterness]," "if she wanted to express it, it would not come out like that; she has neither the intelligence nor the cold candor of self-accounting necessary for such a formulation." This passage is instead Flaubert's narrator "bestow[ing] the power of mature expression upon the material which [Emma Bovary] affords. . . . If Emma could do this herself, she would no longer be what she is, she would have outgrown herself and thereby saved herself" (484). Because Auerbach noted that we do not encounter here a "representation of the content of Emma's consciousness, of *what* she feels *as* she feels it," he then understandably concluded that this should *not* be understood as an instance of *"erlebte Rede"* (i.e., free indirect discourse) (485). My point, though, is that free indirect discourse *should* be understood in a broader sense, as giving voice to any forces that impinge upon consciousness.

37. Zola, *La Bête Humaine*, 60 [1050].

38. Deleuze, *The Logic of Sense*, 321, 324–25.

39. My thanks to Robert Fellman for pointing out that *The Masterpiece* illustrates this point better than does *The Belly of Paris*.

40. For a helpful discussion of the political and judicial institutions against which *The Human Beast* was directed, see Roger Pearson's introduction to Zola, *La Bête Humaine*, xxiv–xxix.

41. Casey Finch and Peter Bowen, in "'The Tittle-Tattle of Highbury': Gossip and the Free Indirect Style in *Emma*," *Representations* 31 (1990): 1–18, connect Austen's use of free indirect discourse to something that determines consciousness—namely, ideology—arguing that free indirect discourse channels not the thoughts of individual characters but is rather the novelistic parallel to "gossip." They connect ideology, gossip, and free indirect discourse by arguing that both gossip and free indirect discourse "function as forms par excellence of surveillance, and both serve ultimately to locate the subject—characterological or political—within a seemingly benign but ultimately coercive narrative or social matrix" (3–4). I briefly return to Finch and Bowen's approach to free indirect discourse and Frances Ferguson's critique of this approach briefly in what follows.

42. On Balzac's use of free indirect discourse, see especially Auerbach's analysis of *Le Père Goriot* (1834) in *Mimesis*, 468–74. Auerbach makes a compelling case that the collective sentiment, rendered through free indirect discourse, that the boarding-house owner Madame Vauquer should be pitied because she is of that class of *"women who have had troubles"* and was not treated well by her husband is actually the consequence of Madame Vauquer's ability to manipulate a "harmony between her person and what we (and Balzac too, occasionally) call her milieu" (470). For George Eliot on the necessity of inferences for civilization, see what are apparently Dorothea's thoughts about Casaubon ("Here was a man who could understand the higher inward life, and with whom there could be some spiritual communion; nay,

one who could illuminate principle with the widest knowledge: a man whose learning almost amounted to a proof of whatever he believed!"), which are followed by the narrator's claim that "Dorothea's inferences may seem large; but really life could never have gone on at any period but for this liberal allowance of conclusions, which has facilitated marriage under the difficulties of civilization." *Middlemarch: An Authoritative Text, Backgrounds, Criticism*, ed. Bert G. Hornsback (New York: Norton, 2000), 15. On the role of free indirect discourse in this passage, see Violeta Sotirova, "Historical Transformations of Free Indirect Style," in *Stylistics: Prospect & Retrospect*, ed. D. L. Hoover and S. Lattig (Amsterdam: Rodopi, 2007), 129–41.

43. Arendt, *The Human Condition*, 178.

44. Bruno Latour, *Politics of Nature: How to Bring the Sciences into Democracy* (Cambridge, MA: Harvard University Press, 2004), 65, 66.

45. For an example of the latter claim, see Franco Moretti's assertion that "not much happens as long as free indirect style remains confined to Western Europe; at most, we have the gradual, entropic drift from 'reflective' to 'non-reflective' consciousness: that is to say, from the sharp punctual utterances like those in *Mansfield Park*, to Flaubert's all-encompassing moods, where the character's inner space is unknowingly colonized by the commonplaces of public opinion." Franco Moretti, *Graphs, Maps, Trees: Abstract Models for a Literary History* (New York: Verso, 2005), 82.

46. On Balzac's interest in milieu theory, see Auerbach, *Mimesis*, 474–82. Flaubert wrote in his December 15–16, 1866, letter to George Sand that he "believe[d] that great Art is scientific and impersonal. What you have to do is to transport yourself, by an intellectual effort, into your Characters—not attract them to yourself." *Flaubert–Sand: The Correspondence*, trans. Francis Steegmuller and Barbara Bray (New York: Knopf, 1993), 49. On Eliot's interest in evolutionary sciences, see Gillian Beer, *Darwin's Plots: Evolutionary Narrative in Darwin, George Eliot, and Nineteenth-Century Fiction* (Cambridge: Cambridge University Press, 2009).

47. Miller, *The Novel and the Police*, 25; Finch and Bowen, "'The Tittle-Tattle of Highbury,'" 14.

48. One can fairly argue that nineteenth-century novels, by creating multiple population models, thereby naturalized the idea of population itself. However, as I discuss more explicitly in the previous chapter, population is an extraordinarily flexible concept and one that is moreover arguably antinormative, since the point of using population concepts is generally to *alter* some aspect of the population.

49. My thanks to Amanda Jo Goldstein for this suggestion in her response to an earlier version of this chapter.

50. Foucault described this active turn to passivity in volume 1 of *The History of Sexuality* (New York: Pantheon, 1978), noting that with the rise of biopolitics, "one might say that the ancient right to take life or let live was replaced by a power to foster life or disallow it to the point of death" (138).

51. For Dickens's and Eliot's engagements with statistics—and, by implication, population—see Steinlight, "Dickens's 'Supernumeraries'"; Rosenthal, "The Large Novel and the Law of Large Numbers"; and Caroline Levine, "The Enormity Effect: Realist Fiction, Literary Studies, and the Refusal to Count," *Genre* 50, no. 1 (2017): 59–75.

4. Building Beaches: Global Flows, Romantic-Era Terraforming, and the Anthropocene

1. Erasmus Darwin, *The Botanic Garden, a Poem, in Two Parts*, 4th ed. (London: Printed for J. Johnson, St. Paul's Church-Yard, 1799), vol. I, canto IV, p. 208, note to l. 320. *The Botanic Garden* was tremendously popular when it appeared, but interest fell off in the later nineteenth century to the point that, as Erasmus Darwin's grandson Charles Darwin noted in his "Life of Erasmus Darwin" for the translation of Ernst Krause's *Erasmus Darwin* (1879), "notwithstanding the former high estimation of his poetry by men of all kinds in England, no one of the present generation reads, as it appears, a single line of it." Charles Darwin, *Charles Darwin's "The Life of Erasmus Darwin,"* ed. Desmond King-Hele (Cambridge: Cambridge University Press, 2002), 33–34. For helpful accounts of the reasons behind this eclipse of interest, see Noel Jackson, "Rhyme and Reason: Erasmus Darwin's Romanticism," *Modern Language Quarterly* 70, no. 2 (2009): 171–94; and Dahlia Porter, *Science, Form, and the Problem of Induction in British Romanticism* (Cambridge: Cambridge University Press, 2018), 73–112. Devin Griffiths, *The Age of Analogy: Science and Literature between the Darwins* (Baltimore, MD: Johns Hopkins University Press, 2016), suggests that this decline in the popularity of Erasmus Darwin's verse obscures the fact that many of his premises continued to inform both science and literature in the nineteenth century.

2. Paul Foot, *Red Shelley* (London: Sidgwick & Jackson, 1980), 227–73, documents the long afterlife of *Queen Mab* in nineteenth-century British radical labor movements.

3. See Paul J. Crutzen, "Geology of Mankind," *Nature* 415, no. 6867 (2002): 23; for contextualization of Crutzen's seminal article, see Jeremy Davies, *The Birth of the Anthropocene* (Oakland: University of California Press, 2016).

4. Mary A. Favret, "War in the Air," *Modern Language Quarterly* 65, no. 4 (December 2004): 538, 543; also included in *War at a Distance: Romanticism and the Making of Modern Wartime* (Princeton, NJ: Princeton University Press, 2009). See also Fabien Locher and Jean-Baptiste Fressoz, "Modernity's Frail Climate: A Climate History of Environmental Reflexivity," *Critical Inquiry* 38, no. 3 (2012): 579–98. For expanded versions of Favret's concise description of this shift in weather science, see Vladimir Janković, *Reading the Skies: A Cultural History of English Weather, 1650–1820* (Chicago: University of Chicago Press, 2000).

5. John Dalton, *Meteorological Observations and Essays* (London: Printed for W. Richardson, J. Phillips, and W. Pennington, 1793), 76.

6. Darwin, *The Botanic Garden*, vol. I: 414 (note XXXIII).

7. John Williams, *The Climate of Great Britain; or Remarks on the Change It Has Undergone, Particularly within the Last Fifty Years* (London: C. and R. Baldwin, 1806), 334.

8. Thomas Malthus, *An Essay on the Principle of Population, as It Affects the Future Improvement of Society. With Remarks on the Speculations of Mr. Godwin, M. Condorcet, and Other Writers* (London: J. Johnson, 1798), 183–84.

9. In addition to characterizing Romantic science and political theory, the operation of untethering also marked key eighteenth-century British transformations of

labor and land. The processes of enclosure and "improvement," for example, meant—despite what a term such as "enclosure" initially seems to imply—dissolving local land rights associated with traditional commons so that large tracts of land could be dealt with as homogenous units and untethering production from local communities so that an abstract "laborer" could move "freely" between countryside and cities. See, e.g., H. C. Darby, *A New Historical Geography of England* (Cambridge: Cambridge University Press, 1973), 302–89. In even more explicit fashion, the slave trade meant forcibly removing African peoples from local contexts so that they could be inserted into global networks. See, e.g., Eric Eustace Williams, *Capitalism and Slavery* (Chapel Hill: University of North Carolina Press, 1994). For compelling accounts of the ways that these processes of untethering modified the meaning and mission of Romantic-era literature, see Alan Liu, *Wordsworth: The Sense of History* (Stanford, CA: Stanford University Press, 1989); and Ian Baucom, *Specters of the Atlantic: Finance Capital, Slavery, and the Philosophy of History* (Durham, NC: Duke University Press, 2005).

10. Dalton, *Meteorological Observations*, 90.

11. We might see this Romantic-era interest in light as a force that produces its effects orthogonally—that is, at an angle to the direction of the solar rays themselves—as one of the key points of difference between "Romanticism" and "Enlightenment." As the periodizing term itself suggests, Enlightenment thinkers understood progress as a process that took place in the same plane, or planes, as flows of "light," and thus those institutions or forces that impeded progress—superstition, a conspiring priesthood, etc.—were simply obstacles that prevented full illumination. For the Romantics, by contrast, progress could never bear this kind of straightforward relationship to the light of reason. From this perspective, Malthus's original 1798 essay on population appears as an attempt to introduce the principle of orthogonal drag into the Enlightenment schema of William Godwin's *Of Political Justice*.

12. Dalton, *Meteorological Observations*, 91.

13. Malthus, *An Essay on the Principle of Population*, 14.

14. Immanuel Kant, "Perpetual Peace: A Philosophical Sketch," in *Kant: Political Writings*, ed. Hans Siegbert Reiss, trans. H. B. Nisbet (Cambridge: Cambridge University Press, 1990), 106. My thanks to Evan Gottlieb for drawing my attention to this aspect of Kant's text.

15. As Alan Bewell notes in *Wordsworth and the Enlightenment: Nature, Man, and Society in the Experimental Poetry* (New Haven, CT: Yale University Press, 1989), for late-eighteenth-century scientists, "What was happening in America was little short of astonishing: not only had a relatively small number of ill-equipped human beings radically transformed a landscape, but they had also begun to change its climate" (244).

16. Bruno Latour, *Science in Action: How to Follow Scientists and Engineers through Society* (Cambridge, MA: Harvard University Press, 1987), 215–57.

17. For a discussion of networks and research on Earth's magnetic field, see Tim Fulford, Debbie Lee, and Peter J. Kitson, *Literature, Science, and Exploration in the Romantic Era: Bodies of Knowledge* (Cambridge: Cambridge University Press, 2004), 149–66; and Patricia Fara, *Sympathetic Attractions: Magnetic Practices, Beliefs, and*

Symbolism in Eighteenth-Century England (Princeton, NJ: Princeton University Press, 1996); for weather observation networks, see Favret, "War in the Air," 543. Establishing a viable network depended in part on the standardization of instruments, so that observers had some confidence that measurements obtained in one location were commensurable with measurements obtained in another.

18. Fulford, Lee, and Kitson, *Literature, Science, and Exploration*, 153.

19. Dalton, *Meteorological Observations*, 11–17, 36–38.

20. Janković, *Reading the Skies*, 156, 158.

21. Fulford, Lee, and Kitson, *Literature, Science and Exploration*, 13.

22. As Latour notes in *Science in Action*, this suggests that "knowledge" should not be understood as something "that could be described by itself or by opposition to ignorance or to 'belief'" but rather can be understood only "by considering a whole cycle of accumulation: how to bring things [e.g., measurements or samples] back to a place for someone to see it for the first time so that others might be sent again to bring other things back" (220).

23. Williams, *The Climate of Great Britain*, 349. For a brief discussion of Williams's proposal, see Janković, *Reading the Skies*, 1, 147.

24. Williams, *Climate of Great Britain*, 343–44.

25. Desmond King-Hele's *Erasmus Darwin and the Romantic Poets* (New York: St. Martin's, 1986) provides the classic account of Erasmus Darwin, but I draw also on more recent discussions in Alan Bewell, "Erasmus Darwin's Cosmopolitan Nature," *ELH* 76 (2009): 19–48, reprinted in *Natures in Translation: Romanticism and Colonial Natural History* (Baltimore, MD: Johns Hopkins University Press, 2016), 53–86; and Jackson, "Rhyme and Reason." See also Siobhan Carroll, "Crusades against Frost: Frankenstein, Polar Ice, and Climate Change in 1818," *European Romantic Review* 24, no. 2 (2013): 211–30, which also connects Darwin's discussions of ice and weather to both Percy Bysshe Shelley's *Queen Mab* and Mary Shelley's *Frankenstein* and establishes that debates about global terraforming were not limited to poetry but were engaged—often with explicit reference to Darwin's poetic accounts—in early-nineteenth-century British periodicals, especially in connection with discussions of British government–funded trips to the Arctic. These discussions were encouraged by the 1815 Mount Tambora volcanic explosion, which produced worldwide climatic change, and a "year without summer" in 1816 in Britain and Europe; see both Carroll, "Crusades against Frost," 215–19; and Gillen D'Arcy Wood, *Tambora: The Eruption That Changed the World* (Princeton, NJ: Princeton University Press), 2014.

26. Darwin, *The Botanic Garden*, I: 59–60 (Canto I, ll. 527–31). In this and in subsequent citations from *The Botanic Garden*, I provide the volume and page number, followed by a parenthetical explanation of the canto number and the specific line numbers of the reference, when the latter is applicable.

27. Alan Bewell, *Romanticism and Colonial Disease* (Baltimore, MD: Johns Hopkins University Press, 1999), 31; see also Alan Bewell, "Jefferson's Thermometer: Colonial Biogeographical Constructions of the Climate of America," in *Romantic Science: The Literary Forms of Natural History*, ed. Noah Heringman (Albany: State University of New York Press, 2008), 111–38.

28. Bewell, "Erasmus Darwin's Cosmopolitan Nature," 21.

29. Darwin, *The Botanic Garden*, I: iii.

30. Percy Bysshe Shelley, *Queen Mab; a Philosophical Poem*, ed. Jonathan Words-worth (New York: Woodstock, 1990), 75 (Canto VI), 105 (Canto VIII). In this and the following references to *Queen Mab*, I note canto numbers for each citation parenthetically.

31. Even Shelley's suggestion that the sea could be dotted with convenient islands had some precedent in eighteenth-century science, for Erasmus Darwin had claimed in *The Temple of Nature; or, The Origin of Society: A Poem, with Philosophical Notes* (London: J. Johnson, 1803) that it was probable that "the ocean has decreased in quantity during the short time which human history has existed" and would con-tinue to do so in the future, making it likely that islands could be seeded throughout these shallower seas (24, note to l. 268).

32. I discuss the context of Shelley's interest in science and technology more fully in Robert Mitchell, "'Here Is Thy Fitting Temple': Science, Technology, and Fiction in Shelley's *Queen Mab*," *Romanticism on the Net* 21 (2001).

33. Shelley's notes on vegetarianism in *Queen Mab* served as the basis for his pub-lished pamphlet *A Vindication of Natural Diet*. For a nuanced discussion of Shelley's vegetarianism, see Timothy Morton, "Sustaining Natures: Shelley and Ecocriti-cism," in *Shelley and the Revolution in Taste* (Cambridge: Cambridge University Press, 1994); and Timothy Morton, "Joseph Ritson, Percy Shelley, and the Making of Romantic Vegetarianism," *Romanticism* 12, no. 1 (2006): 52–61.

34. Shelley, *Queen Mab*, 233, 232.

35. Percy Bysshe Shelley, "Mont Blanc; Lines Written in the Vale of Chamouni," in *Shelley's Poetry and Prose*, ed. Donald H. Reiman and Sharon B. Powers (New York: Norton, 1977), 89 (ll. 1–2).

36. Timothy Morton, *The Poetics of Spice: Romantic Consumerism and the Exotic* (Cambridge: Cambridge University Press, 2000), 93.

37. I expand on this sense of excentricity in Robert Mitchell, "Cryptogamia," *European Romantic Review* 21, no. 5 (2010): 631–51.

38. For Thomas Nagel's original account of the "view from nowhere," see *The View from Nowhere* (New York: Oxford University Press, 1986); for a compelling dis-cussion of the history of the concept of "objectivity," see Lorraine Daston and Peter Galison, *Objectivity* (New York: Zone, 2007). Though I described Shelley's image as "cosmic" in the earlier version of this argument developed in "Global Flows: Romantic-Era Terraforming," in *British Romanticism and Early Globalization: Develop-ing the Modern World Picture*, ed. Evan Gottlieb (Lewisburg, PA: Bucknell University Press, 2014), 199–218, my use of Hannah Arendt's work later in this chapter encour-aged me to reconsider how to describe the kind of image that Shelley employs. For Arendt, the consideration of nature from a "cosmic" rather than an earthly stand-point is one of the defining characteristics of the modern sciences, but this stand-point necessarily produces what she describes as earth-alienation. Hannah Arendt, *The Human Condition*, 2nd ed. (Chicago: University of Chicago Press, 1998), 264. While Shelley's image draws for its veracity on the sciences, it is intended to work against earth-alienation by bringing readers back to the globe on which they live.

39. Shelley, *Queen Mab*, 14 (Canto I).

40. Anna Letitia Barbauld, "A Summer Evening's Meditation," in *Poems* (London: Printed for Joseph Johnson, 1773), 136.

41. Barbauld's image is itself connected to Thomas Wright's suggestion in *An Original Theory or New Hypothesis of the Universe, Founded upon the Laws of Nature, and Solving by Mathematical Principles the General Phenomena of the Visible Creation* (London: Printed for the Author, and sold by H. Chapelle, in Grosvenor-Street, 1750) that imagining the complete destruction of some of the millions of inhabitable worlds in the universe, or even "the total Dissolution of a System of Worlds," is in fact a "chearful" idea, since it "must convince [us] of [our] Immortality, and reconcile [us] to all those little Difficulties incident to human Nature, without the least Anxiety" (76). My thanks to Dahlia Porter for bringing both the Wright and Barbauld connections to my attention.

42. Chakrabarty stresses the latter dynamic, noting that for nineteenth-century liberals such as John Stuart Mill, "Indians or Africans were *not yet* civilized enough to rule themselves" but could in principle grow up to the point that they also occupied the present, rather than past, of mankind. Dipesh Chakrabarty, *Provincializing Europe: Postcolonial Thought and Historical Difference* (Princeton, NJ: Princeton University Press, 2000), 8. Domenico Losurdo, *Liberalism: A Counter-History*, trans. Gregory Elliott (New York: Verso, 2011), documents the extent to which stadial histories of mankind underwrote liberal defenses of slavery from Locke onward. Writing from a standpoint much more sympathetic to liberalism, the political theorist Ryan also stresses the extent to which liberalism intrinsically aims to encompass the earth; see Alan Ryan, *The Making of Modern Liberalism* (Princeton, NJ: Princeton University Press, 2012), 107–22.

43. Though the goal of *Provincializing Europe* is clearly to reform Marxist criticism from within, Chakrabarty nevertheless stresses repeatedly that one should not simply dismiss liberalism (see, for example, 4, 8, 13, 14, 23, 250).

44. Ian Baucom, "History 4°: Postcolonial Method and Anthropocene Time," *Cambridge Journal of Postcolonial Literary Inquiry* 1, no. 1 (2014): 140. The other, perhaps more common, critique of both Chakrabarty and the concept of the Anthropocene more generally is that blaming humans in general for global warming ignores the very specific kinds of human relations that have led to these threats; see, e.g., Jason W. Moore, "The Capitalocene, Part I: On the Nature and Origins of Our Ecological Crisis," *Journal of Peasant Studies* 44, no. 3 (2017): 594–63.

45. Dipesh Chakrabarty, "Baucom's Critique: A Brief Response," *Cambridge Journal of Postcolonial Literary Inquiry* 1, no. 2 (2014): 250. For a useful contextualization of Chakrabarty's essay within the tradition of literary-critical approaches to relationships between literature and the sciences, see Devin Griffiths, "Romantic Planet: Science and Literature within the Anthropocene," *Literature Compass* 14, no. 1 (2017).

46. The debate between Chakrabarty and Baucom is complicated by the uncertain referent of "extinction" for each. In "The Climate of History: Four Theses," *Critical Inquiry* 35, no. 2 (2009): 197–222, Chakrabarty oscillates between claiming that the *human species* is threatened with extinction and the claim that at stake is "the survival of human life *as developed in the Holocene period*"; that is, human societies organized around institutions such as agriculture, cities, durable architecture, and

the arts (213; see also Chakrabarty's stress on the importance of "parametric [that is, boundary] conditions for the existence of institutions central to our idea of modernity and the meanings we derive from them," 217). Baucom reads Chakrabarty as focused solely on the extinction of the human species ("History 4°," 140–41). Yet Baucom's disinclination to advocate for any specific "content" of freedom threatens to render the latter an inherently formal category and confuses the question of whether Baucom is also committed to the survival of institutions such as agriculture, cities, durable architecture, and the arts.

47. As critics of the concept of the Anthropocene have pointed out, though discussions of the Anthropocene begin with acknowledgment of the extraordinary complexity of ecological processes, they nevertheless often lead to desires for a technocratic fix. This latter is exemplified in the conclusion of Crutzen's seminal article of 2002, in which he argued that a "daunting task lies ahead for scientists and engineers to guide society towards environmentally sustainable management during the era of the Anthropocene. This will require appropriate human behaviour at all scales, and may well involve internationally accepted, large-scale geo-engineering projects, for instance to 'optimize' climate" ("Geology of Mankind," 23). On geo-engineering, see Davies, *The Birth of the Anthropocene*, 52–56; and Clive Hamilton, *Earthmasters: The Dawn of the Age of Climate Engineering* (New Haven, CT: Yale University Press, 2013).

48. Shelley's vision of global transformation also avoids the primitivist or Rousseauvian premise that humans should "return" to some presumably better past state of human relationships with one another and their natural environments. As a consequence, Shelley avoids the "fall from grace" paradigm that characterizes many versions of the Anthropocene, that is, the premise that humans have become a force of global environmental transformation only by transgressing the virtuous limits within which all other plants and animals are contained. For a discussion of this dimension of many versions of the Anthropocene, see Davies, *The Birth of the Anthropocene*, 7, 25, 108. Dipesh Chakrabarty's version of the fall is what he describes as "ecological overshoot": see, e.g., "The Politics of Climate Change Is More Than the Politics of Capitalism," *Theory, Culture, & Society* 34, no. 2–3 (2017): 27, 32–34.

49. Marjorie Levinson, "A Motion and a Spirit: Romancing Spinoza," *Studies in Romanticism* 46, no. 4 (2007): 386. For Shelley's interest in Spinoza, see Fazel Abroon, "Necessity and the Origin of Evil in the Thought of Spinoza and Shelley," *Keats-Shelley Review* 14 (2000): 56–70; and Colin Jager, "Shelley after Atheism," *Studies in Romanticism* 49, no. 4 (2010): 611–31.

50. Shelley stressed in *A Defence of Poetry*, in *Shelley's Poetry and Prose: A Norton Critical Edition*, ed. D. H. Reiman and N. Fraistat (New York: Norton, 1977), that poetry enables redemptive joy: "Poetry thus makes immortal all that is best and most beautiful in the world; it arrests the vanishing apparitions which haunt the interlunations of life, and veiling them, or in language or in form, sends them forth among mankind, bearing sweet news of kindred joy to those with whom their sisters abide—abide, because there is no portal of expression from the caverns of the spirit which they inhabit into the universe of things. Poetry redeems from decay the visitations of the divinity in man" (505).

51. Arendt, *The Human Condition*, 1.

52. Melinda Cooper, *Life as Surplus: Biotechnology and Capitalism in the Neoliberal Era* (Seattle: University of Washington Press, 2008), 15–18.

53. Donella H. Meadows, Dennis L. Meadows, Jørgen Randers, and William W. Behrens III, *The Limits to Growth: A Report for the Club of Rome's Project on the Predicament of Mankind* (New York: Universe, 1972), 86; cited in Cooper, *Life as Surplus*, 16.

54. Cooper, *Life as Surplus*, 18.

55. Julian Lincoln Simon, *The Ultimate Resource 2*, 2nd ed. (Princeton, NJ: Princeton University Press, 1996), 66; cited in Cooper, *Life as Surplus*, 18.

56. Kim Stanley Robinson's most well-known "emigration-from-earth" novels are *Red Mars* (New York: Bantam, 1993), *Green Mars* (New York: Bantam, 1994), and *Blue Mars* (New York: Bantam, 1994). At one level, the Mars trilogy is a scientifically plausible account of how humans might terraform Mars over the course of several centuries to make it habitable for humans. However, at a more fundamental level, it is the story of the shipping network that links the politics, economics, and ecology of Mars and Earth. Fredric Jameson's powerful readings of Robinson's trilogy in *Archaeologies of the Future: The Desire Called Utopia and Other Science Fictions* (New York, Verso, 2005), 393–416, are helpful, though Jameson's emphasis on the limits of utopian thinking does not encourage him to attend closely to either the question of flows or to fundamental transformations of human beings in Robinson's series.

57. "Generation Starships," *Encyclopedia of Science Fiction*, http://www.sf-encyclopedia.com/entry/generation_starships.

58. Though Robinson has the voyage last 160 years because of the plausible maximum speed such a ship could reach, it is no doubt not coincidental that seven generations is also a timeframe that plays an important role in contemporary ecological thinking. See, e.g., Stewart Brand, *Whole Earth Discipline: Why Dense Cities, Nuclear Power, Transgenic Crops, Restored Wildlands, and Geoengineering Are Necessary* (New York: Penguin, 2010), which endorses the "'seven generations' approach to future responsibility long credited to the Iroquois League" (79).

59. For example, phosphorus, which the inhabitants require for farming, has become increasingly scarce, yet it is not clear where the "leak" in the phosphorus cycle might be located. Devi, the ship's chief engineer, notes that "everyone gets recycled into the system. There's a lot of phosphorus in our bones that has to be retrieved. In fact I wonder if the missing phosphorus is in people's cremation ashes! You're only allowed to keep a pinch, but maybe it's adding up." Kim Stanley Robinson, *Aurora* (New York: Orbit, 2015), 102.

60. Malthus, *An Essay on the Principle of Population*, 369.

61. I discuss further the importance of the category of "moral restraint" for Malthus's text in Chapter 6.

62. Chris Otter, "The Technosphere: A New Concept for Urban Studies," *Urban History* 44, no. 1 (2017): 151–52.

63. The unnamed narrator of the last part of *Aurora* makes the same point, noting that "the many virtual, simulated, and indoor spaces that so many Terrans seem happy to inhabit" mean that these humans are "in effect occupying spaceships on the land" (469).

64. See "Regeneration," in Ephraim Chambers, *Cyclopaedia: Or, an Universal Dictionary of Arts and Sciences*, 4 vols. (London: Rivington et al., 1778).

65. This closing scene forms a pair with the water scene that opens the novel, which describes Freya's much more placid childhood sailing trip with her father on a lake inside the spaceship.

66. As Amitav Ghosh notes in *The Great Derangement: Climate Change and the Unthinkable* (Chicago: University of Chicago Press, 2016), "through much of human history, people regarded the ocean with great wariness," and even those peoples who "made their living from the sea, through fishing or trade, generally did not build large settlements on the water's edge" but rather situated cities in areas "protected from the open ocean by bays, estuaries, or deltaic river systems" (37). In *The Lure of the Sea: The Discovery of the Seaside in the Western World, 1750–1840* (Cambridge: Polity, 1994), Alain Corbin documents the fairly recent European discovery of the pleasures of seaside beaches. Robinson's representation of the beach as a site of ecstatic learning, rather than a place for habitation, links up well with this history of human relationships to the sea.

67. Bill McKibben, *Eaarth: Making a Life on a Tough New Planet* (New York: Time, 2010).

68. This is another way of approaching what Kathryn Yusoff captures, in the title of her book, as the need for *A Billion Black Anthropocenes (or None)* (Minneapolis: University of Minnesota Press, 2018).

69. McKibben, *Eaarth*.

70. For further real-world examples, see the Anthropocene ToolKit website: https://cissct.duke.edu/teaching-learning.

71. Brand, *Whole Earth Discipline*.

5. Liberalism and the Concept of the Collective Experiment

1. John Stuart Mill, *On Liberty*, in *Collected Works of John Stuart Mill*, ed. John M. Robson (Toronto: University of Toronto Press, 1963–1991), 18:281. Mill also used the phrase "experiments of living" (261).

2. Louis Lasagna, "A Plea for the 'Naturalistic' Study of Medicines," *European Journal of Clinical Pharmacology* 7 (1974): 153. More specifically, Lasagna proposed that doctors would inform patients—now thought of more as autonomous consumers—of the possible risks of such treatments, and each patient-consumer would "make his own judgment" about whether to take the experimental drug. Louis Lasagna, "Consensus among Experts: The Unholy Grail," *Perspectives in Biology and Medicine* 19, no. 4 (1976): 547. On the neoliberal background of Lasagna's proposal, see Edward Nik-Khah, "Neoliberal Pharmaceutical Science and the Chicago School of Economics," *Social Studies of Science* 44, no. 4 (2014): 489–517.

3. On these developments, see Nikolas S. Rose, *The Politics of Life Itself: Biomedicine, Power, and Subjectivity in the Twenty-First Century* (Princeton, NJ: Princeton University Press, 2007), esp. 131–54. On population-level tissue- and data-gathering protocols, see Robert Mitchell and Catherine Waldby, "National Biobanks: Clinical Labour, Risk Production, and the Creation of Biovalue," *Science, Technology, and*

Human Values 35, no. 3 (2010): 330–55; and Robert Mitchell, "US Biobanking Strategies and Biomedical Immaterial Labor," *Biosocieties* 7, no. 3 (2012): 224–44. On the more general development of population-level data-gathering techniques, see Orit Halpern, Robert Mitchell, and Bernard Dionysius Geoghagen, "The Smartness Mandate: Notes toward a Critique," *Grey Room* 68 (2017): 106–29.

4. John Arbuthnot, *Mr. Maitland's Account of Inoculating the Smallpox Vindicated, from Dr. Wagstaffe's Misrepresentations of That Practice, with Some Remarks on Mr. Massey's Sermon* (London: Printed and sold by J. Peele, at Lock's Head in Paternoster-Row, 1722), 2.

5. For discussion of Arbuthnot's background and interest in the smallpox inoculation debate, see Genevieve Miller, *The Adoption of Inoculation for Smallpox in England and France* (Philadelphia: University of Pennsylvania Press, 1957), 106–11; and Andrea Alice Rusnock, *Vital Accounts: Quantifying Health and Population in Eighteenth-Century England and France* (Cambridge: Cambridge University Press, 2002), 43–70.

6. Arbuthnot cites Mather's use of the term "experiment" (36) and reproduces Mather's March 10, 1721/1722, "Letter from Boston in New England" (58–61). He also cites Nettleton's reference to smallpox inoculation as an experiment (56) and reproduces Nettleton's "A Letter from Dr. Nettleton, at Halifax in Yorkshire, to Dr. Jurin, R. S. Secretary" (54–58).

7. Arbuthnot contended that "if the Doctor's Aphorism, laid down . . . That an *Experiment, to make it useful, always must be nearly uniform*; there must be no such Thing as the Practice of Physick; unless by the Word *nearly* he allows a very great Latitude" (14).

8. See Chapter 2 for more on political arithmetic.

9. Arbuthnot also suggests that "the same Odds wou'd be a sufficient prudential Motive to any private Person to proceed upon, abstracting from the more occult and abstruse Causes which seem to favour this Operation" (21).

10. Arbuthnot was relatively uninterested in why individuals might make different decisions about whether to be inoculated. He implied that differing judgments were based on differing assessments of the "Odds" (i.e., probability) of the success of smallpox inoculation in preventing this disease. However, since he also claimed that the ratios he provided in his text would convince any rational person to be inoculated—"the same Odds wou'd be a sufficient prudential Motive to any private Person to proceed upon" (21)—he implied that equivalently rational thinkers would make the same decisions. Arbuthnot here exemplified a wider tendency of eighteenth-century authors interested in probability to assume that all rational thinkers would, when presented with the same evidence, draw the same conclusions; see Lorraine Daston, *Classical Probability in the Enlightenment* (Princeton, NJ: Princeton University Press, 1995), 49–58.

11. On the construction of what Rusnock calls a "correspondence network" in the early eighteenth century for disseminating information about smallpox inoculation, see *Vital Accounts*, 55–70.

12. John Green, *A Sermon Preached before His Grace George, Duke of Marlborough, President, the Vice-Presidents, the Treasurer, &C. Of the Hospitals for the Small-Pox. On*

Tuesday, April 26, 1763. By the Right Reverend Father-in-God John Lord Bishop of Lincoln (London: Printed by H. Woodfall, in Paternoster-Row, 1763), 14.

13. As I document in Chapter 1, Green's stress on the link between smallpox inoculation and stable commerce was commonplace in the series of yearly sermons that commemorated the founding of the smallpox hospital.

14. For Arbuthnot's references to Newgate, see *Mr. Maitland's Account*, 23–25. On the importance of Newgate prison tests for the British inoculation effort, see Miller, *The Adoption of Inoculation*, 80–91; and Rusnock, *Vital Accounts*, 30. Colonial slaves were also among the early test subjects, which further complicates the question of rights and choice; Miller, *The Adoption of Inoculation*, 93, 125, 164.

15. Miller, *The Adoption of Inoculation*, 23, 267–76, argues that the fact that the British aristocracy was much more supportive of smallpox inoculation efforts than the French aristocracy was a key reason for the early adoption of smallpox inoculation in Britain and its much later adoption in France.

16. On Burke as a conservative, see Isaac Kramnick, *The Rage of Edmund Burke: Portrait of an Ambivalent Conservative* (New York: Basic Books, 1977); and Daniel I. O'Neill, *Edmund Burke and the Conservative Logic of Empire* (Oakland: University of California Press, 2016); on Burke as a liberal, see J. G. A. Pocock, "The Political Economy of Burke's Analysis of the French Revolution," in *Virtue, Commerce, and History: Essays on Political Thought and History, Chiefly in the Eighteenth Century* (Cambridge: Cambridge University Press, 1985), 193–212; Yuval Levin, *The Great Debate: Edmund Burke, Thomas Paine, and the Birth of Right and Left* (New York: Basic Books, 2014); and Domenico Losurdo, *Liberalism: A Counter-History*, trans. Gregory Elliott (New York: Verso, 2011). As O'Neill points out, readings of Burke as a liberal generally focus on his support for the American colonists and on his strictures against government overreach, while readings of him as a conservative tend to focus on his claims in *Reflections on the Revolution in France* for the importance of tradition and stable social hierarchies (8). Yet most of these readings of Burke as primarily a liberal or conservative nevertheless also stress the difficulty of applying these categories disjunctively. For Kramnick, for example, Burke is an "ambivalent conservative," while for Levin, he is a "conservative liberal." From the perspective that I develop in this chapter, Losurdo provides the most useful approach to the question of Burke's political allegiances, for he clarifies that liberalism has always presumed a hierarchical division between the small number of those who are worthy of freedom (and who must thus be protected from government overreach) and the much greater number of uncivilized humans who are not worthy of freedom (and who must be under direct and often violent government control).

17. Edmund Burke, *Reflections on the Revolution in France, and on the Proceedings in Certain Societies in London Relative to That Event. In a Letter Intended to Have Been Sent to a Gentleman in Paris* (London: Printed for J. Dodsley, in Pall-Mall, 1790), 44–45.

18. Burke did not dispute the existence of "rights of men" but argued that they should never be considered abstractly: "These metaphysic rights entering into common life, like rays of light which pierce into a dense medium, are, by the laws of nature, refracted from their strait line. Indeed, in the gross and complicated mass of human passions and concerns, the primitive rights of men undergo such a variety of

refractions and reflections, that it becomes absurd to talk of them as if they continued in the simplicity of their original direction" (*Reflections*, 90–91). For an acute
analysis of Burke's account of the rights of man, see James K. Chandler, *Wordsworth's
Second Nature: A Study of the Poetry and Politics* (Chicago: University of Chicago
Press, 1984), 32–35.

19. Though my focus is different than Pocock's, my argument resonates with his
claim in "The Political Economy of Burke's Analysis of the French Revolution" that
though Burke, like Hume and Smith, promoted a liberal, Whig order of commercial
relations, Burke saw the latter as dependent upon a more primary foundation of
"manners."

20. This is highlighted by the importance of Burke for nineteenth-century liberals; see Losurdo, *Liberalism: A Counter-History*, 37–38, 54, 59, 62–63, 130–33.

21. Or, as Mill put it on the first page of his text (*On Liberty*, 217), he sought to
establish "the nature and limits of the power which can be legitimately exercised by
society over the individual."

22. Mill was equally concerned with legal constraints and the stultifying effects
of "opinion," contending "protection . . . against the tyranny of the magistrate is not
enough: there needs protection also against the tyranny of the prevailing opinion
and feeling; against the tendency of society to impose, by other means than civil
penalties, its own ideas and practices as rules of conduct on those who dissent from
them; to fetter the development, and, if possible, prevent the formation, of any individuality not in harmony with its ways, and compel all characters to fashion themselves upon the model of its own. There is a limit to the legitimate interference of
collective opinion with individual independence: and to find that limit, and maintain it against encroachment, is as indispensable to a good condition of human
affairs, as protection against political despotism." *On Liberty*, 219–20.

23. Though Mill did not explain his choice of the word "experiment" in *On Liberty*, he could have been certain, given his important earlier work on the philosophy
of science in *A System of Logic* (1843), that contemporary readers would have understood his use of the term as having a quasi-scientific sense. On Mill's debate with
William Whewell over the nature of the scientific method and the progress of
science, see Laura J. Snyder, *Reforming Philosophy: A Victorian Debate on Science and
Society* (Chicago: University of Chicago Press, 2006).

24. Or, as Mill wrote in *On Liberty*, "A person whose desires and impulses are
his own—are the expression of his own nature, as it has been developed and modified by his own culture—is said to have a character" (264). For reflections on the
relationship between liberalism and Mill's concept of character, see Elaine Hadley,
Living Liberalism: Practical Citizenship in Mid-Victorian Britain (Chicago: University of
Chicago Press, 2010), 70–106.

25. Mill, *On Liberty*, 260–61.

26. Mill (*On Liberty*, 224) stressed that he foregoes "any advantage which could
be derived to my argument from the idea of abstract right, as a thing independent of
utility. I regard utility as the ultimate appeal on all ethical questions; but it must be
utility in the largest sense, grounded on the permanent interests of man as a progressive being."

27. As a professed utilitarian, Mill held that progress resulted when collective happiness increased. However, that answer raises the question of what enabled happiness to increase, and the answer to that latter question seems to have been an increase in knowledge and individuality.

28. Mill thus lamented the fact that, in the past, it has more often been the case that "one partial and incomplete truth" has "substitute[d] . . . for another," with "improvement" then being limited to the fact that "the new fragment of truth is more wanted, more adapted to the needs of the time, than that which it displaces." *On Liberty*, 252–53.

29. Mill, *On Liberty*, 215; see Wilhelm von Humboldt, *The Limits of State Action* (Indianapolis, IN: Liberty Fund, 1993), 48. Mill referred to this text as *Sphere and Duties of Government*. Though Humboldt composed *Ideen zu einem Versuch die Grenzen der Wirksamkeit des Staats zu bestimmen* (Ideas toward an attempt to determine the limits of the activity of the state) in 1791–1792, only short parts of the text were published in 1792 in the *Berlinische Monatsschrift*, and the text as a whole did not appear before Humboldt's death in 1835. The full version of the text first appeared in the 1852 German collected works of Humboldt and was translated into English in 1854, and this latter was the text that Mill consulted. For accounts of the composition and publishing history of Humboldt's text, see J. W. Burrow's introduction to Humboldt, *The Limits of State Action*, xvii–lviii; and David Sorkin, "Wilhelm von Humboldt: The Theory and Practice of Self-Formation (Bildung), 1791–1810," *Journal of the History of Ideas* 44, no. 1 (1983): 55–73.

30. Mill, *On Liberty*, 261; Mill's quotations are drawn from Chapter II ("Of the individual man, and the highest ends of his existence") of Humboldt's *The Limits of State Action* (10, 12).

31. Johann Gottfried Herder, "On the Cognition and Sensation of the Human Soul (1778)," in *Philosophical Writings* (Cambridge: Cambridge University Press, 2002), 236.

32. On the importance of both Leibniz and Pietism for Humboldt, see Ernst Lichtenstein, *Zur Entwicklung des Bildungsbegriffs von Meister Eckhart bis Hegel* (Heidelberg: Quelle & Meyer, 1966), 22–25; Sorkin, "Wilhelm von Humboldt," 59–68; and Paul R. Sweet, "Young Wilhelm von Humboldt's Writings (1789–93) Reconsidered," *Journal of the History of Ideas* 34, no. 3 (1973): 471. In *The German Tradition of Self-Cultivation: "Bildung" from Humboldt to Thomas Mann* (London: Cambridge University Press, 1975), Walter Horace Bruford also stresses the importance of the Stoic ideal of self-sufficiency for Humboldt (1, 14). While Humboldt's interest in Leibniz and the Stoics encouraged his interest in self-development (*Bildung*), neither source explains Humboldt's equal emphasis on the need to pursue *Bildung* through engagement with others. As Sorkin puts it, for Humboldt "self-formation . . . requires social bonds," yet both Leibniz's monadology and Stoicism rejected precisely these kinds of bonds (59). For Leibniz, monads are "windowless" (the "Monads have no windows through which anything can come in or go out"), and so each monad strove toward its perfection or entelechy alone, with any apparent coordination and cooperation among monads the consequence of a harmony among monads preestablished by God. Gottfried Wilhelm Leibniz, "Monadology,"

trans. George Montgomery, in *Discourse on Metaphysics; Correspondence with Arnauld; Monadology* (La Salle, IL: Open Court, 1902), 252, 262. For the Stoics, the goal of self-sufficiency required that social bonds be eliminated as much as possible. Humboldt's emphasis on the necessity of social bonds for self-formation is thus better explained through his uptake of the Pietist vision that I explain in what follows.

33. On the importance of individual diversity and conversations for the Pietists, see James Daryl Clowes, "'Of Art and Women I Had No Knowledge': The Development of Schleiermacher's Understanding of Cognition, Self Identity, Community and Gender," PhD diss., University of Washington, 1996, 38–44; and Koppel S. Pinson, *Pietism as a Factor in the Rise of German Nationalism* (New York: Columbia University Press, 1934), 60–90.

34. Mill, *On Liberty*, 267.

35. Mill's invocation of the language of genius here is in tension with his rejection, in other texts, of any kind of innate, unalterable differences among individuals. He contended in "Utility of Religion" (1874), for example, that "the power of education is almost boundless: there is not one natural inclination which is not strong enough to coerce, and if needful, to destroy by disuse." John Stuart Mill, "Utility of Religion," in *Collected Works of John Stuart Mill*, ed. John M. Robson (Toronto: University of Toronto Press, 1963–1991), 10:409. Mill was in this text a typical political economist, and his "views especially echo those of Adam Smith," who had argued in *The Wealth of Nations* that "differences in talent" were not innate but the result of the division of labor. Diane B. Paul and Benjamin Day, "John Stuart Mill, Innate Differences, and the Regulation of Reproduction," *Studies in History and Philosophy of Biological and Biomedical Sciences* 39 (2008): 223. For Smith's original articulation of this claim, see Adam Smith, *An Inquiry into the Nature and Causes of the Wealth of Nations*, ed. R. H. Campbell and A. S. Skinner (Indianapolis, IN: Liberty Fund, 1981), 1:28. From this perspective, diversity among individuals was a consequence of the necessary difference in circumstances among individuals; that is, "the real effective education of a people is given them by the circumstances by which they are surrounded": John Stuart Mill, "The Condition of Ireland [20]," in *Collected Works of John Stuart Mill*, ed. John M. Robson (Toronto: University of Toronto Press, 1963–1991), 24:955. Hayek adopts and makes central to his theory of knowledge the nearly identical claim that individual diversity is a consequence of the physical separation of individuals from one another.

36. While both Hayek and von Mises originally argued this claim in articles intended for other economists, Hayek also made it the centerpiece of his manifesto-like defense of liberalism, *The Road to Serfdom*, which became a rallying point for neoliberals throughout the twentieth century.

37. Friedrich A. von Hayek, *The Road to Serfdom: Text and Documents* (Chicago: University of Chicago Press, 2007), 100.

38. Friedrich A. von Hayek, "The Use of Knowledge in Society," *American Economic Review* 35, no. 4 (1945): 519.

39. That is, competition is the most "efficient" means of planning, since it makes the "full[est] use of the existing knowledge." Hayek, "The Use of Knowledge in Society," 521.

40. Hayek made essentially the same claim in *The Road to Serfdom*, 95.

41. As I noted in Chapter 2, Hayek explicitly connected his claims about the wisdom of markets to evolutionary biology and to Ernst Mayr in the 1980s.

42. In *The Road to Serfdom*, Hayek's ambivalent neo-Burkean description of the market—it was a tradition, the essence of which must be respected but could also be consciously optimized—led to an equally ambivalent neo-Burkean concept of experimentation. Hayek tended to align "experimentation" with government planning and suggested that this led to totalitarianism. See *The Road to Serfdom*, 45, 51, 14, 196. Yet as I have noted, Hayek also presented the market as itself the product of something like Burke's mode of unconscious, collective, long-term experimentation.

43. Hayek, *The Road to Serfdom*, 46.

44. Winch astutely describes neoliberalism as "the belief that an harmonious relationship can be established between Smithian economic liberalism and Burkean conservatism," in the sense that, "by combining the two positions one arrives at a spontaneous economic order that is the unintended outcome of individual choices, and a legal and governmental regime that respects custom and tradition while being protective of those 'little platoons'—the family, the Church, and other voluntary associations—that are thought to be essential to social cohesion and even nationhood," and points to Hayek as "the most influential exponent of this view." Donald Winch, *Riches and Poverty: An Intellectual History of Political Economy in Britain, 1750–1834* (Cambridge: Cambridge University Press, 1996), 11, 12n23. Though I agree with both this description of neoliberalism and the importance of Hayek, my account seeks to explain how Hayek, by taking seriously the importance of the role of information for the concept of collective experimentation, entwined the Smithian and Burkean positions in a way that was not simply arbitrary or contradictory.

45. Hayek, "The Use of Knowledge in Society," 526.

46. In the introduction to *The Road to Serfdom*, Hayek argued against the belief that German intellectual history was intrinsically antiliberal and oriented toward the sort of authoritarian state exemplified by National Socialism. Such a position "overlooks the fact that, when eighty years ago John Stuart Mill was writing his great essay *On Liberty*, he drew his inspiration, more than from any other men, from two Germans—Goethe and Wilhelm von Humboldt" (61). Hayek proposed that the German authoritarian vision first emerged in philosophers such as Fichte (182–83), which suggests that Hayek's neoliberalism should be understood as an attempt to reclaim a "liberal" mode of German Romanticism or, at least, German late-eighteenth-century philosophy. For a contrary reading, in which Humboldt and Fichte are *both* part of the German liberal tradition, which latter is opposed to German Romanticism proper, see Frederick C. Beiser, *Enlightenment, Revolution, and Romanticism: The Genesis of Modern German Political Thought, 1790–1800* (Cambridge, MA: Harvard University Press, 1992).

47. Mill, *On Liberty*, 292–93.

48. Hayek, *The Road to Serfdom*, 125, 126.

49. Foucault captured this aspect in his suggestion that neoliberalism demands that each individual become an "entrepreneur of himself" (*The Birth of Biopolitics:*

Lectures at the Collège de France, 1978–79, trans. Graham Burchell, ed. Michel Senellart [New York: Palgrave Macmillan, 2008], 226), a phrase that captured the attention of many critics. See, e.g., Philip Mirowski, *Never Let a Serious Crisis Go to Waste: How Neoliberalism Survived the Financial Meltdown* (New York: Verso, 2014), 93–102; and Wendy Brown, *Undoing the Demos: Neoliberalism's Stealth Revolution* (New York: Zone, 2015), 79–11. Foucault's description is accurate, but I would stress the close connection between Mill's earlier concept of experiments in living and the neoliberal concept of an entrepreneurship of the self. The latter is, in essence, what happens to the concept of experiments in living when these are understood as necessarily market based.

50. Even if Mill's stress on the rarity of genius underscored aspects of individuality that were not the result of conscious choice, his concept of the genius nevertheless presumed that an individual chose to work hard to express his or her innate endowments.

51. See, for example, Mitchell and Waldby, "National Biobanks," and Mitchell, "US Biobanking Strategies."

52. Ulrich Beck, *Risk Society: Towards a New Modernity*, trans. Mark Ritter (London: Sage, 1992), 19.

53. In subsequent work, Beck addressed the fact that risks are often inequitably distributed, though he reiterated the necessarily global dimension of modern risks; see Ulrich Beck, *World at Risk*, trans. Ciaran Cronin (Malden, MA: Polity, 2009), 160–86.

54. See Lasagna, "Consensus among Experts." This emphasis on disunity among experts was also stressed by Arbuthnot.

55. Beck stressed that risk determinations function in this way not when groups oppose science but rather when they link science to their own values and aspirations: "Risk consciousness is neither a traditional nor a lay person's consciousness, but is essentially determined by and oriented toward science. For, in order to recognize risks at all and make them the reference point of one's own thought and action, it is necessary on principle that invisible causality relationships between objectively, temporally, and spatially very divergent conditions, as well as more or less speculative projections, be *believed*, that they be *immunized* against the objections that are always possible. . . . One no longer ascends merely from personal experience to general judgments, but rather general knowledge devoid of personal experience becomes the central determinant of personal experience." *Risk Society*, 72.

56. For a chilling account of the neoliberal use of the lack of consensus among experts to pursue specific promarket agendas, see Mirowski, *Never Let a Serious Crisis Go to Waste*, 223–30. Mirowski builds on Robert Proctor's concept of "agnotology" (the uses of ignorance) developed in Robert Proctor and Londa L. Schiebinger, *Agnotology: The Making and Unmaking of Ignorance* (Stanford, CA: Stanford University Press, 2008); and Robert Proctor, *Golden Holocaust: Origins of the Cigarette Catastrophe and the Case for Abolition* (Berkeley: University of California Press, 2011). See also Naomi Oreskes and Erik M. Conway, *Merchants of Doubt: How a Handful of Scientists Obscured the Truth on Issues from Tobacco Smoke to Global Warming* (New York: Bloomsbury, 2010).

57. To put this another way, what remains unclear in Beck's account is whether the final values established by communities must ultimately be subordinated to scientific knowledge, such as research about global warming. Though Beck addressed this point in *World Risk Society*, his advocacy of what he describes there alternately as "reflexive realism" and "realist constructivism" (88–89) does not solve the problem. However, insofar as Beck suggested that it "requires crass ignorance or decidedly selective vision to overlook the link between an ominously rising temperature curve and increasing greenhouse gas emissions" (92), he seemed to assume that ultimate values about how a group wishes to live must be subordinated to climate science research. My thanks to Jamie Lorimer for discussion about this point.

58. *The Road to Serfdom* was Hayek's most influential effort to prove that failure to hew narrowly to liberalism led to National Socialist– or Stalinist-style totalitarianism. Foucault discussed National Socialism briefly in *The History of Sexuality* (New York: Pantheon, 1978), 149–50, and noted the obsessive neoliberal focus on totalitarianism in *The Birth of Biopolitics*, 101–21.

59. Mill, *On Liberty*, 217–18.

60. Roberto Esposito, *Bíos: Biopolitics and Philosophy*, trans. Timothy Campbell (Minneapolis: University of Minnesota Press, 2008), 71.

61. For helpful discussion of the relationship of Esposito's approach to immunity to those of other contemporary theorists such as Niklas Luhmann, Peter Sloterdijk, Donna Haraway, and Jacques Derrida, see Timothy C. Campbell, *Improper Life: Technology and Biopolitics from Heidegger to Agamben* (Minneapolis: University of Minnesota Press, 2011); and Cary Wolfe, *Before the Law: Humans and Other Animals in a Biopolitical Frame* (Chicago: University of Chicago Press, 2013), esp. 38, 90–94.

62. Arbuthnot, *Mr. Maitland's Account*, 35. While Arbuthnot's text implies that in cases of Pestilence, the state compels individuals to act in ways that facilitate the survival of most members of the population, the state could presumably also legitimately choose other criteria, such as ensuring the survival of the "most valuable" members of the population. In *Third Person: Politics of Life and Philosophy of the Impersonal*, trans. Zakiya Hanafi (Malden: Polity, 2012), Roberto Esposito—building on Foucault's brief account of the intersection of race and biopolitics in *The History of Sexuality*—tracks the emergence of precisely this kind of racial hygienic logic in nineteenth-century European biology, philosophy, anthropology, and linguistics (20–59).

63. Beck, *Risk Society*, 58.

64. Or, as Beck puts it in *Risk Society*, "acceptable values [for contaminants] make possible a *permanent ration of collective standardized poisoning*," and in this sense, one is "no longer concerned with questions of ethics at all but with how far one of the most minimal rules of social life—not to poison each other—may be *violated*" (65).

65. On neoliberalism as a constructivism rather than a naturalism, see Mirowski, *Never Let a Serious Crisis Go to Waste*, 53–57.

66. On this link between the concepts of experience and experiment, see the entry on "Experience" in Raymond Williams, *Keywords: A Vocabulary of Culture and Society*, rev. ed. (New York: Oxford University Press, 2015), 126–29. I also discuss the relationship between these terms in Robert Mitchell, *Experimental Life: Vitalism*

in Romantic Science and Literature (Baltimore, MD: Johns Hopkins University Press, 2014).

67. Hannah Arendt, *The Human Condition*, 2nd ed. (Chicago: University of Chicago Press, 1998), 57.

68. Michel Foucault, "The Risks of Security," in *The Essential Works of Foucault, 1954–1984*, vol. 3: *Power*, ed. James D. Faubion (New York: The New Press, 2000), 366.

6. Life, Self-Regulation, and the Liberal Imagination

1. Michel Foucault, *Society Must Be Defended: Lectures at the Collège de France, 1975–76*, trans. David Macey (New York: Picador, 2003), 239–64; Michel Foucault, *Security, Territory, Population: Lectures at the Collège de France, 1977–78*, trans. Graham Burchell, ed. Michel Senellart (New York: Palgrave Macmillan, 2007), 29–53, 333–61; Michel Foucault, *The Birth of Biopolitics: Lectures at the Collège de France, 1978–79*, trans. Graham Burchell, ed. Michel Senellart (New York: Palgrave Macmillan, 2008), 129–57.

2. Foucault, *Security, Territory, Population*, 29–49.

3. Antoine Laurent Lavoisier and Armand Seguin, *Premier mémoire sur la transpiration des animaux*, in *Œuvres de Lavoisier: Publiées par les soins de son excellence le Ministre de l'instruction publique et des cultes* (Paris: Imprimerie impériale, 1862), 2:713. On the Regulator movement, see William S. Powell, ed., *The Regulators in North Carolina: A Documentary History, 1759–1776* (Raleigh: State Department of Archives and History, 1971). Mary Wollstonecraft, *A Vindication of the Rights of Woman: With Strictures on Political and Moral Subjects* (London: Printed for J. Johnson, No. 72, St. Paul's Church Yard, 1792), 36; Mary Wollstonecraft Shelley, *Frankenstein, or, the Modern Prometheus*, 3rd ed., ed. David Lorne Macdonald and Kathleen Dorothy Scherf (Peterborough: Broadview, 2012), 52.

4. Jane Austen, *Pride and Prejudice* (New York: Penguin, 1972), 198.

5. As a consequence of space considerations, I consider only "theoretical" (e.g., economic and philosophical) Romantic-era texts in this chapter. For discussion of ways that the discourse of regulation bears on Romantic-era poems and novels, see Robert Mitchell, "Regulating Life: Romanticism, Science, and the Liberal Imagination," *European Romantic Review* 29, no. 3 (2018): 275–93; and Robert Mitchell, "*Frankenstein* and the Sciences of Self-Regulation," forthcoming.

6. Jon Mee, *Romanticism, Enthusiasm, and Regulation: Poetics and the Policing of Culture in the Romantic Period* (New York: Oxford University Press, 2005); D. A. Miller, *The Novel and the Police* (Berkeley: University of California Press, 1988). See also Alan Richardson's discussion in *Literature, Education, and Romanticism: Reading as Social Practice, 1780–1832* (Cambridge: Cambridge University Press, 1994), 185–202, of Romantic novelists' aspirations to engender "well-regulated minds" by means of the domestic novel.

7. Georges Canguilhem, *Ideology and Rationality in the History of the Life Sciences* (Cambridge, MA: MIT Press, 1988), 81–102; Randy E. Barnett, *Restoring the Lost Constitution: The Presumption of Liberty* (Princeton, NJ: Princeton University Press,

2004), 302–13; Michael Friedman, "Regulative and Constitutive," *Southern Journal of Philosophy* 30 (1991): 73–102. On the biological sense of regulation, see also E. F. Adolph, "Early Concepts of Physiological Regulation," *Physiology Reviews* 41 (1961): 737–70. See as well Canguilhem's helpful effort to distinguish biological from social self-regulation in *The Normal and the Pathological*, trans. Carolyn R. Fawcett in collaboration with Robert S. Cohen (New York: Zone, 1991), 237–56.

8. René Descartes, *The Philosophical Writings of Descartes*, trans. John Cottingham, Robert Stoothoff, and Dugald Murdoch (Cambridge: Cambridge University Press, 1984), 1:77–78. John Harris's *Lexicon Technicum: or, an universal English dictionary of arts and sciences: explaining not only the terms of art, but the arts themselves* (London: Printed for Dan. Brown et al., 1708) defines a regulator as "a small Spring belonging to the *Ballance* in the new Pocket-Watches" (vol. 1, s.v. "Regulator"), while the entry for "Regulator" in Ephraim Chambers's *Cyclopaedia: Or, an Universal Dictionary of Arts and Sciences* (London: Rivington et al., 1778) notes that for a time-keeping device, this is "a small SPRING belonging to the balance, serving to adjust the going, and to make it go either faster or slower."

9. Henry III, c. 25, cited in Julian Hoppit, "Reforming Britain's Weights and Measures, 1660–1824," *English Historical Review* 108, no. 426 (1993): 82.

10. For accounts of the history of measures in sixteenth- through nineteenth-century England, see Hoppit, "Reforming Britain's Weights and Measures"; Witold Kula, *Measures and Men* (Princeton, NJ: Princeton University Press, 1986); Ronald Edward Zupko, *Revolution in Measurement: Western European Weights and Measures since the Age of Science* (Philadelphia: American Philosophical Society, 1990); and Aashish Velkar, *Markets and Measurements in Nineteenth-Century Britain* (Cambridge: Cambridge University Press, 2012).

11. The term "regulating" often appeared in seventeenth-century legal proclamations, as the following examples, chosen randomly from Early English Books Online, underscore: *By the King, a proclamation for the better regulating lotteries within the kingdoms of Great Britain and Ireland* (1665); *By the King, a proclamation for regulating the colours to be worn on merchant ships* (1683); and *An act for the better regulating of measures in and throughout this kingdom* (1695).

12. Gottfried Wilhelm Leibniz, *Essais de théodicée sur la bonté de dieu, la liberté de l'homme et l'origine du mal: Nouvelle édition, augmentée de l'histoire de la vie & des ouvrages de l'auteur, par M. L. de Neufville* (Amsterdam: F. Changuion, 1734), 1:xxxii; Canguilhem, *Ideology and Rationality*, 84. For Clarke's description of the way that God allows his creation "to move regularly," see Samuel Clarke, Gottfried Wilhelm Leibniz, and Isaac Newton, *The Leibniz-Clarke Correspondence, Together with Extracts from Newton's "Principia" and "Opticks"* (Manchester: Manchester University Press, 1956), 13–14.

13. The link between divine action and political sovereignty was explicit. Clarke claimed that if Leibniz's God had no need to intervene in the universe once he had created it, this was "merely a nominal kingdom . . . wherein all things would continually go on without [the king's] government or interposition." Clarke, Leibniz, and Newton, *The Leibniz-Clarke Correspondence*, 14. Leibniz responded that this was equivalent to saying that "a king, who should originally have taken care to have his

subjects so well educated, and should, by his care in providing for their substance, preserve them so well in their fitness for their several stations . . . as that he should have no occasion ever to be amending any thing amongst them; would be only a nominal king" (19–20). For helpful discussion of this debate in the context of the history of the concept of biological regulation, see Canguilhem, *Ideology and Rationality*, 83–87.

14. Canguilhem (*Ideology and Rationality*, 84) argued that Leibniz's image of divine regulation drew on the image of automatic adjustment provided by Christiaan Huygens's new watch regulator, while Clarke opposed this notion of automatic adjustment. McLaughlin proposes that Leibniz's and Clarke's opposed understandings of divine regulation map more precisely to the difference between a hand-held watch regulator, which operated automatically, and the "governor" of a building-mounted clock (i.e., a human being who at least once a day adjusted the clock). As McLaughlin notes, for seventeenth-century authors, clocks (as opposed to watches) were often not models for precision but rather of created objects that required constant supervision. Peter McLaughlin, "Regulation, Assimilation, and Life: Kant, Canguilhem, and Beyond," 2007, http://www.philosophie.uni-hd.de/md/philsem/personal/mcl_regulation.pdf, 3–5.

15. On the context of Locke's writing, see Kelly's general introduction to John Locke, *Locke on Money*, ed. P. H. Kelly (Oxford: Oxford University Press, 1991); and James Thompson, *Models of Value: Eighteenth-Century Political Economy and the Novel* (Durham, NC: Duke University Press, 1996), 41–72.

16. John Locke, *Some Considerations of the Consequences of the Lowering of Interest, and Raising the Value of Money*, in *Locke on Money*, 210.

17. Because of Canguilhem's focus in *Ideology and Rationality* on the prehistory of the physiological concept of regulation, this second model of regulation is absent from his account. Though, given Canguilhem's focus, this is understandable, it means that his emphasis on eighteenth-century concepts of regulation as committed to the premise of "conservation or restoration of a closed system" (87) is true of physics and physiology but not more generally true of all eighteenth-century approaches to regulation.

18. Though Alessia Pannese focuses on the mid-eighteenth to late nineteenth century and does not consider the authors on which I focus here, her reflections in "The Non-Orientability of the Mechanical in Thomas Carlyle's Early Essays," *Journal of Interdisciplinary History of Ideas* 6, no. 11 (2017): 3:1–3:19, on relationships among machines, mechanical causality, automatic behavior, and the will is helpful for parsing out different ways that authors in these periods understood the relationship of automatic behavior to willing. Pannese stresses that some eighteenth-century authors understood "automatism" as a virtuous *accomplishment* that resulted from the earlier exercise of the will. Pannese notes that David Hartley, for example, stressed in *Observations on Man* (1749) that the child's willed effort to walk eventually resulted in automatic walking (3:8–10).

19. I employ here the formulation "he or she" in order to capture the fact that Locke includes some women—namely, widows—as market participants, at least in *Some Considerations* (212, 219).

20. This then leads to the tricky—and unanswerable—question of whether freedom should be understood as an attribute of individuals or of the market. In *Some Considerations*, freedom is attributed to the market, rather than to individuals—for example, the "free" value of gold or silver is equivalent to its "Market value" (325, 327). In Locke's discourses on government, freedom is an attribute of individuals, both in the sense that the natural state of humans (i.e., the "state [that] all men are naturally in") is "a state of perfect freedom to order their actions and dispose of their possessions and persons, as they think fit, within the bounds of the law of nature," while the "freedom of men under government is, to have a standing rule to live by, common to every one of the society, and made by the legislative power erected in it; a liberty to follow my own will in all things, where the rule prescribes not; and not to be subject to the inconstant, uncertain, unknown, arbitrary will of another man." John Locke, *Two Treatises of Government: And a Letter Concerning Toleration* (New Haven, CT: Yale University Press, 2003), 101, 110. While markets are instances of "freedom of men under government," Locke links individual differences, freedom, and markets in a way that makes it impossible to assign freedom either solely to individuals or to markets; rather, freedom is for Locke what emerges when individual differences are bound to markets.

21. James Steuart, *An Inquiry into the Principles of Political Oeconomy: Being an Essay on the Science of Domestic Policy in Free Nations* (London: Printed for A. Millar and T. Cadell, 1767), 1:24. Keith Tribe—and, following Tribe, Taylor—have argued that Steuart could not understand the economy as self-regulating because "the Sovereign or Statesman is essential in the structure" of Steuart's account, in the sense that the Sovereign or Statesman "is the sole expression of a unity which is otherwise dispersed among individual units or the categories which articulate those units." Keith Tribe, *Genealogies of Capitalism* (Atlantic Highlands, NJ: Humanities Press, 1981), 48; see also his *Land, Labour, and Economic Discourse* (London: Routledge & Kegan Paul, 1978), 80–109. Taylor helpfully glosses Tribe's claim: "For Steuart, the statesman's attention is necessary because economic relations lack an autonomous logic by which they would regulate themselves." Christopher Taylor, *Empire of Neglect: The West Indies in the Wake of British Liberalism* (Durham, NC: Duke University Press, 2018), 41. By contrast, Taylor contends, Adam Smith subscribes to the view of self-regulating economic processes. While I acknowledge significant differences between Steuart and Smith, neither Tribe nor Taylor account for Steuart's claim that his Sovereign/Statesman is a fictional perspective that *each* reader of the text is supposed to take on. Hence, Steuart writes that though his book "seems addressed to a statesman, the real object of the inquiry is to influence the spirit of those whom he governs" (1:xiv; see also 3–4). In other words, if each member of the polity acts *as though* a statesman/sovereign organizes economic affairs, this will result in a state of economic affairs that looks as though it had been organized by such a figure. Though Steuart's is not an especially complex account of self-regulation, it is nevertheless *not* an account that requires an actual sovereign.

22. Adam Smith, *An Inquiry into the Nature and Causes of the Wealth of Nations*, ed. R. H. Campbell and A. S. Skinner (Indianapolis, IN: Liberty Fund, 1981), 1:I.v (47–64). Smith found an additional way to valorize the limited perspectives of individuals,

arguing in *The Wealth of Nations* that improvements in the process of production were possible only because the division of labor focused an individual's attention on one part of the production process; i.e., "Men are much more likely to discover easier and readier methods of attaining any object, when the whole attention of their minds is directed towards that single object, than when it is dissipated among a great variety of things" (I:I.i.8 [20]). The differences in individual perspective upon which Smith focused here were not innate but instead a function of the position of individuals within a market and within a production process.

23. In *On the Principles of Political Economy, and Taxation*, 3rd ed. (London: John Murray, 1821), for example, David Ricardo defined the "principal problem in Political Economy" as that of "determin[ing] the laws which regulate" the distribution of "the produce of the earth" among the three "classes" of land-owners, owners of capital, and laborers (v).

24. Some mid-eighteenth-century authors explicitly connected "the standard of taste" to questions of political economy and population. In *An Inquiry into the Principles of Political Economy*, for example, Steuart contended that the standard of taste was essential in enabling an increase of total population. "It is not," Steuart wrote, "in the most fruitful countries of the world, nor in those which are the best calculated for nourishing great multitudes, that we find the most inhabitants. It is in climates less favoured by nature, and where the soil only produces to those who labour, and in proportion to the industry of everyone, where we may expect to find great multitudes; and even these will be found greater or less, in proportion as the turn of the inhabitants is directed to ingenuity and industry." That is, population and trade increase in those countries in which there are more "useful manufactures, which, being refined by the ingenious, will determine what is called the standard of taste; this taste will increase consumption, which will again multiply workmen, and these will encourage the production of food for their nourishment" (1:35).

25. Edmund Burke, *A Philosophical Enquiry into the Origin of Our Ideas of the Sublime and Beautiful* (London: Printed for R. and J. Dodsley, in Pall-Mall, 1759), 6.

26. David Hume, "Of the Standard of Taste," in *Four Dissertations* (London: Printed for A. Millar, 1757), 203, 204.

27. Burke, *A Philosophical Enquiry*, 14.

28. Hume, "Of the Standard of Taste," 203.

29. I draw especially on the following discussions of Malthus: Kenneth Smith, "The Malthusian Controversy," thesis, University of London, 1951; Frances Ferguson, "Malthus, Godwin, Wordsworth, and the Spirit of Solitude," in *Literature and the Body: Essays on Populations and Persons*, ed. Elaine Scarry (Baltimore, MD: Johns Hopkins University Press, 1988), 106–24; Mitchell Dean, *The Constitution of Poverty: Toward a Genealogy of Liberal Governance* (New York: Routledge, 1991), 87–105; Donald Winch, *Riches and Poverty: An Intellectual History of Political Economy in Britain, 1750–1834* (Cambridge: Cambridge University Press, 1996), 221–405; Mary Poovey, *A History of the Modern Fact: Problems of Knowledge in the Sciences of Wealth and Society* (Chicago: University of Chicago Press, 1998), 278–95; and David Collings, *Monstrous Society: Reciprocity, Discipline, and the Political Uncanny, c. 1780–1848* (Lewisburg, PA: Bucknell University Press, 2009), 161–92. Whenever possible, I cite from Thomas

Malthus, *An Essay on the Principle of Population: Or a View of Its Past and Present Effects on Human Happiness, with an Inquiry into Our Prospects Respecting the Future Removal or Mitigation of the Evils Which It Occasions*, ed. Donald Winch and Patricia James (Cambridge: Cambridge University Press, 1992), noting this parenthetically as "1992." However, Winch and James's edition, based on the 1803 edition of Malthus's text, does not appear to include several passages from the 1798 edition that were deleted in later editions. When citing these passages, I cite directly from Thomas Malthus, *An Essay on the Principle of Population, as It Affects the Future Improvement of Society. With Remarks on the Speculations of Mr. Godwin, M. Condorcet, and Other Writers* (London: J. Johnson, 1798), noted parenthetically as "1798."

30. See Joseph J. Spengler, *French Predecessors of Malthus: A Study in Eighteenth-Century Wage and Population Theory* (Durham, NC: Duke University Press, 1942); Foucault, *Security, Territory, Population*, 29–86.

31. See Genevieve Miller, *The Adoption of Inoculation for Smallpox in England and France* (Philadelphia: University of Pennsylvania Press, 1957); and Foucault, *Security, Territory, Population*, 55–86.

32. See, e.g., Malthus's claim in the 1798 edition that when population increases, "the number of labourers . . . being above the proportion of the work in the market, the price of labour must tend toward a decrease; while the price of provisions would at the same time tend to rise" (1798: 30; 1992: 25).

33. For helpful accounts of Malthus's Romanticism, see Ferguson, "Malthus, Godwin, Wordsworth"; Tim Fulford, "Apocalyptic Economics and Prophetic Politics: Radical and Romantic Responses to Malthus and Burke," *Studies in Romanticism* 40, no. 3 (2001): 345–68; and Collings, *Monstrous Society*, 161–92.

34. Malthus uses "regulation" as a synonym for political measures when he refers to "tyrannical regulations" (1798: 134) and as a synonym for natural processes when he claims that "population bears a regular proportion to the food that the earth is made to produce" (1798: 55). Malthus employs both senses of the term in his critique of Godwin's future ideal society, arguing that where Godwin attributes misery to "political regulations" (1798: 176), misery is actually a consequence of "laws inherent in the nature of man" that are "absolutely independent of all human regulation" (1798: 191).

35. William Hazlitt, *A Reply to the Essay on Population, by the Rev. T. R. Malthus. In a Series of Letters. To Which Is Added, Extracts from the Essay, with Notes* (London: Printed for Longman, Hurst, Rees, and Orme, Paternoster-Row, 1807), 44–45.

36. As Winch notes, Malthus in fact introduced the category of moral restraint in 1803; see 1992: xiii.

37. Hazlitt, *A Reply to the Essay on Population*, 46.

38. See esp. Roberto Esposito, *Bíos: Biopolitics and Philosophy*, trans. Timothy Campbell (Minneapolis: University of Minnesota Press, 2008); and Roberto Esposito, *Immunitas: The Protection and Negation of Life*, trans. Zakiya Hanafi (Malden: Polity, 2011).

39. Esposito, *Bíos*, 52–53. Esposito claims that immunity is the key to understanding the connection between biopolitics and modernity, for "only when biopolitics is linked to the immunitary dynamic of the negative protection of life does biopolitics reveal its specifically modern genesis" (9).

40. Esposito, *Immunitas*, 8.

41. Hazlitt, *A Reply to the Essay on Population*, 44.

42. Kant uses "regulative" (*regulativ*) and "constitutive" (*konstitutiv*) throughout his critical works; see especially *Critique of Pure Reason*, trans. Norman Kemp Smith (New York: St. Martin's, 1965), 449–54 (A508–15; B536–43), 514–18 (A615–20; B642–48), 532–49 (A642–68; B670–96); and *Critique of Judgment*, trans. Werner S. Pluhar (Indianapolis, IN: Hackett, 1987), 4–5 (Ak. 167–68), 251–55 (Ak. 372–75), 283–94 (Ak. 401–10). When referring to the *Critique of Pure Reason*, I use the abbreviation *CPR* and note parenthetically the standard German pagination for the "A" and "B" versions of this text; when referring to *Critique of Judgment*, I use the abbreviation *CJ* and note parenthetically, via "Ak.," the pagination for the standard *Akademie Edition* of Kant's works. (The *Kritik der Urteilskraft* is in section 1, volume 5 of this edition.) For concise discussions of Kant's constitutive/regulative distinction, see Friedman, "Regulative and Constitutive"; and Frederick Rauscher, "The Regulative and Constitutive in Kant's and Hegel's Theories of History," *Idealistic Studies* 32, no. 2 (Summer 2002): 121–42.

43. Kant, *CPR* 450 (A509; B537).

44. Gilles Deleuze, *Difference and Repetition*, trans. Paul Patton (New York: Columbia University Press, 1995), 258. For Kant's account of the regulative idea of reason in terms of posing "problems," see *CPR* 449–50 (A508; B536).

45. Hazlitt, *A Reply to the Essay on Population*, 58–59.

46. For Malthus's invocations of pressure, see 1992: 113, 116, 133, 164.

47. Hannah Arendt, *Lectures on Kant's Political Philosophy* (Chicago: University of Chicago Press, 1982), 29. On the likely origins for Kant's distinctive approach to philosophy, see John H. Zammito, *Kant, Herder, and the Birth of Anthropology* (Chicago: University of Chicago Press, 2002).

48. Immanuel Kant, "Idea for a Universal History with a Cosmopolitan Purpose," trans. H. B. Nisbet, in *Kant: Political Writings*, ed. Hans Siegbert Reiss (Cambridge: Cambridge University Press, 1990), 44.

49. Immanuel Kant, "Perpetual Peace: A Philosophical Sketch," trans. H. B. Nisbet, in *Kant: Political Writings*, ed. Hans Siegbert Reiss (Cambridge: Cambridge University Press, 1990), 106.

50. Arendt, *Lectures on Kant's Political Philosophy*, 27.

51. For useful accounts of Kant's racialized claims about standards of beauty, see Simon Gikandi, *Slavery and the Culture of Taste* (Princeton, NJ: Princeton University Press, 2011), 103–7; Pauline Kleingeld, "Kant's Second Thoughts on Race," *Philosophical Quarterly* 57, no. 229 (2007): 573–92; and John Hoffmann, "Kant's Aesthetic Categories: Race in the *Critique of Judgment*," *Diacritics* 44, no. 2 (2016): 54–81. Hoffmann provides an especially nuanced reading of this section of Kant's *Critique of Judgment*, largely because he acknowledges that Kant is divided on the question of race (for example, precisely because Kant "believes that nature intends diversity, he longs to keep races apart": 73). Following Zerilli's lead, I stress Kant's commitment to diversity, rather than Kant's attempt to keep diversity channeled within purported racial groups. Linda M. G. Zerilli, *A Democratic Theory of Judgment* (Chicago: University of Chicago Press, 2016); Linda M. G. Zerilli, *Feminism and the Abyss of Freedom* (Chicago: University of Chicago Press, 2005).

52. See especially *CJ* 55–64 (Ak. 212–19).

53. The following paragraphs owe much to Arendt's stress on the social dimension of thinking in Kant in *Lectures on Kant's Political Philosophy* (esp. 40–41) and Zerilli's compelling extension, in *Feminism and the Abyss of Freedom* and *A Democratic Theory of Judgment*, of Arendt's approach to Kant.

54. The principle of purposiveness can also be described as "the principle of nature's being commensurate with our cognitive power" (*CJ* 28 [Ak. 188]).

55. Or, as Kant put it, "the basis of this pleasure is found in the universal, though subjective, condition of reflective judgments, namely, the purposive harmony of an object (whether a product of nature or of art) with the mutual relation of the cognitive powers (imagination and understanding) that are required for every empirical cognition" (*CJ* 31 [Ak. 191]).

56. See Gilles Deleuze's discussion in "The Idea of Genesis in Kant's Esthetics," in *Desert Islands and Other Texts, 1953–1974* (Cambridge, MA: Semiotext(e), 2004), 56–71, of the "free indeterminate agreement" among the faculties in judgments of taste. Vivasvan Soni, "Playing at Judgment: Aporias of Liberal Freedom in Kant's *Critique of Judgment*," in *Literary/Liberal Entanglements: Toward a Literary History for the Twenty-First Century*, ed. Corinne Harol and Mark Simpson (Toronto: University of Toronto Press, 2017), 151–91, provides a critical account of the conceptual problems entailed by Kant's equation of "freedom" with "play" and argues that these are also the aporias of liberalism (151–52). Soni contends that Kant's concept of "free play" cannot provide a workable model for politics, because Kant's concept of freedom as play "is only free *because and for as long as* it refuses to be directed toward any ends" (177) and hence necessarily leads to the model of "an infinite conversation . . . that can never properly conclude" (152). As Soni notes, this dimension of liberalism was criticized by critics such as Leo Strauss and Carl Schmitt, who stressed that "an infinite conversation" is "incapable of providing a ground for decision." Against Kant's model of free play, Soni suggests that rule-oriented games provide a better point of orientation for politics (179–81). While I agree with Soni that Kant is remarkably tight-lipped concerning the content of his concept of "play" (169), his representation of the regular, or self-regulated, play (*regelmäßiges Spiel*) of the imagination seems to open up Kant's work to Soni's preferred model of games.

57. Kant contended in the *Critique of Judgment* that "only in *society* is the beautiful of empirical interest. And if we grant that the urge to society is natural to man but that his fitness and propensity for it, i.e., *sociability*, is a requirement of man as a creature with a vocation for society and hence is a property pertaining to his *humanity*, then we must also inevitably regard taste as an ability to judge whatever allows us to communicate even our *feeling* to everyone else, and hence regard taste as a means of furthering something that everyone's natural inclination demands" (*CJ* 163 [Ak 296–97]).

58. Though my use of the term "regulative play" seems to align Arendt's and Zerilli's approaches with Friedrich Schiller's account of play in *On the Aesthetic Education of Man, in a Series of Letters* (New York: Frederick Ungar, 1965), both Arendt and Zerilli focus on actual differences among humans in a way that is absent in Schiller.

59. Zerilli, *A Democratic Theory of Judgment*, esp. 28–40.

60. Hannah Arendt, *The Human Condition*, 2nd ed. (Chicago: University of Chicago Press, 1998), 57.

61. On the connection between the common world and aspirations for immortality, see Arendt, *The Human Condition*, 17–21.

62. Malthus 1992: 369.

63. As Arendt notes in *The Human Condition*, it is "characteristic of all natural processes that they come into being without the help of man," and we describe as "automatic" "all courses of movement which are self-moving and therefore outside the range of willful and purposeful interference" (150–51).

64. Following Lavoisier's—and, more directly, Claude Bernard's—lead, the early-twentieth-century physiologist Walter B. Cannon proposed the concept of regulation as the catch-all term for the ability of parts and systems of animal bodies to persist in certain states despite changes in the external environment. See Walter B. Cannon, "The Body Physiologic and the Body Politic," *Science* 93, no. 2401 (1941): 1–10; Walter B. Cannon, "Organization for Physiological Homeostasis," *Physiological Reviews* 9, no. 3 (1929): 399–431; and Walter B. Cannon, *The Wisdom of the Body*, 1st ed. (New York: Norton, 1932). My reflections on Cannon and his legacy owe much to Benjamin J. Murphy, "Not So New Materialism: Homeostasis Revisited," *Configurations* 27, no. 1 (2019): 1–36.

65. Sharon E. Kingsland, *The Evolution of American Ecology, 1890–2000* (Baltimore, MD: Johns Hopkins University Press, 2005), 165.

66. See also Frank B. Golley, *A History of the Ecosystem Concept in Ecology: More Than the Sum of the Parts* (New Haven, CT: Yale University Press, 1993), 24.

67. Kingsland, *The Evolution of American Ecology*, 189. See also Joel B. Hagen, *An Entangled Bank: The Origins of Ecosystem Ecology* (New Brunswick, NJ: Rutgers University Press, 1992), 100–121.

68. Lawrence K. Frank, "Foreword," *Annals of the New York Academy of Sciences* 50 (1948): 189–96. For discussion of this conference, see Kingsland, *The Evolution of American Ecology*, 186–89.

69. G. Evelyn Hutchinson, "Circular Causal Systems in Ecology," *Annals of the New York Academy of Sciences* 50 (1948): 221. On Hutchinson, see Hagen, *An Entangled Bank*, 62–76.

70. Kingsland, *The Evolution of American Ecology*, 194.

71. Howard T. Odum, *Environment, Power, and Society* (New York: Wiley-Interscience, 1971). For discussion of Odum's effort to "root ethics in the laws of energy and the principles of ecology," see Kingsland, *The Evolution of American Ecology*, 201–3.

72. See esp. Robert Van Horn and Philip Mirowski, "The Rise of the Chicago School of Economics and the Birth of Neoliberalism," in *The Road from Mont Pèlerin: The Making of the Neoliberal Thought Collective*, ed. Philip Mirowski and Dieter Plehwe (Cambridge, MA: Harvard University Press, 2009), 139–78; Edward Nik-Khah, "George Stigler, the Graduate School of Business, and the Pillars of the Chicago School," in *Building Chicago Economics: New Perspectives on the History of America's Most Powerful Economics Program*, ed. Robert Van Horn, Philip Mirowski, and Thomas A. Stapleford (Cambridge: Cambridge University Press, 2011), 116–47; and Edward

Nik-Khah and Robert Van Horn, "Inland Empire: Economics Imperialism as an Imperative of Chicago Neoliberalism," *Journal of Economic Methodology* 19, no. 3 (2012): 259–82.

73. Nik-Khah and Van Horn, "Inland Empire," 263–67.

74. See George J. Stigler and Clare Friedland, "What Can Regulators Regulate? The Case of Electricity," *Journal of Law & Economics* 5 (1962): 1–16.

75. See George J. Stigler, "The Theory of Economic Regulation," *Bell Journal of Economics and Management Science* 2, no. 1 (1971): 3–21.

76. See, e.g., George J. Stigler, "Regulation: The Confusion of Means and Ends," in *Regulating New Drugs*, ed. Richard L. Landau (Chicago: University of Chicago Center for Policy Study, 1973), 10–19.

77. Stigler, "The Theory of Economic Regulation," 10. Stigler's description of purchasing as "voting" was less a metaphor than Stigler's attempt to express what he understood as the true meaning of voting. Since for Stigler all social relations are economic in nature, the *political* sense of "voting" is in fact a metaphor for what is fundamentally an economic act. Stigler thus used the term "voting" to describe economic purchases so that his readers would understand the political sense of voting in economic terms (rather than, for example, expanding their understanding of economic terms).

78. Nik-Khah, "George Stigler," 141; see also Nik-Khah and Van Horn, "Inland Empire," 270.

79. The literature on the intersection of enclosure and Romantic aesthetics is vast, but important early discussions include John Barrell, *The Dark Side of the Landscape: The Rural Poor in English Painting, 1730–1840* (Cambridge: Cambridge University Press, 1980); Ann Bermingham, *Landscape and Ideology: The English Rustic Tradition, 1740–1860* (Berkeley: University of California Press, 1986); and Alan Liu, *Wordsworth: The Sense of History* (Stanford, CA: Stanford University Press, 1989).

80. See, for example, The Ecologist, *Whose Common Future? Reclaiming the Commons* (Philadelphia: New Society Publishers, 1993); Michael Hardt and Antonio Negri, *Empire* (Cambridge, MA: Harvard University Press, 2000), 300–3; Michael Hardt and Antonio Negri, *Multitude: War and Democracy in the Age of Empire* (New York: Penguin, 2004), 196–208; and Michael Hardt and Antonio Negri, *Commonwealth* (Cambridge, MA: Harvard University Press, 2009), viii, 120–125, 135–200, 387n1.

81. Neoliberal theorists also underscore the "classical liberal" provenance of an author such as Elinor Ostrom, who received the Nobel Prize in Economics for her research on the commons and who has also been claimed by the left. In his Introduction to Elinor Ostrom, *The Future of the Commons: Beyond Market Failure and Government Regulation*, with contributions by Christina Chang, Mark Pennington, and Vlad Tarko (London: Institute of Economic Affairs, 2010)—a book drawn from Ostrom's lecture for the Institute of Economic Affairs's annual Friedrich Hayek lecture—the business school professor Philip Booth insists that "in no sense do Professor Ostrom's ideas conflict with the idea of a free economy" but rather they "si[t] firmly within the classical liberal tradition of political economy" (11, 15). See also Mark Pennington, "Elinor Ostrom, Common-Pool Resources, and the Classical Liberal Tradition," in *The Future of the Commons*, 21–47.

82. Bruno Latour, *Politics of Nature: How to Bring the Sciences into Democracy* (Cambridge, MA: Harvard University Press, 2004), 8.

83. Ecologist, *Whose Common Future?*, 7. Hardt and Negri's account of the commons, though emphatically far left, is nevertheless more difficult to situate, since they are interested in contemporary commons produced half within, half outside the most modern formations of capitalist relations (e.g., immaterial labor).

84. Pennington, "Elinor Ostrom," 40.

85. See Bruno Latour, *We Have Never Been Modern* (Cambridge, MA: Harvard University Press, 1993); Bruno Latour, *Pandora's Hope: Essays on the Reality of Science Studies* (Cambridge, MA: Harvard University Press, 1999); and Latour, *Politics of Nature*.

86. Bruno Latour, "'It's the Development, Stupid!' Or: How to Modernize Modernization," 2007, http://www.bruno-latour.fr/node/153, p. 5.

87. On asbestos, see Latour, *Politics of Nature*, 23–24; on air conditioning, see Bruno Latour, *Facing Gaia: Eight Lectures on the New Climatic Regime*, trans. Catherine Porter (Cambridge: Polity, 2017).

88. See Latour, *Politics of Nature*; and Latour, *Facing Gaia*.

89. As I noted in Chapter 3, this premise underwrites Bruno Latour's actor-network theory and is most convincingly exemplified in his *The Pasteurization of France*, trans. Alan Sheridan and John Law (Cambridge, MA: Harvard University Press, 1988), in which he contends that it was by means of the alliances that Louis Pasteur established among his laboratory, other researchers, farmers, cows, and the anthrax bacillus itself that the latter became visible and could be treated medically.

90. Bruno Latour discusses his distinction between matters of fact and matters of concern extensively in *Politics of Nature* and provides a brief introduction in "Why Has Critique Run out of Steam? From Matters of Fact to Matters of Concern," *Critical Inquiry* 30 (2004): 225–48.

91. Latour, *Politics of Nature*, 8.

92. Latour, *Politics of Nature*, 133.

93. Latour, *Facing Gaia*, esp. 111–45.

94. Latour, *Politics of Nature*, 235; see also 151, 197.

95. Bruno Latour, "On Some of the Affects of Capitalism," lecture given at the Royal Academy, Copenhagen, February 26, 2014, http://www.bruno-latour.fr/node/550, pp. 10–11.

96. Philip Mirowski, "What Is Science Critique? Lessig, Latour," in *The Routledge Handbook of the Political Economy of Science*, ed. David Tyfield, Rebecca Lave, Samuel Randalls, and Charles Thorpe (New York: Routledge, 2017), 446. Mirowski notes that Latour's proximity to neoliberalism is less evident if one considers only the Chicago School, for those economists indeed theorize self-regulating markets. However, there are "two factions" of the neoliberal movement—"one, the Chicago School, which upheld neoclassical theory as the gospel of the neoliberal movement; and the other, the Austrian variant, which rejected neoclassical economics as having any validity whatsoever" (446)—and Latour is much closer to the second wing (which includes Hayek).

97. In "On Some of the Affects of Capitalism," Latour contends that "capitalism" is not a useful concept but rather a means of encouraging simultaneous optimism and pessimism, and he turns instead to locutions such as "the wide expansion of the reach of 'market organizations' along metrological chains" (11). Drawing on Ferdinand Braudel, Latour distinguishes between market relations and capitalism, arguing that "capitalism . . . feeds on, parasitizes and distorts marketplaces" (4) and suggests that revealing the conceptual incoherence of the concept of "the economy" and its aspiration of self-regulation will promote the weaker, and for him less problematic, form of market relations (which also aligns his position with that of neoliberal advocates of the commons).

98. See Mirowski's "What Is Science Critique?" for a discussion of this lacuna within Latour's work.

99. For a critical overview of the debate between "limits of growth" advocates and their procapitalist critics, see Melinda Cooper, *Life as Surplus: Biotechnology and Capitalism in the Neoliberal Era* (Seattle: University of Washington Press, 2008), 15–50. Many of the most well-researched recent proposals for solving the problems of climate change—e.g., Stewart Brand, *Whole Earth Discipline: Why Dense Cities, Nuclear Power, Transgenic Crops, Restored Wildlands, and Geoengineering Are Necessary* (New York: Penguin, 2010); George Monbiot, *Heat: How to Stop the Planet from Burning* (Brooklyn: South End, 2009); Thomas L. Friedman, *Hot, Flat, and Crowded: Why We Need a Green Revolution—and How It Can Renew America*, 2nd ed. (New York: Farrar, Straus and Giroux, 2009); and Steven Stoft, *Carbonomics: How to Fix the Climate and Charge It to OPEC* (Nantucket, MA: Diamond, 2008)—seem to me liberal insofar as they seek to mediate among ecological limits, capitalism, and individual choice. See also David Collings's astute point in *Stolen Future, Broken Present: The Human Significance of Climate Change* (Ann Arbor, MI: Open Humanities Press, 2014) that even minimal proposals for legislative regulation have failed in the United States precisely because, in the view of "free-market dogmatists . . . capitalism is more *real* than the biosphere" (81). On hopes that "smart" technologies can deliver ecological salvation, see Orit Halpern, Robert Mitchell, and Bernard Dionysius Geoghagen, "The Smartness Mandate: Notes toward a Critique," *Grey Room* 68 (2017): 106–29.

100. On Arendt's critique of the modern sciences, see my "Romanticism and the Experience of Experiment," *Wordsworth Circle* 46, no. 3 (2015): 132–42.

101. Bruno Latour discusses the relationship of habit and automatism in *An Inquiry into Modes of Existence: An Anthropology of the Moderns*, trans. Catherine Porter (Cambridge, MA: Harvard University Press, 2013), 265–80.

102. Latour, *Politics of Nature*, 199.

103. On the Greek polis, see Arendt, *The Human Condition*; on the ward system, councils, and *Räte*, see Hannah Arendt, *On Revolution* (New York: Penguin, 1973), 215–82. For a helpful discussion of the connection between Arendt's understanding of the connection between local council systems and larger republican and federal units, see Matthew Block, "The Federalism of Hannah Arendt," MA thesis, Duke University, March 2020.

Works Cited

Abrams, M. H. *Doing Things with Texts: Essays in Criticism and Critical Theory*. Ed. Michael Fischer. New York: Norton, 1989.

———. *The Mirror and the Lamp: Romantic Theory and the Critical Tradition*. New York: Oxford University Press, 1953.

Abroon, Fazel. "Necessity and the Origin of Evil in the Thought of Spinoza and Shelley." *Keats-Shelley Review* 14 (2000): 56–70.

Addison, Joseph, and Richard Steele. *The Spectator*. Ed. Donald Frederic Bond. 5 vols. Oxford: Clarendon, 1965.

Adolph, E. F. "Early Concepts of Physiological Regulation." *Physiology Reviews* 41 (1961): 737–70.

Agamben, Giorgio. *Homo Sacer: Sovereign Power and Bare Life*. Trans. Daniel Heller-Roazen. Stanford, CA: Stanford University Press, 1998.

———. *Remnants of Auschwitz: The Witness and the Archive*. New York: Zone, 1999.

Altick, Richard D. *The English Common Reader: A Social History of the Mass Reading Public, 1800–1900*. 2nd ed. Columbus: Ohio State University Press, 1998.

Anderson, Amanda. *Bleak Liberalism*. Chicago: University of Chicago, 2016.

Arbuthnot, John. *Mr. Maitland's Account of Inoculating the Smallpox Vindicated, from Dr. Wagstaffe's Misrepresentations of That Practice, with Some Remarks on Mr. Massey's Sermon*. London: Printed and sold by J. Peele, at Lock's Head in Paternoster-Row, 1722.

Arendt, Hannah. *The Human Condition*. 2nd ed. Chicago: University of Chicago Press, 1998.

———. "Introduction *into* Politics." In *The Promise of Politics*, 93–200. New York: Schocken, 2005.

———. *Lectures on Kant's Political Philosophy*. Chicago: University of Chicago Press, 1982.

———. *On Revolution*. New York: Penguin, 1973.

Armstrong, Nancy. *Desire and Domestic Fiction: A Political History of the Novel*. New York: Oxford University Press, 1987.

Armstrong, Nancy, and Len Tennenhouse. *Novels in the Time of Democratic Writing: The American Example*. Philadelphia: University of Pennsylvania Press, 2018.

———. "The Problem of Population and the Form of the American Novel." *American Literary History* 20, no. 4 (2008): 667–85.

Auerbach, Erich. *Mimesis: The Representation of Reality in Western Literature*. Princeton, NJ: Princeton University Press, 1953.

Austen, Jane. *Emma: An Authoritative Text, Contexts, Criticism*. Ed. George Justice. New York, Norton, 2012.

———. *Pride and Prejudice.* New York: Penguin, 1972.

Bacon, Francis. *The Essays, or Councils, Civil and Moral, of Sir Francis Bacon, Lord Verulam, Viscount St. Alban with a Table of the Colours of Good and Evil, and a Discourse of the Wisdom of the Ancients.* London: Printed for H. Herringman, R. Scot, R. Chiswell, A. Swalle, and R. Bentley, 1696.

Banfield, Ann. *Unspeakable Sentences: Narration and Representation in the Language of Fiction.* Boston: Routledge & Kegan Paul, 1982.

Barbauld, Anna Letitia. *Poems.* London: Printed for Joseph Johnson, 1773.

Barnett, Randy E. *Restoring the Lost Constitution: The Presumption of Liberty.* Princeton, NJ: Princeton University Press, 2004.

Barrell, John. *The Dark Side of the Landscape: The Rural Poor in English Painting, 1730–1840.* Cambridge: Cambridge University Press, 1980.

Barrows, Susanna. *Distorting Mirrors: Visions of the Crowd in Late Nineteenth-Century France.* New Haven, CT: Yale University Press, 1981.

Baucom, Ian. "History 4°: Postcolonial Method and Anthropocene Time." *Cambridge Journal of Postcolonial Literary Inquiry* 1, no. 1 (2014): 123–42.

———. *Specters of the Atlantic: Finance Capital, Slavery, and the Philosophy of History.* Durham, NC: Duke University Press, 2005.

Beck, Ulrich. *Risk Society: Towards a New Modernity.* Trans. Mark Ritter. London: Sage, 1992.

———. *World at Risk.* Trans. Ciaran Cronin. Malden, MA: Polity, 2009.

Beer, Gillian. *Darwin's Plots: Evolutionary Narrative in Darwin, George Eliot, and Nineteenth-Century Fiction.* Cambridge: Cambridge University Press, 2009.

Beiser, Frederick C. *Enlightenment, Revolution, and Romanticism: The Genesis of Modern German Political Thought, 1790–1800.* Cambridge, MA: Harvard University Press, 1992.

Bender, John B. *Imagining the Penitentiary: Fiction and the Architecture of Mind in Eighteenth-Century England.* Chicago: University of Chicago Press, 1987.

Berlin, Isaiah. *Liberty.* Oxford: Oxford University Press, 2002.

Bermingham, Ann. *Landscape and Ideology: The English Rustic Tradition, 1740–1860.* Berkeley: University of California Press, 1986.

Bernoulli, Daniel. "Essai d'une nouvelle analyse de la mortalité causée par la petite vérole." In *Histoire de l'Académie Royale des Sciences*, 2–45. Paris: De l'imprimerie royale, 1766.

Bewell, Alan. "Erasmus Darwin's Cosmopolitan Nature." *ELH* 76 (2009): 19–48.

———. "Jefferson's Thermometer: Colonial Biogeographical Constructions of the Climate of America." In *Romantic Science: The Literary Forms of Natural History*, ed. Noah Heringman, 111–38. Albany: State University of New York Press, 2008.

———. *Natures in Translation: Romanticism and Colonial Natural History.* Baltimore, MD: Johns Hopkins University Press, 2016.

———. *Romanticism and Colonial Disease.* Baltimore, MD: Johns Hopkins University Press, 1999.

———. *Wordsworth and the Enlightenment: Nature, Man, and Society in the Experimental Poetry.* New Haven, CT: Yale University Press, 1989.

Blanchard, Elodie Vieille. "Modelling the Future: An Overview of the 'Limits to Growth' Debate." *Centaurus* 52 (2010): 91–116.

Block, Matthew. "The Federalism of Hannah Arendt." MA thesis, Duke University. March 2020.

Bloom, Harold. "The Internalization of Quest-Romance." In *Romanticism and Consciousness: Essays in Criticism*, ed. Harold Bloom, 3–24. New York: Norton, 1970.

Boltanski, Luc, and Eve Chiapello. *The New Spirit of Capitalism*. Trans. Gregory Elliott. New York: Verso, 2005.

Boyle, James. *Shamans, Software, and Spleens: Law and the Construction of the Information Society*. Cambridge, MA: Harvard University Press, 1996.

Brand, Stewart. *Whole Earth Discipline: Why Dense Cities, Nuclear Power, Transgenic Crops, Restored Wildlands, and Geoengineering Are Necessary*. New York: Penguin, 2010.

Broglio, Ron. *Beasts of Burden: Biopolitics, Labor, and Animal Life in British Romanticism*. Albany: State University of New York Press, 2017.

Brooks, Cleanth. *The Well Wrought Urn: Studies in the Structure of Poetry*. New York: Harcourt, Brace, Jovanovich, 1975.

Brown, Wendy. *Undoing the Demos: Neoliberalism's Stealth Revolution*. New York: Zone, 2015.

Bruford, Walter Horace. *The German Tradition of Self-Cultivation: "Bildung" from Humboldt to Thomas Mann*. London: Cambridge University Press, 1975.

Burke, Edmund. *A Philosophical Enquiry into the Origin of Our Ideas of the Sublime and Beautiful*. London: Printed for R. and J. Dodsley, in Pall-Mall, 1759.

———. *Reflections on the Revolution in France, and on the Proceedings in Certain Societies in London Relative to That Event. In a Letter Intended to Have Been Sent to a Gentleman in Paris*. London: Printed for J. Dodsley, in Pall-Mall, 1790.

Campbell, Timothy C. *Improper Life: Technology and Biopolitics from Heidegger to Agamben*. Minneapolis: University of Minnesota Press, 2011.

Canguilhem, Georges. *Ideology and Rationality in the History of the Life Sciences*. Cambridge, MA: MIT Press, 1988.

———. *The Normal and the Pathological*. Trans. Carolyn R. Fawcett in collaboration with Robert S. Cohen. New York: Zone, 1991.

Cannon, Walter B. "The Body Physiologic and the Body Politic." *Science* 93, no. 2401 (1941): 1–10.

———. "Organization for Physiological Homeostasis." *Physiological Reviews* 9, no. 3 (1929): 399–431.

———. *The Wisdom of the Body*. 1st ed. New York: Norton, 1932.

Carroll, Siobhan. "Crusades against Frost: Frankenstein, Polar Ice, and Climate Change in 1818." *European Romantic Review* 24, no. 2 (2013): 211–30.

Chakrabarty, Dipesh. "Baucom's Critique: A Brief Response." *Cambridge Journal of Postcolonial Literary Inquiry* 1, no. 2 (2014): 245–64.

———. "The Climate of History: Four Theses." *Critical Inquiry* 35, no. 2 (2009): 197–222.

———. "The Politics of Climate Change Is More Than the Politics of Capitalism." *Theory, Culture, & Society* 34, no. 2–3 (2017): 25–37.

—. *Provincializing Europe: Postcolonial Thought and Historical Difference.* Princeton, NJ: Princeton University Press, 2000.

Chambers, Ephraim. *Cyclopaedia: Or, an Universal Dictionary of Arts and Sciences.* 4 vols. London: Rivington et al., 1778.

Chandler, James K. *Wordsworth's Second Nature: A Study of the Poetry and Politics.* Chicago: University of Chicago Press, 1984.

Christensen, Jerome. *Romanticism at the End of History.* Baltimore, MD: Johns Hopkins University Press, 2000.

Clarke, Samuel, Gottfried Wilhelm Leibniz, and Isaac Newton. *The Leibniz-Clarke Correspondence, Together with Extracts from Newton's "Principia" and "Opticks."* Manchester: Manchester University Press, 1956.

Clowes, James Daryl. "'Of Art and Women I Had No Knowledge': The Development of Schleiermacher's Understanding of Cognition, Self Identity, Community and Gender." PhD diss., University of Washington, 1996.

Cohn, Dorrit. *Transparent Minds: Narrative Modes for Presenting Consciousness in Fiction.* Princeton, NJ: Princeton University Press, 1978.

Coleridge, Samuel Taylor, and William Wordsworth. *Lyrical Ballads, with a Few Other Poems.* London: Printed for J. & A. Arch, Gracechurch-Street, 1798.

Collings, David. "After the Covenant: Romanticism, Secularization, and Disastrous Transcendence." *European Romantic Review* 21, no. 3 (2010): 345–61.

—. *Monstrous Society: Reciprocity, Discipline, and the Political Uncanny, c. 1780–1848.* Lewisburg, PA: Bucknell University Press, 2009.

—. *Stolen Future, Broken Present: The Human Significance of Climate Change.* Ann Arbor, MI: Open Humanities Press, 2014.

Collini, Stefan. *Liberalism and Sociology: L. T. Hobhouse and Political Argument in England, 1880–1914.* Cambridge: Cambridge University Press, 1979.

Connell, Philip. *Romanticism, Economics, and the Question of "Culture."* Oxford: Oxford University Press, 2001.

Constant, Benjamin. "The Liberty of the Ancients Compared with That of the Moderns." In *Political Writings,* 308–28. Cambridge: Cambridge University Press, 1988.

—. "The Spirit of Conquest and Usurpation and Their Relation to European Civilization." In *Political Writings,* 44–169. Cambridge: Cambridge University Press, 1988.

Cooper, Melinda. *Family Values: Between Neoliberalism and the New Social Conservatism.* New York: Zone, 2017.

—. *Life as Surplus: Biotechnology and Capitalism in the Neoliberal Era.* Seattle: University of Washington Press, 2008.

Corbin, Alain. *The Lure of the Sea: The Discovery of the Seaside in the Western World, 1750–1840.* Cambridge: Polity, 1994.

Crutzen, Paul J. "Geology of Mankind." *Nature* 415, no. 6867 (2002): 23.

Dalton, John. *Meteorological Observations and Essays.* London: Printed for W. Richardson, J. Phillips, and W. Pennington, 1793.

Darby, H. C. *A New Historical Geography of England.* Cambridge: Cambridge University Press, 1973.

Darwin, Charles. *Charles Darwin's "The Life of Erasmus Darwin."* Ed. Desmond King-Hele. Cambridge: Cambridge University Press, 2002.

Darwin, Erasmus. *The Botanic Garden, a Poem, in Two Parts*. 4th ed. 2 vols. London: Printed for J. Johnson, St. Paul's Church-Yard, 1799.

———. *The Temple of Nature; or, The Origin of Society: A Poem, with Philosophical Notes*. London: J. Johnson, 1803.

Daston, Lorraine. *Classical Probability in the Enlightenment*. Princeton, NJ: Princeton University Press, 1995.

Daston, Lorraine, and Peter Galison. *Objectivity*. New York: Zone, 2007.

Davies, Jeremy. *The Birth of the Anthropocene*. Oakland: University of California Press, 2016.

Dean, Mitchell. *The Constitution of Poverty: Toward a Genealogy of Liberal Governance*. New York: Routledge, 1991.

Defoe, Daniel. *Robinson Crusoe: An Authoritative Text, Contexts, Criticism*. Ed. Michael Shinagel. New York: Norton, 1994.

Deleuze, Gilles. *Difference and Repetition*. Trans. Paul Patton. New York: Columbia University Press, 1995.

———. "The Idea of Genesis in Kant's Esthetics." In *Desert Islands and Other Texts, 1953–1974*, 56–71. Cambridge, MA: Semiotext(e), 2004.

———. *The Logic of Sense*. Trans. Mark Lester with Charles Stivale, ed. Constantin V. Boundas. New York: Columbia University Press, 1990.

Descartes, René. *The Philosophical Writings of Descartes*. Trans. John Cottingham, Robert Stoothoff, and Dugald Murdoch. 3 vols. Cambridge: Cambridge University Press, 1984.

Desrosières, Alain. *The Politics of Large Numbers: A History of Statistical Reasoning*. Trans. Camille Naish. Cambridge, MA: Harvard University Press, 1998.

Dreiser, Theodore. *The Financier*. New York: Penguin, 2008.

Du Bois, W. E. B. "The Talented Tenth." In Booker T. Washington et al., *The Negro Problem: A Series of Articles by Representative American Negroes of Today*, 33–75. New York: J. Pott & Company, 1903.

Eagleton, Terry. *Literary Theory: An Introduction*. Anniversary ed. Malden, MA: Blackwell, 2008.

The Ecologist. *Whose Common Future? Reclaiming the Commons*. Philadelphia: New Society Publishers, 1993.

Eliot, George. *Middlemarch: An Authoritative Text, Backgrounds, Criticism*. Ed. Bert G. Hornsback. New York: Norton, 2000.

Empson, William. *Some Versions of Pastoral*. New York: New Directions, 1974.

Esposito, Roberto. *Bíos: Biopolitics and Philosophy*. Trans. Timothy Cambell. Minneapolis: University of Minnesota Press, 2008.

———. *Immunitas: The Protection and Negation of Life*. Trans. Zakiya Hanafi. Malden: Polity, 2011.

———. *Third Person: Politics of Life and Philosophy of the Impersonal*. Trans. Zakiya Hanafi. Malden: Polity, 2012.

Eyre, Richard. *A Sermon, Preached, at St. Andrew's, Holborn, April the 25th 1765. On the Anniversary Meeting, of the Governors, of the Small-Pox Hospitals*. London: Printed by H. Woodfall, 1765.

Fara, Patricia. *Sympathetic Attractions: Magnetic Practices, Beliefs, and Symbolism in Eighteenth-Century England*. Princeton, NJ: Princeton University Press, 1996.

Favret, Mary A. *War at a Distance: Romanticism and the Making of Modern Wartime.* Princeton, NJ: Princeton University Press, 2009.

———. "War in the Air." *Modern Language Quarterly* 65, no. 4 (December 2004): 531–59.

Ferguson, Frances. "Jane Austen, *Emma*, and the Impact of Form." *Modern Language Quarterly* 61, no. 1 (2000): 157–80.

———. "Malthus, Godwin, Wordsworth, and the Spirit of Solitude." In *Literature and the Body: Essays on Populations and Persons*, ed. Elaine Scarry, 106–24. Baltimore, MD: Johns Hopkins University Press, 1988.

Finch, Casey, and Peter Bowen. "'The Tittle-Tattle of Highbury': Gossip and the Free Indirect Style in *Emma*." *Representations* 31 (1990): 1–18.

Flaubert, Gustave. *Flaubert–Sand: The Correspondence.* Trans. Francis Steegmuller and Barbara Bray. New York: Knopf, 1993.

Foot, Paul. *Red Shelley.* London: Sidgwick & Jackson, 1980.

Forster, E. M. *Aspects of the Novel.* New York: Harvest, 1956.

Foucault, Michel. *The Birth of Biopolitics: Lectures at the Collège de France, 1978–79.* Trans. Graham Burchell. Ed. Michel Senellart. New York: Palgrave Macmillan, 2008.

———. *The Courage of the Truth: Lectures at the Collège de France, 1983–1984.* Trans. Graham Burchell. Ed. Frédéric Gros. New York: Palgrave Macmillan, 2011.

———. *The Essential Works of Foucault, 1954–1984.* Vol. 1: *Ethics.* Ed. Paul Rabinow. New York: The New Press, 1997.

———. *The Essential Works of Foucault, 1954–1984.* Vol. 3: *Power.* Ed. James D. Faubion. New York: The New Press, 2000.

———. *The Government of Self and Others: Lectures at the Collège de France, 1982–83.* Trans. Graham Burchell. Ed. Frédéric Gros. New York: Palgrave Macmillan, 2011.

———. "Governmentality." In Foucault, *Essential Works*, 3:201–22.

———. *Hermeneutics of the Subject: Lectures at the Collège de France, 1981–82.* Trans. Graham Burchell. Ed. Frédéric Gros. New York: Palgrave Macmillan, 2005.

———. *The History of Sexuality.* Vol. 1: *An Introduction.* New York: Pantheon, 1978.

———. "'Omnes et Singulatim': Toward a Critique of Political Reason." In Foucault, *Essential Works*, 3:298–325.

———. *On the Government of the Living: Lectures at the Collège de France, 1979–80.* Trans. Graham Burchell. Ed. Michel Senellart. New York: Palgrave Macmillan, 2014.

———. "The Risks of Security." In Foucault, *Essential Works*, 3:365–81.

———. *Security, Territory, Population: Lectures at the Collège de France, 1977–78.* Trans. Graham Burchell. Ed. Michel Senellart. New York: Palgrave Macmillan, 2007.

———. "Sexuality and Solitude." In Foucault, *Essential Works*, 1:175–84.

———. *Society Must Be Defended: Lectures at the Collège de France, 1975–76.* Trans. David Macey. New York: Picador, 2003.

Frank, Lawrence K. "Foreword." *Annals of the New York Academy of Sciences* 50 (1948): 189–96.

Frederickson, Kathleen. *The Ploy of Instinct: Victorian Sciences of Nature and Sexuality in Liberal Governance.* New York: Fordham University Press, 2014.

Freeden, Michael. *The New Liberalism: An Ideology of Social Reform.* Oxford: Oxford University Press, 1986.

Frey, Anne. *British State Romanticism: Authorship, Agency, and Bureaucratic Nationalism.* Stanford, CA: Stanford University Press, 2010.

Friedman, Michael. "Regulative and Constitutive." *Southern Journal of Philosophy* 30 (1991): 73–102.

Friedman, Milton, and Rose D. Friedman. *Free to Choose: A Personal Statement.* San Diego, CA: Harcourt Brace Jovanovich, 1990.

Friedman, Thomas L. *Hot, Flat, and Crowded: Why We Need a Green Revolution—and How It Can Renew America.* 2nd ed. New York: Farrar, Straus and Giroux, 2009.

Fulford, Tim. "Apocalyptic Economics and Prophetic Politics: Radical and Romantic Responses to Malthus and Burke." *Studies in Romanticism* 40, no. 3 (2001): 345–68.

Fulford, Tim, Debbie Lee, and Peter J. Kitson. *Literature, Science, and Exploration in the Romantic Era: Bodies of Knowledge.* Cambridge: Cambridge University Press, 2004.

Gallagher, Catherine. *The Body Economic: Life, Death, and Sensation in Political Economy and the Victorian Novel.* Princeton, NJ: Princeton University Press, 2006.

Gerard, Alexander. *An Essay on Genius.* London: Printed for W. Strahan; T. Cadell in the Strand, 1774.

Geulen, Eva. "Response and Commentary." In *Romanticism and Biopolitics*, ed. Alastair Hunt and Matthis Rudolf. Romantic Circles Praxis Series. 2012. https://www.rc.umd.edu/praxis/biopolitics/HTML/praxis.2012.geulen.html.

Ghosh, Amitav. *The Great Derangement: Climate Change and the Unthinkable.* Chicago: University of Chicago Press, 2016.

Gigante, Denise. "The Monster in the Rainbow: Keats and the Science of Life." *PMLA* 117 (2002): 433–48.

Gikandi, Simon. *Slavery and the Culture of Taste.* Princeton, NJ: Princeton University Press, 2011.

Gilmartin, Kevin. *Print Politics: The Press and Radical Opposition in Early-Nineteenth-Century England.* Cambridge: Cambridge University Press, 1996.

Glass, D. V. "The Population Controversy in Eighteenth-Century England. Part I. The Background." *Population Studies* 6, no. 1 (1952): 69–91.

Glimp, David. *Increase and Multiply: Governing Cultural Reproduction in Early Modern England.* Minneapolis: University of Minnesota Press, 2003.

Godwin, William. *The Enquirer, Reflections on Education, Manners, and Literature. In a Series of Essays.* London: Printed for G. G. and J. Robinson, Paternoster-Row, 1797.

———. *Enquiry Concerning Political Justice, and Its Influence on Modern Morals and Happiness.* 3rd ed. Ed. Isaac Kramnick. New York: Penguin, 1976.

———. *Of Population: An Enquiry Concerning the Power of Increase in the Numbers of Mankind.* London: Printed for Longman, Hurst, Rees, Orme and Brown, 1820.

———. "Of the Sources of Genius." In *The Enquirer, Reflections on Education, Manners, and Literature. In a Series of Essays,* 12–28. London: Printed for G. G. and J. Robinson, Paternoster-Row, 1797.

Golley, Frank B. *A History of the Ecosystem Concept in Ecology: More Than the Sum of the Parts.* New Haven, CT: Yale University Press, 1993.

Goodlad, Lauren M. E. *Victorian Literature and the Victorian State: Character and Governance in a Liberal Society.* Baltimore, MD: Johns Hopkins University Press, 2003.

Gray, Thomas. *An Elegy Written in a Country Church Yard.* 5th ed. London: Printed for R. Dodsley in Pall-Mall, 1751.

Green, John. *A Sermon Preached before His Grace George, Duke of Marlborough, President, the Vice-Presidents, the Treasurer, &C. Of the Hospitals for the Small-Pox. On Tuesday, April 26, 1763. By the Right Reverend Father-in-God John Lord Bishop of Lincoln.* London: Printed by H. Woodfall, in Paternoster-Row, 1763.

Greene, John C. "Aristotle to Darwin: Reflections on Ernst Mayr's Interpretation in *The Growth of Biological Thought.*" *Journal of the History of Biology* 25, no. 2 (1992): 257–84.

Griffiths, Devin. *The Age of Analogy: Science and Literature between the Darwins.* Baltimore, MD: Johns Hopkins University Press, 2016.

———. "Romantic Planet: Science and Literature within the Anthropocene." *Literature Compass* 14, no. 1 (2017).

Gross, Jonathan David. *Byron: The Erotic Liberal.* Lanham, MD: Rowman & Littlefield, 2001.

Guillory, John. *Cultural Capital: The Problem of Literary Canon Formation.* Chicago: University of Chicago Press, 1993.

Guyer, Sara Emilie. *Reading with John Clare: Biopoetics, Sovereignty, Romanticism.* New York: Fordham University Press, 2015.

Hacking, Ian. *The Taming of Chance.* New York: Cambridge University Press, 1990.

Hadley, Elaine. *Living Liberalism: Practical Citizenship in Mid-Victorian Britain.* Chicago: University of Chicago Press, 2010.

Hagen, Joel B. *An Entangled Bank: The Origins of Ecosystem Ecology.* New Brunswick, NJ: Rutgers University Press, 1992.

Hale, Piers J. "Finding a Place for the Anti-Malthusian Tradition in the Victorian Evolution Debates." In *New Perspectives on Malthus*, ed. Robert J. Mayhew, 182–207. Cambridge: Cambridge University Press, 2016.

Halpern, Orit, Robert Mitchell, and Bernard Dionysius Geoghagen. "The Smartness Mandate: Notes toward a Critique." *Grey Room* 68 (2017): 106–29.

Hamilton, Clive. *Earthmasters: The Dawn of the Age of Climate Engineering.* New Haven, CT: Yale University Press, 2013.

Hardin, Garrett. "The Tragedy of the Commons." *Science* 162, no. 3859 (1968): 1243–48.

Hardin, Russell. *Liberalism, Constitutionalism, and Democracy.* Oxford: Oxford University Press, 1999.

Hardt, Michael, and Antonio Negri. *Commonwealth.* Cambridge, MA: Harvard University Press, 2009.

———. *Empire.* Cambridge, MA: Harvard University Press, 2000.

———. *Multitude: War and Democracy in the Age of Empire.* New York: Penguin, 2004.

Harris, John. *Lexicon Technicum: or, an universal English dictionary of arts and sciences: explaining not only the terms of art, but the arts themselves.* London: Printed for Dan. Brown et al., 1708.

Harvey, David. *A Brief History of Neoliberalism.* New York: Oxford University Press, 2005.

Havens, George R. "Rousseau, Melon, and Sir William Petty." *Modern Language Notes* 55, no. 7 (1940): 499–503.

Hayek, Friedrich A. von. "Evolution and Spontaneous Order." 1983. https://www.youtube.com/watch?v=yQhqZ-iWMRM.

————. *The Fatal Conceit: The Errors of Socialism*. Chicago: University of Chicago Press, 1989.

————. *The Road to Serfdom: Text and Documents*. Chicago: University of Chicago Press, 2007.

————. "The Use of Knowledge in Society." *American Economic Review* 35, no. 4 (1945): 519–30.

Hazlitt, William. *A Reply to the Essay on Population, by the Rev. T. R. Malthus. In a Series of Letters. To Which Is Added, Extracts from the Essay, with Notes*. London: Printed for Longman, Hurst, Rees, and Orme, Paternoster-Row, 1807.

Herder, Johann Gottfried. "On the Cognition and Sensation of the Human Soul (1778)." In *Philosophical Writings*, 187–243. Cambridge: Cambridge University Press, 2002.

Hey, Jody. "Regarding the Confusion between the Population Concept and Mayr's 'Population Thinking.'" *Quarterly Review of Biology* 86, no. 4 (2011): 253–64.

Hirschman, Albert O. *The Passions and the Interests: Political Arguments for Capitalism before Its Triumph*. 20th anniversary ed. Princeton, NJ: Princeton University Press, 1997.

Hitchcock, Susan Tyler. *Frankenstein: A Cultural History*. New York: Norton, 2007.

Hobhouse, L. T. *Liberalism and Other Writings*. Cambridge: Cambridge University Press, 1994.

Hoffmann, John. "Kant's Aesthetic Categories: Race in the *Critique of Judgment*." *Diacritics* 44, no. 2 (2016): 54–81.

Hoppit, Julian. "Reforming Britain's Weights and Measures, 1660–1824." *English Historical Review* 108, no. 426 (1993): 82–104.

Humboldt, Wilhelm von. *The Limits of State Action*. Indianapolis, IN: Liberty Fund, 1993.

Hume, David. *An Enquiry Concerning the Principles of Morals*. London: Printed for A. Millar, 1751.

————. *Essays, Moral, Political, and Literary*. Ed. Eugene F. Miller. Indianapolis, IN: Liberty Classics, 1987.

————. *Four Dissertations*. London: Printed for A. Millar, 1757.

————. "Of the Populousness of Antient Nations." In *Essays and Treatises on Several Subjects*, 155–262. Edinburgh: Printed for A. Kincaid, and A. Donaldson, 1753.

————. "Of the Standard of Taste." In *Four Dissertations*, 203–40.

————. *Writings on Economics*. Madison: University of Wisconsin Press, 1955.

Hutchinson, G. Evelyn. "Circular Causal Systems in Ecology." *Annals of the New York Academy of Sciences* 50 (1948): 221–46.

Hunt, Alastair, and Matthis Rudolf, eds. "Romanticism and Biopolitics." Special issue of *Romantic Circles Praxis Series*. 2012.

Jackson, Noel. "Rhyme and Reason: Erasmus Darwin's Romanticism." *Modern Language Quarterly* 70, no. 2 (2009): 171–94.

Jager, Colin. "Shelley after Atheism." *Studies in Romanticism* 49, no. 4 (2010): 611–31.

Jameson, Fredric. *Archaeologies of the Future: The Desire Called Utopia and Other Science Fictions*. New York, Verso, 2005.

————. *The Political Unconscious: Narrative as a Socially Symbolic Act*. Ithaca, NY: Cornell University Press, 1981.

Janković, Vladimir. *Reading the Skies: A Cultural History of English Weather, 1650–1820.* Chicago: University of Chicago Press, 2000.

Kant, Immanuel. *Critique of Judgment.* Trans. Werner S. Pluhar. Indianapolis, IN: Hackett, 1987.

———. *Critique of Pure Reason.* Trans. Norman Kemp Smith. New York: St. Martin's, 1965.

———. *Gesammelte Schriften.* 23 vols. Berlin: Hrsg. von der Koeniglich-Preussischen Akademie der Wissenschaften zu Berlin, 1902–.

———. "Idea for a Universal History with a Cosmopolitan Purpose." Trans. H. B. Nisbet. In *Kant: Political Writings,* ed. Hans Siegbert Reiss, 41–53. Cambridge: Cambridge University Press, 1990.

———. "Perpetual Peace: A Philosophical Sketch." Trans. H. B. Nisbet. In *Kant: Political Writings,* ed. Hans Siegbert Reiss, 93–130. Cambridge: Cambridge University Press, 1990.

Kantor, Jamison. "Immortality, Romanticism, and the Limit of the Liberal Imagination." *PMLA* 133, no. 3 (2018): 508–25.

Kenyon-Jones, Christine. *Kindred Brutes: Animals in Romantic-Period Writing.* Aldershot: Ashgate, 2001.

King-Hele, Desmond. *Erasmus Darwin and the Romantic Poets.* New York: St. Martin's, 1986.

Kingsland, Sharon E. *The Evolution of American Ecology, 1890–2000.* Baltimore, MD: Johns Hopkins University Press, 2005.

Kingstone, Helen. "Human-Animal Elision: A Darwinian Universe in George Eliot's Novels." *Nineteenth-Century Contexts* 40, no. 1 (2018): 87–103.

Kittler, Friedrich. "Über die Sozialisation Wilhelm Meisters." In *Dichtung als Sozialisationsspiel: Studien zu Goethe und Gottfried Keller,* ed. Gerhard Kaiser and Friedrich Kittler, 13–124. Göttingen: Vandenhoeck & Ruprecht, 1978.

Klancher, Jon. "Godwin and the Republican Romance: Genre, Politics, and Contingency in Cultural History." *Modern Language Quarterly* 56, no. 2 (1995): 145–65.

———. *The Making of English Reading Audiences, 1790–1832.* Madison: University of Wisconsin Press, 1987.

Klein, Naomi. *The Shock Doctrine: The Rise of Disaster Capitalism.* 1st ed. New York: Metropolitan, 2007.

Kleingeld, Pauline. "Kant's Second Thoughts on Race." *Philosophical Quarterly* 57, no. 229 (2007): 573–92.

Kramnick, Isaac. *The Rage of Edmund Burke: Portrait of an Ambivalent Conservative.* New York: Basic Books, 1977.

———. *Republicanism and Bourgeois Radicalism: Political Ideology in Late Eighteenth-Century England and America.* Ithaca, NY: Cornell University Press, 1990.

Kreilkamp, Ivan. "Dying Like a Dog in *Great Expectations.*" In Morse and Danahay, *Victorian Animal Dreams,* 81–94.

Kula, Witold. *Measures and Men.* Princeton, NJ: Princeton University Press, 1986.

Lasagna, Louis. "Consensus among Experts: The Unholy Grail." *Perspectives in Biology and Medicine* 19, no. 4 (1976): 537–48.

———. "A Plea for the 'Naturalistic' Study of Medicines." *European Journal of Clinical Pharmacology* 7 (1974): 153–54.

Latour, Bruno. *Facing Gaia: Eight Lectures on the New Climatic Regime*. Trans. Catherine Porter. Cambridge: Polity, 2017.

———. *An Inquiry into Modes of Existence: An Anthropology of the Moderns*. Trans. Catherine Porter. Cambridge, MA, Harvard University Press, 2013.

———. "'It's the Development, Stupid!' Or: How to Modernize Modernization." 2007. http://www.bruno-latour.fr/node/153.

———. "On Some of the Affects of Capitalism." Lecture given at the Royal Academy, Copenhagen, February 26, 2014. http://www.bruno-latour.fr/node/550.

———. *Pandora's Hope: Essays on the Reality of Science Studies*. Cambridge, MA: Harvard University Press, 1999.

———. *The Pasteurization of France*. Trans. Alan Sheridan and John Law. Cambridge, MA: Harvard University Press, 1988.

———. *Politics of Nature: How to Bring the Sciences into Democracy*. Cambridge, MA: Harvard University Press, 2004.

———. *Science in Action: How to Follow Scientists and Engineers through Society*. Cambridge, MA: Harvard University Press, 1987.

———. *We Have Never Been Modern*. Cambridge, MA: Harvard University Press, 1993.

———. "Why Has Critique Run out of Steam? From Matters of Fact to Matters of Concern." *Critical Inquiry* 30 (2004): 225–48.

Lavoisier, Antoine Laurent, and Armand Seguin. *Premier mémoire sur la respiration des animaux*. 6 vols. Vol. 2. In *Œuvres de Lavoisier: Publiées par les soins de son excellence le Ministre de l'instruction publique et des cultes*. Paris: Imprimerie impériale, 1862.

———. *Premier mémoire sur la transpiration des animaux*. 6 vols. Vol. 2. In *Œuvres de Lavoisier: Publiées par les soins de son excellence le Ministre de l'instruction publique et des cultes*. Paris: Imprimerie impériale, 1862.

Leibniz, Gottfried Wilhelm. *Essais de théodicée sur la bonté de dieu, la liberté de l'homme et l'origine du mal: Nouvelle édition, augmentée de l'histoire de la vie & des ouvrages de l'auteur, par M. L. de Neufville*. 2 vols. Vol. 1. Amsterdam: F. Changuion, 1734.

———. "Monadology." Trans. George Montgomery. In *Discourse on Metaphysics; Correspondence with Arnauld; Monadology*, 249–72. La Salle, IL: Open Court, 1902.

Lemke, Thomas. "Beyond Foucault: From Biopolitics to the Government of Life." In *Governmentality: Current Issues and Future Challenges*, ed. Ulrich Bröckling, Susanne Krasmann, and Thomas Lemke, 165–82. New York: Routledge, 2011.

———. "The Risks of Security: Liberalism, Biopolitics, and Fear." In Lemm and Vatter, eds., *The Government of Life*, 59–74.

Lemm, Vanessa, and Miguel Vatter, eds. *The Government of Life: Foucault, Biopolitics, and Neoliberalism*. New York: Fordham University Press, 2014.

Letwin, William. *The Origins of Scientific Economics: English Economic Thought, 1660–1776*. London: Methuen, 1963.

Levin, Yuval. *The Great Debate: Edmund Burke, Thomas Paine, and the Birth of Right and Left*. New York: Basic Books, 2014.

Levine, Caroline. "The Enormity Effect: Realist Fiction, Literary Studies, and the Refusal to Count." *Genre* 50, no. 1 (2017): 59–75.

Levine, George Lewis, and U. C. Knoepflmacher. *The Endurance of Frankenstein: Essays on Mary Shelley's Novel*. Berkeley: University of California Press, 1979.

Levinson, Marjorie. "A Motion and a Spirit: Romancing Spinoza." *Studies in Romanticism* 46, no. 4 (2007): 367–408.

Lichtenstein, Ernst. *Zur Entwicklung des Bildungsbegriffs von Meister Eckhart bis Hegel.* Heidelberg: Quelle & Meyer, 1966.

Linley, Margaret. "The Living Transport Machine: George Eliot's *Middlemarch.*" In *Transport in British Fiction: Technologies of Movement, 1840–1940,* ed. Adrienne E. Gavin and Andrew E. Humphries, 84–100. Basingstoke: Palgrave, 2015.

Liu, Alan. *Wordsworth: The Sense of History.* Stanford, CA: Stanford University Press, 1989.

Locher, Fabien, and Jean-Baptiste Fressoz. "Modernity's Frail Climate: A Climate History of Environmental Reflexivity." *Critical Inquiry* 38, no. 3 (2012): 579–98.

Locke, John. *Locke on Money.* Ed. P. H. Kelly. 2 vols. Oxford: Oxford University Press, 1991.

———. *Some Considerations of the Consequences of the Lowering of Interest, and Raising the Value of Money.* In *Locke on Money,* 203–342.

———. *Two Treatises of Government: And a Letter Concerning Toleration.* New Haven, CT: Yale University Press, 2003.

Lonsdale, Roger, Thomas Gray, William Collins, and Oliver Goldsmith. *The Poems of Thomas Gray, William Collins, Oliver Goldsmith.* New York: Norton, 1972.

Losurdo, Domenico. *Liberalism: A Counter-History.* Trans. Gregory Elliott. New York: Verso, 2011.

Lukàcs, Georg. "Narrate or Describe?" In *Writer & Critic and Other Essays,* ed. A. D. Kahn, 110–48. New York: Universal Library, 1971.

———. *Studies in European Realism.* Trans. Edith Bone. New York: Howard Fertig, 2002.

Lynch, Deidre. *The Economy of Character: Novels, Market Culture, and the Business of Inner Meaning.* Chicago: University of Chicago Press, 1998.

MacKenzie, Scott R. *Be It Ever So Humble: Poverty, Fiction, and the Invention of the Middle-Class Home.* Charlottesville: University of Virginia Press, 2013.

Maddox, Isaac. *A Sermon Preached before His Grace Charles, Duke of Marlborough, President, the Vice-Presidents and Governors of the Hospital for the Small-Pox, and for Inoculation, at the Parish-Church of St. Andrew Holburn, on Thursday, March 5, 1752.* London: Printed by H. Woodfall, 1753.

Makdisi, Saree. *William Blake and the Impossible History of the 1790s.* Chicago: University of Chicago Press, 2003.

Mallory-Kani, Amy. "'Contagious Air(s)': Wordsworth's Poetics and Politics of Immunity." *European Romantic Review* 26, no. 6 (2015): 699–717.

Malthus, Thomas. *An Essay on the Principle of Population, as It Affects the Future Improvement of Society. With Remarks on the Speculations of Mr. Godwin, M. Condorcet, and Other Writers.* London: J. Johnson, 1798.

———. *An Essay on the Principle of Population: Or a View of Its Past and Present Effects on Human Happiness, with an Inquiry into Our Prospects Respecting the Future Removal or Mitigation of the Evils Which It Occasions.* Ed. Donald Winch and Patricia James. Cambridge: Cambridge University Press, 1992.

Manent, Pierre. *An Intellectual History of Liberalism.* Princeton, NJ: Princeton University Press, 1994.

Mann, Thomas. *Buddenbrooks*. Frankfurt am Main: Fischer Taschenbuch Verlag, 1989.

———. *Buddenbrooks*. Translated by John E. Woods. New York: Vintage, 1993.

Marx, Karl. *Capital: A Critique of Political Economy*. Vol. 1. Trans. Ben Fowkes. New York: Vintage, 1977.

———. *A Contribution to the Critique of Political Economy*. In Karl Marx and Friedrich Engels, *Collected Works*, 50 vols. New York: International Publishers, 1859.

Mayr, Ernst. *Animal Species and Evolution*. Cambridge, MA: Belknap Press of Harvard University Press, 1965.

———. "Darwin and the Evolutionary Theory in Biology." In *Evolution and Anthropology: A Centennial Appraisal*, ed. Betty J. Meggers, 1–10. Washington, DC: Anthropological Society of Washington, 1959.

———. "Speciation and Selection." *Proceedings of the American Philosophical Society* 93, no. 6 (1949): 514–19.

McCormick, Ted. "Population: Modes of Seventeenth-Century Demographic Thought." In *Mercantilism Reimagined: Political Economy in Early Modern Britain and Its Empire*, ed. Philip J. Stern and Carl Wennerlind, 25–45. Oxford: Oxford University Press, 2014.

———. *William Petty and the Ambitions of Political Arithmetic*. Oxford: Oxford University Press, 2009.

McKeon, Michael. *The Origins of the English Novel, 1600–1740*. Baltimore, MD: Johns Hopkins University Press, 1987.

McKibben, Bill. *Eaarth: Making a Life on a Tough New Planet*. New York: Time, 2010.

McLane, Maureen N. *Romanticism and the Human Sciences: Poetry, Population, and the Discourse of the Species*. Cambridge: Cambridge University Press, 2000.

McLaughlin, Peter. "Regulation, Assimilation, and Life: Kant, Canguilhem, and Beyond." 2007. http://www.philosophie.uni-hd.de/md/philsem/personal/mcl_regulation.pdf.

Meadows, Donella H., Dennis L. Meadows, Jørgen Randers, and William W. Behrens III. *The Limits to Growth: A Report for the Club of Rome's Project on the Predicament of Mankind*. New York: Universe, 1972.

Mee, Jon. *Romanticism, Enthusiasm, and Regulation: Poetics and the Policing of Culture in the Romantic Period*. New York: Oxford University Press, 2005.

———. "'The Use of Conversation': William Godwin's Conversable World and Romantic Sociability." *Studies in Romanticism* 50 (2001): 567–90.

Melville, Herman. *Moby-Dick: A Norton Critical Edition*. Ed. Hershel Parker. New York: Norton, 2018.

Mill, John Stuart. "Autobiography." In *Collected Works of John Stuart Mill*, 1:1–290.

———. *Collected Works of John Stuart Mill*. Ed. John M. Robson. 33 vols. Toronto: University of Toronto Press, 1963–1991.

———. "The Condition of Ireland [20]." In *Collected Works of John Stuart Mill*, 24:955–957.

———. *On Liberty*. In *Collected Works of John Stuart Mill*, 18:213–310.

———. "Utility of Religion." In *Collected Works of John Stuart Mill*, 10:403–28.

Miller, D. A. *The Novel and the Police*. Berkeley: University of California Press, 1988.

Miller, Genevieve. *The Adoption of Inoculation for Smallpox in England and France*. Philadelphia: University of Pennsylvania Press, 1957.

Mirowski, Philip. *Machine Dreams: Economics Becomes a Cyborg Science.* Cambridge: Cambridge University Press, 2002.

———. *More Heat Than Light: Economics as Social Physics; Physics as Nature's Economics.* Cambridge: Cambridge University Press, 1989.

———. *Never Let a Serious Crisis Go to Waste: How Neoliberalism Survived the Financial Meltdown.* New York: Verso, 2014.

———. *Science-Mart: Privatizing American Science.* Cambridge, MA: Harvard University Press, 2011.

———. "What Is Science Critique? Lessig, Latour." In *The Routledge Handbook of the Political Economy of Science,* ed. David Tyfield, Rebecca Lave, Samuel Randalls, and Charles Thorpe. New York: Routledge, 2017.

Mirowski, Philip, and Dieter Plehwe, eds. *The Road from Mont Pèlerin: The Making of the Neoliberal Thought Collective.* Cambridge, MA: Harvard University Press, 2009.

Mitchell, Robert. "Biopolitics and Population Aesthetics." *South Atlantic Quarterly* 115, no. 2 (2016): 367–98.

———. "Cryptogamia." *European Romantic Review* 21, no. 5 (2010): 631–51.

———. *Experimental Life: Vitalism in Romantic Science and Literature.* Baltimore, MD: Johns Hopkins University Press, 2014.

———. "*Frankenstein* and the Sciences of Self-Regulation." Forthcoming.

———. "Global Flows: Romantic-Era Terraforming." In *British Romanticism and Early Globalization: Developing the Modern World Picture,* ed. Evan Gottlieb, 199–218. Lewisburg, PA: Bucknell University Press, 2014.

———. "'Here Is Thy Fitting Temple': Science, Technology, and Fiction in Shelley's *Queen Mab.*" *Romanticism on the Net* 21 (2001).

———. "Regulating Life: Romanticism, Science, and the Liberal Imagination." *European Romantic Review* 29, no. 3 (2018): 275–93.

———. "Romanticism and the Experience of Experiment." *Wordsworth Circle* 46, no. 3 (2015): 132–42.

———. "US Biobanking Strategies and Biomedical Immaterial Labor." *Biosocieties* 7, no. 3 (2012): 224–44.

Mitchell, Robert, and Catherine Waldby. "National Biobanks: Clinical Labour, Risk Production, and the Creation of Biovalue." *Science, Technology, and Human Values* 35, no. 3 (2010): 330–55.

Monbiot, George. *Heat: How to Stop the Planet from Burning.* Brooklyn: South End, 2009.

Montes, Leonidas. "Newton's Real Influence on Adam Smith and Its Context." *Cambridge Journal of Economics* 32 (2008): 555–76.

Montesquieu, Charles de Secondat, baron de. *Persian Letters; Trans. Mr. Ozell.* 2 vols. London: Printed for J. Tonson, 1722.

———. *The Spirit of Laws; Translated from the French of M. De Secondat, Baron De Montesquieu, by Mr. Nugent.* Trans. Thomas Nugent. London: Printed for J. Nourse, and P. Vaillant, 1752.

Moore, Jason W. "The Capitalocene, Part I: On the Nature and Origins of Our Ecological Crisis." *Journal of Peasant Studies* 44, no. 3 (2017): 594–63.

Moretti, Franco. *Atlas of the European Novel, 1800–1900.* New York: Verso, 1998.

————. *Distant Reading*. New York: Verso, 2013.

————. *Graphs, Maps, Trees: Abstract Models for a Literary History*. New York: Verso, 2005.

————. *Signs Taken for Wonders*. Rev. ed. New York: Verso, 1988.

————. "The Slaughterhouse of Literature." *Modern Language Quarterly* 61, no. 1 (2000): 207–27.

————. *The Way of the World: The Bildungsroman in European Culture*. London: Verso, 1987.

Morgen, Sandra. *Into Our Own Hands: The Women's Health Movement in the United States, 1969–1990*. New Brunswick, NJ: Rutgers University Press, 2002.

Morse, Deborah Denenholz, and Martin A. Danahay. *Victorian Animal Dreams: Representations of Animals in Victorian Literature and Culture*. Aldershot: Ashgate, 2007.

Morton, Timothy. "Joseph Ritson, Percy Shelley, and the Making of Romantic Vegetarianism." *Romanticism* 12, no. 1 (2006): 52–61.

————. *The Poetics of Spice: Romantic Consumerism and the Exotic*. Cambridge: Cambridge University Press, 2000.

————. "Sustaining Natures: Shelley and Ecocriticism." In *Shelley and the Revolution in Taste*, 207–40. Cambridge: Cambridge University Press, 1994.

Mossner, Ernest Campbell. "Hume and the Ancient-Modern Controversy, 1725–1752: A Study in Creative Scepticism." *University of Texas Studies in English* 28 (1949): 139–53.

Murphy, Benjamin J. "Not So New Materialism: Homeostasis Revisited." *Configurations* 27, no. 1 (2019): 1–36.

Murray, Julie. "Company Rules: Burke, Hastings, and the Specter of the Modern Liberal State." *Eighteenth-Century Studies* 41, no. 1 (2007): 55–69.

Myers, Victoria. "William Godwin's Enquirer: Between Oratory and Conversation." *Nineteenth-Century Prose* 41, no. 1–2 (2014): 335–78.

Nagel, Thomas. *The View from Nowhere*. New York: Oxford University Press, 1986.

Nelson, Alondra. *Body and Soul: The Black Panther Party and the Fight against Medical Discrimination*. Minneapolis: University of Minnesota Press, 2011.

Nietzsche, Friedrich Wilhelm. *Beyond Good and Evil: Prelude to a Philosophy of the Future*. Trans. Walter Kaufmann. New York: Vintage, 1966.

Nik-Khah, Edward. "George Stigler, the Graduate School of Business, and the Pillars of the Chicago School." In *Building Chicago Economics: New Perspectives on the History of America's Most Powerful Economics Program*, ed. Robert Van Horn, Philip Mirowski, and Thomas A. Stapleford, 116–47. Cambridge: Cambridge University Press, 2011.

————. "Neoliberal Pharmaceutical Science and the Chicago School of Economics." *Social Studies of Science* 44, no. 4 (2014): 489–517.

Nik-Khah, Edward, and Robert Van Horn. "Inland Empire: Economics Imperialism as an Imperative of Chicago Neoliberalism." *Journal of Economic Methodology* 19, no. 3 (2012): 259–82.

Norris, Frank. *The Octopus: A Story of California*. New York: Bantam, 1971.

North, Brownlow. *A Sermon, Preached before His Grace Augustus Henry Duke of Grafton, President, the Vice-Presidents, and Treasurer, &C of the Hospitals for the Small-Pox and Inoculation, on Thursday May, the 6th, 1773, by Brownlow, Lord Bishop of Lichfield and Coventry, and Published at Their Request.* London: Printed by William Woodfall, 1773.

Odum, Eugene P. *Fundamentals of Ecology.* Philadelphia: Saunders, 1954.

Odum, Howard T. *Environment, Power, and Society.* New York: Wiley-Interscience, 1971.

O'Neill, Daniel I. *Edmund Burke and the Conservative Logic of Empire.* Oakland: University of California Press, 2016.

Oreskes, Naomi, and Erik M. Conway. *Merchants of Doubt: How a Handful of Scientists Obscured the Truth on Issues from Tobacco Smoke to Global Warming.* New York: Bloomsbury, 2010.

Ostrom, Elinor. *The Future of the Commons: Beyond Market Failure and Government Regulation,* with contributions by Christina Chang, Mark Pennington, and Vlad Tarko. London: Institute of Economic Affairs, 2010.

Otter, Chris. "Making Liberal Objects: British Techno-Social Relations, 1800–1900." *Cultural Studies* 21, no. 4–5 (2007): 570–90.

———. "The Technosphere: A New Concept for Urban Studies." *Urban History* 44, no. 1 (2017): 145–54.

———. *The Victorian Eye: A Political History of Light and Vision in Britain, 1800–1910.* Chicago: University of Chicago Press, 2008.

Pannese, Alessia. "The Non-Orientability of the Mechanical in Thomas Carlyle's Early Essays." *Journal of Interdisciplinary History of Ideas* 6, no. 11 (2017): 3:1–3:19.

Park, William. "What Was New About the 'New Species of Writing'?" *Studies in the Novel* 2, no. 2 (1970): 112–30.

Patey, Douglas Lane. "The Eighteenth Century Invents the Canon." *Modern Language Studies* 18, no. 1 (1988): 17–37.

Paul, Diane B., and Benjamin Day. "John Stuart Mill, Innate Differences, and the Regulation of Reproduction." *Studies in History and Philosophy of Biological and Biomedical Sciences* 39 (2008): 222–31.

Pennington, Mark. "Elinor Ostrom, Common-Pool Resources, and the Classical Liberal Tradition." In *The Future of the Commons: Beyond Market Failure and Government Regulation,* ed. Elinor Ostrom, 21–47. London: Institute of Economic Affairs, 2012.

Petty, William. *Another Essay in Political Arithmetick, Concerning the Growth of the City of London with the Measures, Periods, Causes, and Consequences Thereof, 1682.* London: Printed by H. H. for Mark Pardoe, 1683.

———. *Five Essays in Political Arithmetick.* London: Printed for Henry Mortlock, 1687.

———. *Political Arithmetic, or a Discourse Concerning, the Extent and Value of Lands, People, Buildings [Etc.].* London: Printed for Robert Clavel at the Peacock and Hen. Mortlock at the Phenix in St. Paul's Church-yard, 1690.

Pinson, Koppel S. *Pietism as a Factor in the Rise of German Nationalism.* New York: Columbia University Press, 1934.

Plotz, John. *The Crowd: British Literature and Public Politics.* Berkeley: University of California Press, 2007.

Pocock, J. G. A. "The Political Economy of Burke's Analysis of the French Revolution." In *Virtue, Commerce, and History: Essays on Political Thought and History, Chiefly in the Eighteenth Century*, 193–212. Cambridge: Cambridge University Press, 1985.

Poovey, Mary. *A History of the Modern Fact: Problems of Knowledge in the Sciences of Wealth and Society*. Chicago: University of Chicago Press, 1998.

Porter, Dahlia. *Science, Form, and the Problem of Induction in British Romanticism*. Cambridge: Cambridge University Press, 2018.

Porter, Theodore M. *The Rise of Statistical Thinking, 1820–1900*. Princeton, NJ: Princeton University Press, 1986.

Powell, William S., ed. *The Regulators in North Carolina: A Documentary History, 1759–1776*. Raleigh: State Department of Archives and History, 1971.

Prendergast, Christopher. "Evolution and Literary History: A Response to Franco Moretti." *New Left Review* 34 (2005): 40–62.

Price, Richard. *An Essay on the Population of England, from the Revolution to the Present Time*. 2nd ed. London: Printed for T. Cadell, 1780.

———. *Observations on Reversionary Payments; on Schemes for Providing Annuities for Widows, and for Persons in Old Age*. 3rd ed. London: T. Cadell, 1773.

Proctor, Robert. *Golden Holocaust: Origins of the Cigarette Catastrophe and the Case for Abolition*. Berkeley: University of California Press, 2011.

Proctor, Robert, and Londa L. Schiebinger. *Agnotology: The Making and Unmaking of Ignorance*. Stanford, CA: Stanford University Press, 2008.

Rajan, Tilottama. *The Supplement of Reading: Figures of Understanding in Romantic Theory and Practice*. Ithaca, NY: Cornell University Press, 1990.

Rauscher, Frederick. "The Regulative and Constitutive in Kant's and Hegel's Theories of History." *Idealistic Studies* 32, no. 2 (Summer 2002): 121–42.

"Review of Frankenstein; or the Modern Prometheus." *Edinburgh Magazine and Literary Miscellany* (1818): 249–53.

Ricardo, David. *On the Principles of Political Economy, and Taxation*. 3rd ed. London: John Murray, 1821.

Richardson, Alan. *Literature, Education, and Romanticism: Reading as Social Practice, 1780–1832*. Cambridge: Cambridge University Press, 1994.

Ritvo, Harriet. *The Animal Estate: The English and Other Creatures in the Victorian Age*. Cambridge, MA: Harvard University Press, 1987.

Robinson, Kim Stanley. *Aurora*. New York: Orbit, 2015.

———. *Blue Mars*. New York: Bantam, 1994.

———. *Green Mars*. New York: Bantam, 1994.

———. *Red Mars*. New York: Bantam, 1993.

Robson, Catherine. *Heart Beats: Everyday Life and the Memorized Poem*. Princeton, NJ: Princeton University Press, 2012.

Roe, Nicholas. *Wordsworth and Coleridge: The Radical Years*. New York: Oxford University Press, 1990.

Roger, Jacques. *Buffon: A Life in Natural History*. Trans. Sarah Lucille Bonnefoi. Ed. L. Pearce Williams. Ithaca, NY: Cornell University Press, 1997.

Rose, Mark. *Authors and Owners: The Invention of Copyright.* Cambridge, MA: Harvard University Press, 1993.

Rose, Nikolas S. *The Politics of Life Itself: Biomedicine, Power, and Subjectivity in the Twenty-First Century.* Princeton, NJ: Princeton University Press, 2007.

Rosenthal, Jesse. "The Large Novel and the Law of Large Numbers; Or, Why George Eliot Hates Gambling." *ELH* 77 (2010): 777–811.

Rousseau, Jean-Jacques. *Basic Political Writings.* Ed. Donald A. Cress. Indianapolis, IN: Hackett, 1988.

———. "Discourse on the Origin and Basis of Inequality among Men [1754]." In *Basic Political Writings*, 23–109.

———. "Discourse on the Sciences and Arts [1750]." In *Basic Political Writings*, xxi–21.

———. *Politics and the Arts: Letter to M. D'Alembert on the Theatre.* Trans. Allan Bloom. Ithaca, NY: Cornell University Press, 1960.

Rusnock, Andrea Alice. *Vital Accounts: Quantifying Health and Population in Eighteenth-Century England and France.* Cambridge: Cambridge University Press, 2002.

Russell, David. "Aesthetic Liberalism: John Stuart Mill as Essayist." *Victorian Studies* 56, no. 1 (2013): 7–30.

Russo, Brent Lewis. "Romantic Liberalism." PhD diss., University of California–Irvine, 2014.

Ryan, Alan. *The Making of Modern Liberalism.* Princeton, NJ: Princeton University Press, 2012.

Schiller, Friedrich. *On the Aesthetic Education of Man, in a Series of Letters.* New York: Frederick Ungar, 1965.

Schumpeter, Joseph A. *History of Economic Analysis.* New York: Oxford University Press, 1954.

Scott, Walter. "Remarks on Frankenstein." *Blackwood's Edinburgh Magazine* 2, no. 12 (1818): 613–20.

Sharpe, William. *A Dissertation upon Genius: Or, an Attempt to Shew, That the Several Instances of Distinction, and Degrees of Superiority in the Human Genius Are Not, Fundamentally, the Result of Nature, but the Effect of Acquisition.* London: Printed for C. Bathurst, 1755.

Shelley, Mary Wollstonecraft. *Frankenstein, or, the Modern Prometheus.* 3rd ed. Ed. David Lorne Macdonald and Kathleen Dorothy Scherf. Peterborough: Broadview, 2012.

———. *The Last Man.* Peterborough: Broadview, 1996.

Shelley, Percy Bysshe. *A Defence of Poetry.* In *Shelley's Poetry and Prose: A Norton Critical Edition*, ed. D. H. Reiman and N. Fraistat, 478–508. New York: Norton, 1977.

———. *The Letters of Percy Bysshe Shelley, Containing Material Never Before Collected.* 2 vols. London: G. Bell and Sons, Ltd., 1914.

———. "Mont Blanc; Lines Written in the Vale of Chamouni." In *Shelley's Poetry and Prose*, ed. Donald H. Reiman and Sharon B. Powers, 89–93. New York: Norton, 1977.

———. *Prometheus Unbound, a Lyrical Drama in Four Acts, with Other Poems.* London: C. and J. Ollier, 1820.

———. *Queen Mab; a Philosophical Poem.* Ed. Jonathan Wordsworth. New York: Woodstock, 1990.

———. *Shelley's Adonais: A Critical Edition.* Ed. Anthony D. Knerr. New York: Columbia University Press, 1984.

———. *A Vindication of Natural Diet: Being One of a Series of Notes to Queen Mab, a Philosophical Poem.* London: Printed for J. Callow, 1813.

Simon, Julian Lincoln. *The Ultimate Resource 2.* 2nd ed. Princeton, NJ: Princeton University Press, 1996.

Smith, Adam. *An Inquiry into the Nature and Causes of the Wealth of Nations.* Ed. R. H. Campbell and A. S. Skinner. 2 vols. Indianapolis, IN: Liberty Fund, 1981.

Smith, Kenneth. "The Malthusian Controversy." Thesis, University of London, 1951.

Snyder, Laura J. *Reforming Philosophy: A Victorian Debate on Science and Society.* Chicago: University of Chicago Press, 2006.

Sober, Elliott. "Evolution, Population Thinking, and Essentialism." *Philosophy of Science* 47, no. 3 (1980): 350–83.

Soni, Vivasvan. "Playing at Judgment: Aporias of Liberal Freedom in Kant's *Critique of Judgment.*" In *Literary/Liberal Entanglements: Toward a Literary History for the Twenty-First Century,* ed. Corinne Harol and Mark Simpson, 151–91. Toronto: University of Toronto Press, 2017.

Sorkin, David. "Wilhelm von Humboldt: The Theory and Practice of Self-Formation (Bildung), 1791–1810." *Journal of the History of Ideas* 44, no. 1 (1983): 55–73.

Sotirova, Violeta. "Historical Transformations of Free Indirect Style." In *Stylistics: Prospect & Retrospect,* ed. D. L. Hoover and S. Lattig, 129–41. Amsterdam: Rodopi, 2007.

Spengler, Joseph J. *French Predecessors of Malthus: A Study in Eighteenth-Century Wage and Population Theory.* Durham, NC: Duke University Press, 1942.

Squire, Samuel. *A Sermon Preached before His Grace Charles, Duke of Marlborough, President, the Vice-Presidents, the Treasurer, &C. Of the Hospitals for the Small-Pox, on Thursday, March 27, 1760.* London: Printed by H. Woodfall, 1760.

Stangeland, Charles Emil. "Pre-Malthusian Doctrines of Population: A Study in the History of Economic Theory." *Studies in History, Economics and Public Law* 21, no. 3 (1904).

Steinlight, Emily. "Dickens's 'Supernumeraries' and the Biopolitical Imagination of Victorian Fiction." *Novel: A Forum on Fiction* 43, no. 2 (2010): 227–50.

Steuart, James. *An Inquiry into the Principles of Political Oeconomy: Being an Essay on the Science of Domestic Policy in Free Nations.* 2 vols. London: Printed for A. Millar and T. Cadell, 1767.

Stigler, George J. "Regulation: The Confusion of Means and Ends." In *Regulating New Drugs,* ed. Richard L. Landau, 10–19. Chicago: University of Chicago Center for Policy Study, 1973.

———. "The Theory of Economic Regulation." *Bell Journal of Economics and Management Science* 2, no. 1 (1971): 3–21.

Stigler, George J., and Clare Friedland. "What Can Regulators Regulate? The Case of Electricity." *Journal of Law & Economics* 5 (1962): 1–16.

Stoft, Steven. *Carbonomics: How to Fix the Climate and Charge It to OPEC*. Nantucket, MA: Diamond, 2008.

Stout, Daniel. *Corporate Romanticism: Liberalism, Justice, and the Novel*. New York: Fordham University Press, 2017.

Sussman, Charlotte. "The Colonial Afterlife of Political Arithmetic: Swift, Demography, and Mobile Populations." *Cultural Critique* 56 (2004): 96–126.

———. *Peopling the World: Representing Human Mobility from Milton to Malthus*. Philadelphia: University of Pennsylvania Press, 2020.

Sweet, Paul R. "Young Wilhelm von Humboldt's Writings (1789–93) Reconsidered." *Journal of the History of Ideas* 34, no. 3 (1973): 469–82.

Taylor, Christopher. *Empire of Neglect: The West Indies in the Wake of British Liberalism*. Durham, NC: Duke University Press, 2018.

Tenger, Zeynep, and Paul Trolander. "Genius versus Capital: Eighteenth-Century Theories of Genius and Adam Smith's *Wealth of Nations*." *MLQ* 55 (1994): 169–89.

Thomas, David Wayne. *Cultivating Victorians: Liberal Culture and the Aesthetic*. Philadelphia: University of Pennsylvania Press, 2004.

Thompson, E. P. *The Making of the English Working Class*. New York: Pantheon, 1964.

Thompson, James. *Models of Value: Eighteenth-Century Political Economy and the Novel*. Durham, NC: Duke University Press, 1996.

Thorslev Jr., Peter L. "Post-Waterloo Liberalism: The Second Generation." *Studies in Romanticism* 28, no. 3 (1989): 437–61.

Thurtle, Phillip. *The Emergence of Genetic Rationality: Space, Time, and Information in American Biological Science, 1870–1920*. Seattle: University of Washington Press, 2007.

Tribe, Keith. *Land, Labour, and Economic Discourse*. London: Routledge & Kegan Paul, 1978.

———. *Genealogies of Capitalism*. Atlantic Highlands, NJ: Humanities Press, 1981.

Trilling, Lionel. *The Liberal Imagination*. New York: NYRB, 2008.

Turner, James. *Reckoning with the Beast: Animals, Pain, and Humanity in the Victorian Mind*. Baltimore, MD: Johns Hopkins University Press, 1980.

Van Horn, Robert, and Philip Mirowski. "The Rise of the Chicago School of Economics and the Birth of Neoliberalism." In Mirowski and Plehwe, *The Road from Mont Pèlerin*, 139–78.

Vatter, Miguel. "Foucault and Hayek: Republican Law and Civil Society." In Lemm and Vatter, *The Government of Life*, 163–84.

Velkar, Aashish. *Markets and Measurements in Nineteenth-Century Britain*. Cambridge: Cambridge University Press, 2012.

Wallerstein, Immanuel Maurice. *The Modern World-System IV: Centrist Liberalism Triumphant, 1789–1914*. Berkeley: University of California Press, 2011.

———. *Unthinking Social Science: The Limits of Nineteenth-Century Paradigms*. Philadelphia: Temple University Press, 2001.

Warsh, David. *Knowledge and the Wealth of Nations: A Story of Economic Discovery*. New York: Norton, 2006.

Watt, Ian. *The Rise of the Novel: Studies in Defoe, Richardson, and Fielding.* Berkeley: University of California Press, 1957.

Weheliye, Alexander G. *Habeas Viscus: Racializing Assemblages, Biopolitics, and Black Feminist Theories of the Human.* Durham, NC: Duke University Press, 2014.

Williams, Eric Eustace. *Capitalism and Slavery.* Chapel Hill: University of North Carolina Press, 1994.

Williams, John. *The Climate of Great Britain; or Remarks on the Change It Has Undergone, Particularly within the Last Fifty Years.* London: C. and R. Baldwin, 1806.

Williams, Raymond. *The Country and the City.* London: Chatto and Windus, 1973.

———. *Keywords: A Vocabulary of Culture and Society.* Rev. ed. New York: Oxford University Press, 2015.

Wilson, Elizabeth. "Underbelly." *differences: A Journal of Feminist Cultural Studies* 21, no. 1 (2010): 194–208.

Winch, Donald. *Riches and Poverty: An Intellectual History of Political Economy in Britain, 1750–1834.* Cambridge: Cambridge University Press, 1996.

Witt, John Fabian. *The Accidental Republic: Crippled Workingmen, Destitute Widows, and the Remaking of American Law.* Cambridge, MA: Harvard University Press, 2004.

Wolfe, Cary. *Before the Law: Humans and Other Animals in a Biopolitical Frame.* Chicago: University of Chicago Press, 2013.

Wollstonecraft, Mary. *A Vindication of the Rights of Woman: With Strictures on Political and Moral Subjects.* London: Printed for J. Johnson, No. 72, St. Paul's Church Yard, 1792.

Woloch, Alex. *The One vs. the Many: Minor Characters and the Space of the Protagonist in the Novel.* Princeton, NJ: Princeton University Press, 2003.

Wood, Gillen D'Arcy. *Tambora: The Eruption That Changed the World.* Princeton, NJ: Princeton University Press, 2014.

Wood, Paul B. *The Aberdeen Enlightenment: The Arts Curriculum in the Eighteenth Century.* Aberdeen: Aberdeen University Press, 1993.

Woodmansee, Martha. *The Author, Art, and the Market: Rereading the History of Aesthetics.* New York: Columbia University Press, 1994.

———. "The Genius and the Copyright: Economic and Legal Conditions of the Emergence of the 'Author.'" *Eighteenth-Century Studies* 17, no. 4 (1984): 425–48.

Woodring, Carl. *Politics in English Romantic Poetry.* Cambridge, MA: Harvard University Press, 1970.

Woolf, Virginia. *A Room of One's Own.* New York: Harvest, 2005.

Wordsworth, William. *Lyrical Ballads, with Other Poems. In Two Volumes.* 1st ed. 2 vols. London: Printed for T. N. Longman and O. Rees, Paternoster-Row, 1800.

———. *Lyrical Ballads, with Other Poems. In Two Volumes.* 3rd ed. 2 vols. London: Printed for T. N. Longman and O. Rees, Paternoster-Row, 1802.

———. *The Prelude, 1799, 1805, 1850: Authoritative Texts, Context and Reception, Recent Critical Essays.* New York: Norton, 1979.

Wright, Thomas. *An Original Theory or New Hypothesis of the Universe, Founded upon the Laws of Nature, and Solving by Mathematical Principles the General Phenomena of the Visible Creation.* London: Printed for the Author, and sold by H. Chapelle, in Grosvenor-Street, 1750.

Young, David B. "Libertarian Demography: Montesquieu's Essay on Depopulation in the *Lettres persanes.*" *Journal of the History of Ideas* 36, no. 4 (1975): 669–82.

Young, Edward. *Conjectures on Original Composition. In a Letter to the Author of Sir Charles Grandison.* London: Printed for A. Millar, in the Strand, and R. and J. Dodsley, in Pall-Mall, 1759.

Young, Ronnie. "James Beattie and the Progress of Genius in the Aberdeen Enlightenment." *Journal for Eighteenth-Century Studies* 36, no. 2 (2013): 245–61.

Yusoff, Kathryn. *A Billion Black Anthropocenes (or None).* Minneapolis: University of Minnesota Press, 2018.

Zammito, John H. *Kant, Herder, and the Birth of Anthropology.* Chicago: University of Chicago Press, 2002.

Zerilli, Linda M. G. *A Democratic Theory of Judgment.* Chicago: University of Chicago Press, 2016.

———. *Feminism and the Abyss of Freedom.* Chicago: University of Chicago Press, 2005.

Zola, Émile. "The Experimental Novel." In *The Experimental Novel and Other Essays,* 1–54. New York: Haskell House, 1964.

———. *Germinal.* In *Les Rougon-Macquart,* 3:1133–591.

———. *Germinal.* Ed. and trans. Roger Pearson. New York: Penguin, 2004.

———. *La bête humaine.* In *Les Rougon-Macquart,* 4:995–1331.

———. *Les Rougon-Macquart, histoire naturelle et sociale d'une famille sous le Second Empire.* 5 vols. Ed. Armand Lanoux and Henri Mitterand. Paris: Bibliothéque de la Pléiade.

———. *La Bête Humaine.* Ed. and trans. Roger Pearson. Oxford: Oxford University Press, 2009.

Zupko, Ronald Edward. *Revolution in Measurement: Western European Weights and Measures since the Age of Science.* Philadelphia: American Philosophical Society, 1990.

Index

Robert Mitchell is Chair of English at Duke University, where he also directs the Center for Interdisciplinary Studies in Science and Cultural Theory. His most recent book, *Experimental Life: Vitalism in Romantic Science and Literature*, won the Michelle Kendrick Memorial Book Prize and the BSLS Book Prize.

Sara Guyer and Brian McGrath, series editors

Lightning Source UK Ltd.
Milton Keynes UK
UKHW041821040321
379729UK00012B/93